QM Library
23 1331829 6

D1389293

LONGITUDINAL DATA ANALYSIS

WILEY SERIES IN PROBABILITY AND STATISTICS

Established by WALTER A. SHEWHART and SAMUEL S. WILKS

A complete list of the titles in this series appears at the end of this volume.

LONGITUDINAL DATA ANALYSIS

Donald Hedeker
Robert D. Gibbons
University of Illinois at Chicago

WILEY-INTERSCIENCE
A JOHN WILEY & SONS, INC., PUBLICATION

Published by John Wiley & Sons, Inc., Hoboken, New Jersey.
Published simultaneously in Canada.

For general information on our other products and services or for technical support, please contact our Customer Care Department within the United States at (800) 762-2974, outside the United States at (317) 572-3993 or fax (317) 572-4002.

Wiley also publishes its books in a variety of electronic formats. Some content that appears in print may not be available in electronic format. For information about Wiley products, visit our web site at www.wiley.com.

Library of Congress Cataloging-in-Publication Data:

Hedeker, Donald R., 1958–
Longitudinal data analysis / Donald Hedeker, Robert D. Gibbons.
p. cm.
Includes bibliographical references and index.
ISBN-13: 978-0-471-42027-9 (cloth)
ISBN-10: 0-471-42027-1 (cloth)
1. Longitudinal method. 2. Medicine—Research—Statistical methods. 3. Medical sciences—Research—Statistical methods. 4. Social sciences—Research—Statistical methods. I. Gibbons, Robert D., 1955—. II. Title.

R853.S7H48 2006
610.72'7—dc22

2005058221

Printed in the United States of America.

10 9 8 7 6 5 4 3 2 1

To Vera, Ava, Lina,
and the memory of Daniel
- D.H.

To Carol, Julie, Jason,
Michael, and the memory of
Rozlyn, Dorothy, and
Burton
- R.D.G.

CONTENTS

PREFACE

The prominence of the longitudinal study has grown tremendously in the past twenty years. The search for causal inference has led researchers in many fields to collect prospective longitudinal data in an effort to draw causal links between interventions and endpoints. In many cases, such studies involve rigorously controlled experiments—for example, prospective randomized single-center and multi-center clinical trials. In other cases, longitudinal data from naturalistic studies are either prospectively collected or retrospectively obtained to explore dynamically changing associations. For example, in a randomized clinical trial, investigators often collect prospective longitudinal data on one or more endpoints in response to a particular intervention relative to a control condition. The focus may be either to determine if there is a significant difference between control and treated individuals at the end of the study, often termed an "endpoint" analysis, or to examine differential rates of change over the course of the study in treated and control conditions. In the case in which subjects are initially randomized to the control and treatment conditions, differences in either the final response or the rate of response over time (*e.g.*, differential linear trends over time) are taken as evidence that the treatment produces an impact on the outcome measure of interest above and beyond chance expectations based on responses in the control condition.

By contrast, in naturalistic studies, subjects who elect to receive a particular treatment are compared to subjects who do not elect to receive that treatment. Often, the outcome of interest is repeatedly measured over the course of the study or in the available data. Furthermore, the timing and intensity of treatment are not controlled, so the treatment or intervention itself may exhibit longitudinal variability. For example, consider a ten-year health services research study in which the total amount of mental health service utilization is the endpoint, and the comparison of interest is between subjects with and without private insurance. At the beginning of the study, a subject may not have private insurance, however, five years into the study, the subject may obtain private insurance. The need to capture the time-varying nature of the intervention of interest is also a distinguishing feature of longitudinal studies. The longitudinal nature of the study and the time-varying treatment variable allow each subject to serve as his or her own control. Of course, in naturalistic longitudinal studies, we must also question whether those subjects who obtained private insurance during the course of the study are comparable to those subjects who did not, in terms of a myriad of other factors that might effect mental health services utilization. For example, individuals who obtain private insurance may do so because they have a family history of mental illness and are concerned that public insurance will not meet their mental health care needs in the future. As such, we would expect that these individuals would have greater mental health service utilization in general, regardless of whether they did or did not obtain private insurance. In this example, family history of mental illness and obtaining private insurance are "confounded," and the process by which this confound has arisen is termed a "selection effect." Longitudinal studies of this kind are widespread in economics, epidemiology, sociology, psychology in specific, and social sciences and medical sciences in general, and tools for reducing or eliminating bias in naturalistic studies have been studied in considerable detail by Cochran [1968], Heckman [1979], Rubin [1974, 1977], Rosenbaum and Rubin [1983], Angrist et al. [1996], and Little and Rubin [2000], to name but a few.

While it would appear that longitudinal studies are now considered foundational for drawing causal inference, they are not without limitations. Perhaps the most dramatic difficulty is the presence of missing data. Stated quite simply, not all subjects remain in the study for the entire length of the study. Reasons for discontinuing the study may be differentially related to the treatment. For example, some subjects may develop side effects to an otherwise effective treatment and must discontinue the study. Alternatively, some subjects might achieve the full benefit of the study early on and discontinue the study because they feel that their continued participation will provide no added benefit. The treatment of missing data in longitudinal studies is itself a vast literature, with major contributions by Laird [1988], Little [1995], Rubin [1976], and Little and Rubin [2002], to name a few. The basic problem is that even in a randomized and well-controlled clinical trial, the subjects who were initially enrolled in the study and randomized to the various treatment conditions may be quite different from those subjects that are available for analysis at the end of the trial. If subjects "drop out" because they already have derived full benefit from an effective treatment, an analysis that only considers those subjects who completed the trial may fail to show that the treatment was beneficial relative to the control condition. This type of analysis is often termed a "completer" analysis. To avoid this type of obvious bias, investigators often resort to an "intent to treat" analysis in which the last available measurement is carried forward to the end of the study as if the subject had actually completed the study. This type of analysis, often termed an "endpoint" analysis, introduces its' own set of problems in that (a) all subjects are treated equally regardless of the actual intensity of their treatment over the course of the study, and (b) the actual responses that would have been observed at the end of the study, if the subject had remained in the study until its' conclusion, may in fact be quite different from the response made at the time of discontinuation. Returning to our example of the study in which subjects discontinue when they feel that they have received full treatment benefit, an endpoint analysis might miss the fact that some of these subjects may have had a relapse had they remained on treatment. Many other objections have been raised about these two simple approaches of handing missing data in longitudinal studies.

In an attempt to provide a more general treatment of longitudinal data, with more realistic assumptions regarding the longitudinal response process and associated missing data mechanisms, statistical researchers have developed a wide variety of more rigorous approaches to the analysis of longitudinal data. Among these the most widely used include mixed-effects regression models [Laird and Ware, 1982], and generalized estimating equation (GEE) models [Liang and Zeger, 1986; Zeger and Liang, 1986]. Variations of these models have been developed for both discrete and continuous outcomes and for a variety of missing data mechanisms. The application of these models to the analysis of clustered and/or longitudinal data is the primary focus of this book. The primary distinction between the two general approaches is that mixed-effects models are "full-likelihood" methods and GEE models are "partial-likelihood" methods. The advantage of statistical models based on partial-likelihood is that (a) they are computationally easier than full-likelihood methods, and (b) they generalize quite easily to a wide variety of outcome measures with quite different distributional forms. The price of this flexibility, however, is that partial likelihood methods are more restrictive in their assumptions regarding missing data than their full-likelihood counterparts. In addition, full-likelihood methods provide estimates of person-specific effects (*e.g.*, person-specific trend lines) that are quite useful in understanding inter-individual variability in the longitudinal response process and in predicting future responses for a given subject or set of subjects from a particular subgroup (*e.g.*, a county, or a hospital, or a community). The distinctions between the various alternative statistical models for analysis of longitudinal data will be fully explored in the following chapters.

The outline of this book is as follows. Chapter 1 provides an overview of some of the history of analysis of longitudinal data and brief discussion of the various methodologies that have been used in analysis of longitudinal data. Chapter 2 presents univariate analysis of variance (ANOVA) models for analysis of repeated measurements. Chapter 3, presents multivariate analysis of variance (MANOVA) methods for analysis of repeated measurements.

Chapter 4 provides the conceptual and statistical foundation for mixed-effects regression models (MRM), which is the primary focus of this book. This chapter presents the general random intercept and trend growth curve model with various extensions including group structures and time-varying covariates. Chapter 5 considers curvilinear and polynomial trend MRMs, including discussion and illustration of the use of orthogonal polynomials. Chapter 6 presents covariance pattern models (CPMs) for analysis of longitudinal data. These models can be viewed as extending the MANOVA approach, while allowing for incomplete data across time and various forms for the error variance–covariance matrix (*i.e.*, autoregressive, Toeplitz, unstructured). Chapter 7 provides a further generalization of the MRM to the case of autocorrelated residuals. This chapter also discusses selection and comparison of the many possible models for the variance–covariance matrix of the repeated measures.

Chapter 8 introduces generalized estimating equations (GEE) models for analysis of longitudinal data. Since this class of models can be viewed as extending generalized linear models (GLMs) to the case of correlated data, both continuous and categorical outcomes are considered. Conceptual and statistical foundations of GEE models are presented, with comparisons to MRMs and CPMs. Chapter 9 presents MRMs for anaylsis of dichotomous longitudinal data. Using ordinary logistic regression as a starting point, the addition of random effects to the model is described, as are the consequences this has on the scale and interpretation of the regression coefficients in the model. Some discussion is paid to the distinction between the population-averaged estimates of the GEE models and the subject-specific estimates of the categorical MRMs. Chapter 10 then extends the model for ordinal outcomes and presents the many varieties of ordinal MRMs, including proportional and nonproportional odds models. Chapter 11 presents MRMs for nominal responses, when the categorized responses are unordered.

Chapter 12 describes models for the analysis of count data using a Poisson process. Such data commonly occur in health services research where the number of service visits is an outcome of interest. The models are then further generalized to the case of a zero-inflated Poisson (ZIP) model, which segments the response process into (a) a logistic regression model for the presence or absence of utilization and (b) a Poisson regression model for the intensity of utilization, conditional on use.

Chapter 13 presents three-level generalization of the previously presented two-level linear and nonlinear MRMs. An example is a multi-center longitudinal clinical trial in which subjects are nested within centers and repeatedly measured over time. Finally, Chapter 14 presents a detailed overview and discussion of the problem of missing data in analysis of longitudinal data. We present an overview of the various approaches to this problem, and we discuss in detail application of selection and pattern mixture models.

Throughout the book, applications of these methods for analysis of longitudinal data are emphasized and extensively illustrated using real examples. However, we have chosen not to focus on software in the book, though some syntax examples are provided. Many programs are available for the analyses presented in this book including SAS, SPSS, STATA, SYSTAT, HLM, MLwiN, MIXREG/MIXOR, and Mplus. To accompany this book, we have decided to post several datasets and computer syntax examples on the website `http://www.uic.edu/~hedeker/long.html`. Our aim is to keep these syntax examples current as new versions of the software programs emerge.

Most of the material from this book grew out of a class on Longitudinal Data Analysis, taught at the University of Illinois at Chicago. This semester-long class typically covers all of the material in Chapters 1 through 9 and 15. Overheads and additional materials from this course are available at `http://www.uic.edu/classes/bstt/bstt513`. The students in this class are diverse, as is their level of statistical background. As a result, this book does not assume a great deal of statistical knowledge. Essentially, one should have a good knowledge of multiple regression and ANOVA modeling, along with some knowledge of matrix algebra, maximum likelihood estimation, and logistic regression.

Several friends, colleagues, and students have helped us with this book. In particular, we thank R. Darrell Bock for teaching us everything that we know in statistics (though he is not responsible for our errors of learning). Our colleague Hakan Demirtas provided extensive help on the chapter on missing data. A big thanks goes to Ann Hohmann at N.I.M.H., who has been an incredible supporter of our work for many years. We are grateful for the support provided by grants MH56146, MH65556, MH66302, and MH01254. Several colleagues and students helped very much in reading over, discussing, and correcting drafts of the chapters. In this regard, thanks go to Michael Berbaum, Richard Campbell, Mark Grant, Zeynep Isgor, Andrew Leon, Julie Tamar Shecter, Matheos Yosef, and an anonymous reviewer. Additionally, we thank Subhash Aryal for help in the Latex preparation of this book. Finally, we thank Steve Quigley and Susanne Steitz of John Wiley & Sons for their assistance throughout the entire process of writing and completing this book.

DONALD HEDEKER AND ROBERT D. GIBBONS

Chicago, Illinois
January, 2006

ACKNOWLEDGMENTS

We have used several datasets from behavioral and medical studies in this text. We are thankful to several people for their graciousness in sharing their data with us: John Davis, Jan Fawcett, Brian Flay, Richard Hough, Michael Hurlburt, Robin Mermelstein, Niels Reisby, and Nina Schooler.

D. H. and R. D. G.

ACRONYMS

AIC	Akaike Information Criterion
ANOVA	Analysis of Variance
ANCOVA	Analysis of Covariance
BIC	Bayesian Information Criterion
CPM	Covariance Pattern Model
CS	Compound Symmetry
DHHS	Department of Health and Human Services
EB	Empirical Bayes
GEE	Generalized Estimating Equations
GLM	Generalized Linear Model
HDRS	Hamilton Depression Rating Scale
HLM	Hierarchical Linear Model
ICC	Intraclass Correlation
IRT	Item Response Theory
IOM	Institute of Medicine
LOCF	Last Observation Carried Forward
MANOVA	Multivariate Analysis of Variance
MAR	Missing at Random

MCAR	Missing Completely at Random
MCMC	Markov Chain Monte Carlo
ML	Maximum Likelihood
MLE	Maximum Likelihood Estimate
MNAR	Missing Not at Random
MRM	Mixed-effects Regression Model
NAS	National Academy of Sciences
OPO	Organ Procurement Organization
OPTN	Organ Procurement Transplantation Network
REML	Restricted Maximum Likelihood
SSCP	Sum of Squares and Cross Products
SSRI	Selective Serotonin Reuptake Inhibitor
TCA	Tricyclic Antidepressant
UN	Unstructured
ZAP	Zero-altered Poisson
ZIP	Zero-inflated Poisson

CHAPTER 1

INTRODUCTION

1.1 ADVANTAGES OF LONGITUDINAL STUDIES

There are numerous advantages of longitudinal studies over cross-sectional studies. First, to the extent that repeated measurements from the same subject are not perfectly correlated, longitudinal studies are more powerful than cross-sectional studies for a fixed number of subjects. Stated in another way, to achieve a similar level of statistical power, fewer subjects are required in a longitudinal study. The reason for this is that repeated observations from the same subject, while correlated, are rarely, if ever, perfectly correlated. The net result is that the repeated measurements from a single subject provide more independent information than a single measurement obtained from a single subject.

Second, in a longitudinal study, each subject can serve as his/her own control. For example, in a crossover study, each subject can receive both experimental and control conditions. In general, intra-subject variability is substantially less than inter-subject variability, so a more sensitive or statistically powerful test is the result. As previously mentioned, in naturalistic or observational studies, the primary intervention of interest may also be time-varying, so that naturalistic intra-subject changes in the intervention can be related to changes in the outcome of interest within individuals. Again, the net result is an exclusion of between-subject variability from measurement error which results in more efficient estimators of treatment-related effects when compared to corresponding cross-sectional designs with the same number and pattern of observations.

Third, longitudinal studies allow an investigator to separate aging effects (*i.e.*, changes over time within individuals), from cohort effects (*i.e.*, differences between subjects at baseline). Such cohort effects are often mistaken for changes occurring within individuals. Without longitudinal data, one cannot differentiate these two competing alternatives.

Finally, longitudinal data can provide information about individual change, whereas cross-sectional data cannot. Statistical estimates of individual trends can be used to better understand heterogeneity in the population and the determinants of growth and change at the level of the individual.

1.2 CHALLENGES OF LONGITUDINAL DATA ANALYSIS

Despite their advantages, longitudinal data are not without their challenges. Observations are not, by definition, independent and we must account for the dependency in data using more sophisticated statistical methods. The appropriate analytical methods are not as well developed, especially for more sophisticated models that permit more general forms of correlation among the repeated measurements. Often, there is a lack of available computer software for application of these more complex statistical models, or the level of statistical sophistication required of the user is beyond the typical level of the practitioner. In certain cases, for example nonlinear models for binary, ordinal, or nominal endpoints, parameter estimation can be computationally intensive due to the need for numerical or Monte Carlo simulation methods to evaluate the likelihood of nonlinear mixed-effects regression models.

An added complication that arises in the context of analysis of longitudinal data is the invariable presence of missing data. In some cases, a subject may be missing one of several measurement occasions; however, it is more likely that there are missing data due to attrition. Attrition, sometimes referred to as "drop-out," refers to a subject removing himself or herself from the study, prior to the end of the study. The data record for this subject therefore prematurely terminates. Several simple approaches to this problem have been proposed, none of which are statistically satisfactory. The simplest approach, termed a "completer analysis," limits the analysis to only those subjects that completed the studymissing data,completer analysis. Unfortunately, the available sample at the end of the study may have little resemblance to the sample initially randomized. Reasons for not completing the study may be confounded with the effects that the study was designed to investigate. For example, in a randomized clinical trial of a new drug versus placebo, only those subjects that did well on the drug may complete the study, giving the potentially false appearance of superiority of drug over placebo. The second simple approach is termed "Last Observation Carried Forward" (LOCF) and involves imputing the last available measurement to all subsequent measurement occasions. While things are somewhat better in the case of LOCF versus completer analyses, in an LOCF analysis, subjects treated in the analysis as if they have had identical exposure to the drug may have quite different exposures in reality or their experience on the drug may be complicated by other factors that led to their withdrawal from the study that are ignored in the analysis. More rigorous statistical alternatives based on mixed-effects regression models with ignorable and nonignorable nonresponse are an important focus of this book. Nevertheless, the presence of missing data, and its treatment in the statistical analysis, is a complicating feature of longitudinal data, making analysis potentially far more complex than analysis of cross-sectional data. The advantage, however, is that all available data from each subject can be used in the analysis, leading to increased statistical power, the ability to estimate subject-specific effects, and decreased bias due to arbitrary exclusion of subjects with incomplete response or the simple imputation of values for the missing responses.

Yet another complicating feature of longitudinal data is that not only does the outcome measure change over time, but the values of the predictors or independent variables can also change over time. For example, in the measurement of the relationship between plasma level of a drug and health status, both plasma level (the predictor) and health status (the outcome) change over time. The goal here is to estimate the dynamic relationship between these two variables over time. Note that this is a relationship that occurs within individuals, and it may vary from individual to individual as well. While our overall objective may be to determine if a relationship exists between drug plasma level and health status in the population, we must be able to model this dynamic relationship within individuals and must reach an overall conclusion regarding whether such a relationship exists in the population. The treatment of time-varying covariates in analysis of longitudinal data permits much stronger statistical inferences about dynamical relationships than can be obtained using cross-sectional data. The price, however, is considerable added complexity to the statistical model.

Finally, in some cases, the repeated measurements involve different conditions that the same subjects are exposed to. A classic example is a crossover design in which two or more treatments are given to the same subject in different orders. In these cases, the statistical inferences may be compromised by order or carry-over effects in which response to a one treatment may be conditional on exposure to a previous treatment. Dealing with carry-over or sequence effects is far from trivial, and is the statistical price paid for the stronger statistical inferences permitted by within-subject experimentation.

1.3 SOME GENERAL NOTATION

To set the stage for the statistical discussion to follow, it is helpful to present a unified notation for the various aspects of the longitudinal design. We index the N subjects in the longitudinal study as

$$i = 1, \ldots, N \text{ subjects.}$$

For a balanced design in which all subjects have complete data, and are measured on the same occasions, we index the measurement occasions as

$$j = 1, \ldots, n \text{ observations.}$$

or in the unbalanced case of unequal numbers of measurements or different time-points for different subjects

$$j = 1, \ldots, n_i \text{ observations for subject } i.$$

The total number of observations are given by

$$\sum_i^N n_i.$$

The repeated responses, or outcomes, or dependent measures for subject i are denoted as the vector

$$\boldsymbol{y}_i = n_i \times 1.$$

The values of the p predictors, or covariates, or independent variables for subject i on occasion j are denoted as (including an intercept term):

$$\boldsymbol{x}_{ij} = p \times 1.$$

For time-invariant predictors (between subject, *e.g.*, sex), the values of $\boldsymbol{x}_{ij}$ are constant for $j = 1, \ldots, n_i$. For time-varying predictors (within-subject, *e.g.*, age), the $\boldsymbol{x}_{ij}$ can take on subject- and timepoint-specific values. To describe the entire matrix of predictors for subject i, we use the notation

$$\boldsymbol{X}_i = n_i \times p.$$

It should be noted that not all of the literature on longitudinal data analysis uses this notation. In other sources, the following notation is sometimes used:

- $i = 1, \ldots, n$ subjects,
- $j = 1, \ldots, t_i$ observations,
- total number of observations = $\sum_i^n t_i$.

1.4 DATA LAYOUT

In fixing ideas for the statistical development to follow, it is also useful to apply this previously described notation to describe a longitudinal dataset as follows.

Subject	Observation	Response	Covariates		
1	1	y_{11}	x_{111}	...	x_{11p}
1	2	y_{12}	x_{121}	...	x_{12p}
.	.	.	.	. .	
1	n_1	y_{1n_1}	x_{1n_11}	...	x_{1n_1p}
.	.	.	.	. .	
.	.	.	.	. .	
.	.	.	.	. .	
.	.	.	.	. .	
N	1	y_{N1}	x_{N11}	...	x_{N1p}
N	2	y_{N2}	x_{N21}	...	x_{N2p}
.	.	.	.	. .	
N	n_N	y_{Nn_N}	x_{Nn_N1}	...	x_{Nn_NP}

In this univariate design, n_i varies by subject and so the number of data lines per subject can vary. In terms of the covariates, if x_r is time-invariant (*i.e.*, a between-subjects variable) then, for a given subject i, the covariate values are the same across time, namely, $x_{i1r} = x_{i2r} = x_{i3r} = \ldots = x_{in_ir}$.

The above layout depicts what is called a 2-level design in the multilevel [Goldstein, 1995] and hierarchical linear modeling [Raudenbush and Bryk, 2002] literatures. Namely, repeated observations at level 1 are nested within subjects at level 2. In some cases, subjects themselves are nested within sites, hospitals, clinics, workplaces, etc. In this case, the design has three levels with level-2 subjects nested within level-3 sites. This book primarily focuses on 2-level designs and models, with Chapter 13 covering 3-level extensions.

1.5 ANALYSIS CONSIDERATIONS

There are several different features of longitudinal studies that must be considered when selecting an appropriate longitudinal analysis. First, there is the form of the outcome or response measure. If the outcome of interest is continuous and normally distributed, much simpler analyses are usually possible (*e.g.*, a mixed-effects linear regression model). By contrast, if the outcome is continuous but does not have a normal distribution (*e.g.*, a count), then alternative nonlinear models (*e.g.*, a mixed-effects Poisson regression model) can be considered. For qualitative outcomes, such as binary (yes or no), ordinal (*e.g.*, sad, neutral, happy), or nominal (republican, democrat, independent), more complex nonlinear models are also typically required.

Second, the number of subjects N is an important consideration for selecting a longitudinal analysis method. The more advanced models (*e.g.*, generalized mixed-effects regression models) that are appropriate for analysis of unbalanced longitudinal data are based on large sample theory and may be inappropriate for analysis of small N studies (*e.g.*, $N < 50$).

Third, the number of observations per subject n_i is also an important consideration when selecting an analytic method. For $n_i = 2$ for all subjects, a simple change score can be computed and the data can be analyzed using methods for cross-sectional data, such as ANCOVA. When $n_i = n$ for all subjects, the design is said to be balanced, and traditional ANOVA or MANOVA models for repeated measurements (*i.e.*, traditional mixed-effects models or multivariate growth curve models) can be used. In the most general case where n_i varies from subject to subject, more general methods are required (*e.g.*, generalized mixed-effects regression models), which are the primary focus of this book.

Fourth, the number and type of covariates is an important consideration for model selection for $E(\boldsymbol{y}_i)$. In the one sample case, we may only have interest in characterizing the rate of change in the population over time. Here, we can use a random-effects regression model, where the parameters of the growth curve are treated as random effects and allowed to vary from subject to subject. In the multiple sample case (*e.g.*, comparison of one or more treatment conditions to control), the model consists of one or more categorical covariates that contrast the various treatment conditions in the design. In the regression case, we may have a mixture of continuous and/or categorical covariates, such as age, sex, and race. When the covariates take on time-specific values (*i.e.*, time-varying covariates), the statistical model must be capable of handling these as well.

Fifth, selection of a plausible variance–covariance structure for the $V(\boldsymbol{y}_i)$ is of critical importance. Different model specifications lead to (a) homogeneous or heterogeneous variances and/or (b) homogeneous or heterogeneous covariances of the repeated measurements over time. Furthermore, residual autocorrelation among the responses may also play a role in modeling the variance–covariance structure of the data.

Each of these factors is important for selecting an appropriate analytical model for analysis of a particular set of longitudinal data. In the following chapters, greater detail on the specifics of these choices will be presented.

1.6 GENERAL APPROACHES

There are several different general approaches to the analysis of longitudinal data. To provide an overview, and to fix ideas for further discussion and more detailed presentation, we present the following outline.

The first approach, which we refer to as the "derived variable" approach, involves the reduction of the repeated measurements into a summary variable. In fact, once reduced, this approach is strictly not longitudinal, since there is only a single measurement per subject. Perhaps the earliest example of the analysis of longitudinal data was presented by Student [1908] in his illustration of the t-test. The objective of the study [Cushny and Peebles, 1905] was to determine changes in sleep as a function of treatment with the hypnotic drug scopolomine. Although hours of sleep were carefully measured by the investigators, day-to-day variability presented statistical challenges in detecting the drug effect using large sample methods available at the time. Student (Gossett) proposed the one sample t-test to test if the average difference between experimental and control conditions was zero.

Examples of derived variables include (a) average across time, (b) linear trend across time, (c) carrying the last observation forward, (d) computing a change score, and (e) computing the area under the curve. A critical problem with all of these approaches is that our uncertainty in the derived variable is proportional to the number of measurements for which it was computed. In the unbalanced case (*e.g.*, drop-outs), different subjects will have different numbers of measurements and hence different uncertainties, therefore violating the commonly made assumption of homoscedasticity. Furthermore, by reducing multiple repeated measurements to a single summary measurement, there is typically a substantial loss of statistical power. Finally, use of time-varying covariates is not possible when the temporal aspect of the data has been removed.

Second, perhaps the simplest but most restrictive model is the ANOVA for repeated measurements [Winer, 1971]. The model assumes compound symmetry which implies constant variances and covariances over time. Clearly such an assumption has little, if any, validity for longitudinal data. Typically, variances increase with time because some subjects respond and others do not, and covariances for proximal occasions are larger than covariances for distal occasions. The model allows each subject to have his or her own trend line, however, the trend lines can only differ in terms of their intercepts, which implies that subjects deviate at baseline, but are consistent thereafter. It is more likely, of course, that subjects will deviate systematically from the overall trend both at baseline and in terms of the rate that they change over time (*i.e.*, their slope).

Third, MANOVA models have also been proposed for analysis of longitudinal data (see Bock [1975]). In the multivariate case, the repeated observations are generally transformed to orthogonal polynomial coefficients, and these coefficients (*e.g.*, constant, linear, quadratic growth rates) are then used as multivariate responses in a MANOVA. The principal disadvantages of this approach is that it does not permit missing data or different measurement occasions for different subjects.

Fourth, generalized mixed-effects regression models, which form the primary emphasis of this book, are now quite widely used for analysis of longitudinal data. These models can be applied to both normally distributed continuous outcomes as well as categorical outcomes and other nonnormally distributed outcomes such as counts that have a Poisson distribution. Mixed-effects regression models are quite robust to missing data and irregularly spaced measurement occasions and can easily handle both time-invariant and time-varying covariates. As such, they are among the most general of the methods for analysis of longitudinal data. They are sometimes called "full-likelihood" methods, because they make full use of all available data from each subject. The advantage is that missing data are ignorable if the missing responses can be explained either by covariates in the model or by the available responses from a given subject. The disadvantage is that full-likelihood methods are more computationally complex than quasi-likelihood methods, such as generalized estimating equations (GEE).

Fifth, covariance pattern models [Jennrich and Schluchter, 1986] can also be used to analyze longitudinal data. Here, the variance–covariance matrix of the repeated outcomes is modeled directly, and there is no attempt at distinguishing within-subjects variance from between-subjects variance, as is the case with mixed-effects regression models. Typically, the variance–covariance matrix is modeled in terms of a relatively small number of parameters, and full-likelihood estimation methods are used.

Sixth, GEE models are often used as a very general and computationally convenient alternative to mixed-effects regression models. They can be used to fit a wide variety of types of outcome measures and do not require complex numerical evaluation of the likelihood for nonlinear models. The disadvantage is that missing data are only ignorable if the missing data are explained by covariates in the model. This is a restrictive assumption in many situations and therefore, GEE models have somewhat limited applicability to incomplete longitudinal data.

1.7 THE SIMPLEST LONGITUDINAL ANALYSIS

The simplest possible longitudinal design consists of a single group and two measurement occassions. A paired t-test can be used to determine if there is significant average change between two timepoints. For this, note that there are N subjects, and the pre-test and post-test measurements for subject i are denoted y_{i1} and y_{i2} respectively. The difference or change score for subject i is denoted $d_i = y_{i2} - y_{i1}$. To test the difference between pre-test and post-test measurements, the null hypothesis can be written as: $H_0 : \mu_{y_1} = \mu_{y_2}$ which is the same as same as writing $H_0 : (\mu_{y_2} - \mu_{y_1}) = 0$. The test statistic is computed as

$$
\begin{aligned}
t &= \bar{d} \,/\, \left(s_d/\sqrt{N}\right) \\
&= \bar{d} \,/\, \left(\sqrt{\left[\sum_i d_i^2 - (\sum_i d_i)^2/N\right] /(N-1)} \,/\, \sqrt{N} \right) \\
&\overset{H_0}{\sim} t_{N-1}
\end{aligned}
$$

and is distributed as Student's t on $N - 1$ degrees of freedom. Notice that we can perform the same test using a regression model, where the difference between pre- and post-test measurements is

$$d_i = \beta_0 + e_i .$$

Assumming normality, we can test $H_0 : \beta_0 = 0$, by computing the ratio of $\hat{\beta}_0$ to its standard error, which also has a t-distribution on $N - 1$ degrees of freedom.

1.7.1 Change Score Analysis

Now consider a slightly more complex situation in which we have randomized subjects into two groups, a treatment and a control group. The groups are designated by $x_i = 0$ for controls and $x_i = 1$ for the treatment group. A regression model for the change score is given by

$$d_i \;=\; \beta_0 \;+\; \beta_1 x_i \;+\; e_i .$$

Note that in this model, β_0 reflects the average change for the control group, and β_1 reflects the difference in average change between the two groups. Hypothesis testing is as follows:

- testing $H_0 : \beta_0 = 0$ tests whether the average change is equal to zero for the control group
- testing $H_0 : \beta_1 = 0$ tests whether the average change is equal for the two groups

Notice that the change score analysis is equivalent to regressing the post-treatment measurement on the treatment variable, using the pre-treatment measurement as a covariate with slope equal to one.

$$\begin{aligned} d_i &= \beta_0 + \beta_1 x_i + e_i, \\ y_{i2} - y_{i1} &= \beta_0 + \beta_1 x_i + e_i, \\ y_{i2} &= y_{i1} + \beta_0 + \beta_1 x_i + e_i. \end{aligned}$$

In many ways, this is an overly restrictive assumption, since there is typically no *a priori* reason why a unit change in the pre-test score should translate to a unit change in the post-test measurement.

1.7.2 Analysis of Covariance of Post-test Scores

When the slope describing the relationship between the pre-test and post-test score is not one (*i.e.*, $\beta_2 \neq 1$), then we have an ANCOVA model for the post-test score, *i.e.*,

$$y_{i2} = \beta_0 + \beta_1 x_i + \beta_2 y_{i1} + e_i.$$

In terms of hypothesis testing we have

- testing $H_0 : \beta_0 = 0$ tests whether the average post-test is equal to zero for the control group subjects with zero pre-test
- testing $H_0 : \beta_1 = 0$ tests whether the post-test is equal for the two groups, given the same value on the pre-test (*i.e.*, conditional on pre-test)
- testing $H_0 : \beta_2 = 0$ tests whether the post-test is related to the pre-test, conditional on group

Note that change score analysis and ANCOVA answer different questions. Change score analysis tests if the average change is the same between the groups, whereas ANCOVA tests if the post-test average is the same between groups for sub-populations with the same pre-test values (*i.e.*, is the conditional average the same between the groups). The choice of which to use depends on the question of interest. The two models often yield similar conclusions for a test of the group effect. If subjects are randomized to group, then ANCOVA is more efficient (*i.e.*, more powerful), however, one must be careful using ANCOVA in nonrandomized settings, where groups are not necessarily similar in terms of pre-test scores (Lord's paradox, see Allison [1990]; Bock [1975]; Maris [1998]; Wright [2005]).

1.7.3 ANCOVA of Change Scores

Many practitioners have argued whether it is better to use the post-treatment ANCOVA (adjusting for pre-treatment) or to use ANCOVA on change scores adjusting for pre-treatment. With a bit of algebra, it is easy to show that the two approaches are identical for testing the null hypothesis of no treatment effect (*i.e.*, $H_0 : \beta_1 = 0$).

$$
\begin{aligned}
d_i &= \beta_0 + \beta_1 x_i + \beta_2 y_{i1} + e_i, \\
y_{i2} - y_{i1} &= \beta_0 + \beta_1 x_i + \beta_2 y_{i1} + e_i, \\
y_{i2} &= \beta_0 + \beta_1 x_i + (1 + \beta_2) y_{i1} + e_i.
\end{aligned}
$$

As can be seen from the equations above, there is no difference whatsoever between the two alternative models in terms of the treatment effect.

1.7.4 Example

To illustrate application of these simple models for the analysis of pre-test versus post-test change, we applied them to the Television School and Family Smoking Prevention and Cessation Project [Flay et al., 1988]. This study was designed to increase knowledge of the effects of tobacco use in school-age children. Characteristics of the sample are as follows:

- *sample* - 1600 7th-graders - 135 classrooms - 28 LA schools
  - between 1 to 13 classrooms per school
  - between 2 to 28 students per classroom
- *outcome* - knowledge of the effects of tobacco use
- *timing* - students tested at pre- and post-intervention
- *design* - schools randomized to
  - a social-resistance classroom curriculum (CC)
  - a media (television) intervention (TV)
  - CC combined with TV
  - a no-intervention control group

Here, we will ignore the clustering of students within classrooms and schools (see Chapter 13 for a description of methods to deal with this) and will concentrate on the potential change across the two study timepoints.

The first hypothesis to be tested is whether there was any overall change across time. The mean pre-intervention score is 2.069 and the mean post-intervention score is 2.662. Thus, there was an overall increase in knowledge scores of 0.59 units. A simple change score analysis (*i.e.*, paired t-test) reveals that this difference is significant ($t_{1599} = 15.01, p <$

.0001). Alternatively, a regression model treating the change score as the dependent variable, with an intercept and no regressors, yields identical results: $\hat{\beta}_0 = .5925$, standard error = .0395, $t_{1599} = 15.01, p < .0001$.

Next, we examine the effect of the CC and TV interventions on change in knowledge. Summary statistics for the 2 × 2 design are given below.

Tobacco and Health Knowledge Scale (THKS)
Subgroup Descriptive Statistics
Pre-Intervention, Post-Intervention, and Difference

	CC = no		CC = yes	
	TV = no	TV = yes	TV = no	TV = yes
N	421	416	380	383
Pre- Int mean	2.152	2.087	2.050	1.979
sd	1.182	1.288	1.285	1.286
Post-Int mean	2.361	2.539	2.968	2.823
sd	1.296	1.437	1.405	1.312
Difference	0.209	0.452	0.918	0.844

Does change across time vary by CC, TV, or both? To test this hypothesis, we begin by graphically displaying the post-intervention THKS means for the four groups in Figure 1.1.

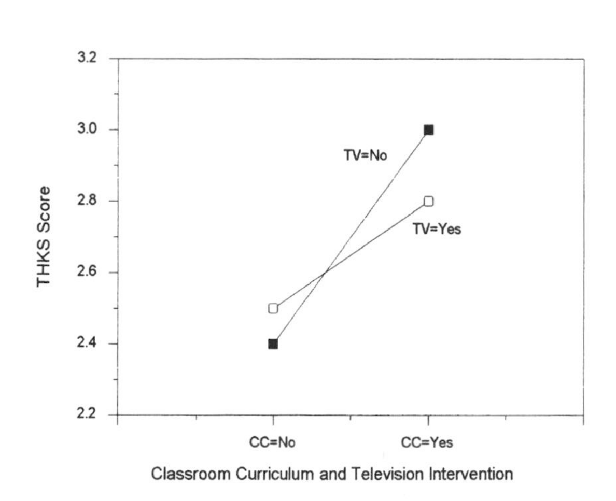

Figure 1.1. Post-intervention THKS means.

Next, we compute a regression analysis for post-intervention knowledge scores. Here, we code the treatment variables CC and TV as simple dummy-codes with 0 indicating no treatment and 1 indicating treatment exposure. The model with main effects and interaction of CC and TV effects yields a significant main effect of CC and a CC by TV interaction. The TV effect approaches significance ($p < .06$).

Variable	Parameter Estimate	Standard Error	t Value	Pr > | t |
Intercept	2.3611	0.0665	35.52	<.0001
CC	0.6074	0.0965	6.29	<.0001
TV	0.1774	0.0943	1.88	0.0600
CC by TV	−0.3234	0.1365	−2.37	0.0180

From this model, we can obtain the estimated post-intervention means of the four groups, based on our coding of the variables CC and TV:

- CC no TV no = $\hat{\beta}_0 = 2.3611$
- CC yes TV no = $\hat{\beta}_0 + \hat{\beta}_1 = 2.3611 + 0.6074 = 2.9685$
- CC no TV yes = $\hat{\beta}_0 + \hat{\beta}_2 = 2.3611 + 0.1774 = 2.5385$
- CC yes TV yes = $\hat{\beta}_0 + \hat{\beta}_1 + \hat{\beta}_2 + \hat{\beta}_3 = 2.3611 + 0.6074 + 0.1774 - 0.3234 = 2.8225$

These agree with the observed means, within rounding error, since the model is a saturated one in terms of the four groups. The ANCOVA model, adjusting for pre-intervention knowledge scores, yields significant main effects and interactions.

Variable	Parameter Estimate	Standard Error	t Value	Pr > | t |
Intercept	1.6613	0.0844	19.69	<.0001
PRETHKS	0.3252	0.0259	12.58	<.0001
CC	0.6406	0.0921	6.95	<.0001
TV	0.1987	0.0900	2.21	0.0273
CC by TV	−0.3216	0.1303	−2.47	0.0136

Here, using the sample average on the pre-intervention scores (= 2.069), we can calculate the adjusted means for the four groups as

- CC no TV no = $\hat{\beta}_0 + 2.069 \times \hat{\beta}_1 = 1.6613 + 2.069 \times 0.3252 = 2.3341$
- CC yes TV no = $\hat{\beta}_0 + 2.069 \times \hat{\beta}_1 + \hat{\beta}_2 = 1.6613 + 2.069 \times 0.3252 + 0.6406 = 2.9747$
- CC no TV yes = $\hat{\beta}_0 + 2.069 \times \hat{\beta}_1 + \hat{\beta}_3 = 1.6613 + 2.069 \times 0.3252 + 0.1987 = 2.5328$
- CC yes TV yes = $\hat{\beta}_0 + 2.069 \times \hat{\beta}_1 + \hat{\beta}_2 + \hat{\beta}_3 + \hat{\beta}_4 = 1.6613 + 2.069 \times 0.3252 + 0.6406 + 0.1987 - 0.3216 = 2.8518$

These are not very different from the estimated marginal means, which is not surprising since these four groups did not differ much in terms of their pre-intervention scores.

Regression on difference scores yields quite similar results to the baseline adjusted post-intervention regression model.

Variable	Parameter Estimate	Standard Error	t Value	Pr > | t |
Intercept	0.2090	0.0757	2.76	0.0058
CC	0.7094	0.1100	6.45	<.0001
TV	0.2429	0.1074	2.26	0.0239
CC by TV	−0.3180	0.1556	−2.04	0.0411

Finally, adding pre-intervention knowledge scores as a covariate, produces identical intervention effects to the model for post-intervention scores.

Variable	Parameter Estimate	Standard Error	t Value	Pr > | t |
Intercept	1.6613	0.0844	19.69	<.0001
PRETHKS	−0.6748	0.0259	−26.10	<.0001
CC	0.6406	0.0921	6.95	<.0001
TV	0.1987	0.0900	2.21	0.0273
CC by TV	−0.3216	0.1303	−2.47	0.0136

These analyses reveal that both CC and TV interventions impact knowledge scores, however, their effects are nonadditive. CC increases knowledge gains overall, however the TV intervention decreases the gain in knowledge when CC is present and increases the gain in knowledge when CC is absent.

1.8 SUMMARY

Longitudinal studies represent enormous advantages over cross-sectional studies in terms of providing foundations for causal inference. It is not surprising that the required level of statistical sophistication required for the analysis of longitudinal data is more advanced as well. Despite their advantages, longitudinal data are not without their challenges. The treatment of missing data plays a far greater role in longitudinal studies than it does in cross-sectional studies and analysis of naturalistic or observational longitudinal data is complicated by numerous sources of bias due to selection effects. In the following chapters, we provide a variety of approaches to the analysis of different types of longitudinal data, present their strengths and limitations, and illustrate their use with real examples.

CHAPTER 2

ANOVA APPROACHES TO LONGITUDINAL DATA

There are two classical approaches to the analysis of longitudinal data. The first is variably called univariate mixed-model, split-plots, or repeated measures ANOVA, and the second is based on multivariate ANOVA (MANOVA). Both models assume interval measurement and normally distributed errors that are homogeneous across groups. In some cases, normality and homogeneity of variance can be brought about through transformation (*e.g.*, natural log transformation). For both models, the primary focus is on comparison of group means, and neither model is informative about individual growth curves (*i.e.*, subject-specific trends). Furthermore, the timepoints are assumed to be fixed across subjects (either evenly or unevenly spaced) and are treated as a classification variable in the ANOVA or MANOVA model. This precludes analysis of unbalanced designs in which different subjects are measured on different occasions. Both models are based on least-squares estimation and are therefore adversely affected by outliers and missing data. While the ANOVA model can handle some missing data (*i.e.*, there are methods for unbalanced ANOVA), the MANOVA model cannot handle any missing data. In terms of the variance–covariance structure for the responses ($\boldsymbol{y}_i$), the ANOVA model assumes compound symmetry (*i.e.*, equal variances and covariances over time), whereas the MANOVA model makes no assumption regarding the specific form of the variance–covariance structure. While this is an important advantage of MANOVA over ANOVA, it is tempered by the larger limitation of requiring complete data for all subjects in the MANOVA model. As such, application of MANOVA must follow deletion of all subjects without complete data, which is essentially a completer analysis, and is prone to substantial bias in that the composition of subjects that complete the study can be quite different from the composition of the subjects at the time of randomization.

Longitudinal Data Analysis. By Donald Hedeker and Robert D. Gibbons

In this chapter, we describe the univariate ANOVA model for within-subject designs in detail. In the social and behavioral sciences literature, this model and extensions of it are often referred to as a "Repeated Measures ANOVA." In Chapter 3, we present the alternative multivariate approach. While these approaches are no longer recommended for routine application (if at all), they are important in that they fix ideas for the development of the more modern and advanced methods that are the primary focus of this book.

2.1 SINGLE-SAMPLE REPEATED MEASURES ANOVA

In the single-sample case, the model is referred to as the randomized blocks ANOVA. In this case, we have no intervention or group effects, but are simply using the model to characterize rates of change over time. With $i = 1, \ldots, N$ subjects and $j = 1, \ldots, n$ measurement occasions, the randomized blocks ANOVA is given by the linear model,

$$y_{ij} = \mu + \pi_i + \tau_j + e_{ij} \ , \qquad (2.1)$$

where μ = the grand mean, π_i = the individual difference component for subject i, which is assumed to be constant over time, τ_j = the effect of time, assumed to be the same for all subjects, and e_{ij} = the error for subject i on occasion j. We also assume that the random components are distributed as $\pi_i \sim N(0, \sigma_\pi^2)$, where σ_π^2 is the between-subjects variance, and $e_{ij} \sim N(0, \sigma_e^2)$, with σ_e^2 as the within-subjects variance. Notice that this is a mixed model because it includes both random (π_i) and fixed (τ_j) parameters.

2.1.1 Design

The data for the model above can be represented by the following two-factor design of subjects crossed with timepoints:

	Timepoint			
Subject	1	2	...	n
1	y_{11}	y_{12}	...	y_{1n}
2	y_{21}	y_{22}	...	y_{2n}
.	.	.	...	.
.	.	.	...	.
N	y_{N1}	y_{N2}	...	y_{Nn}

In this design there is one observation per cell. In other words, each subject is observed once at each timepoint. The design is similar to a randomized blocks design with subjects as blocks. In the simple case of $n = 2$, the design and subsequent analysis is identical to a paired t-test, in terms of testing of the time effect.

In terms of model assumptions, we assume that

$$\begin{aligned}
\sum_{j=1}^{n} \tau_j &= 0, \\
E(y_{ij}) &= \mu + \tau_j, \\
V(y_{ij}) &= V(\mu + \tau_j + \pi_i + e_{ij}) = \sigma_\pi^2 + \sigma_e^2, \\
C(y_{ij}, y_{i'j}) &= 0 \text{ for } i \neq i', \\
C(y_{ij}, y_{ij'}) &= \sigma_\pi^2 \text{ for } j \neq j'.
\end{aligned}$$

Here, $E(\cdot)$, $V(\cdot)$, and $C(\cdot)$ denote expectation, variance, and covariance respectively. Note that the first covariance statement indicates that subjects are independent of each other, whereas the second covariance statement indicates that the covariance is σ_π^2 for any two repeated measures within the same subject. This covariance can be expressed as a correlation, which reflects the magnitude of the within-subject association in a more interpretable metric:

$$Corr(y_{ij}, y_{ij'}) = \frac{\sigma_\pi^2}{\sigma_\pi^2 + \sigma_e^2}. \tag{2.2}$$

This correlation is termed the intraclass correlation. Note that it is the same for all longitudinal pairs of measures and therefore represents the average correlation of y from any two timepoints. In the above formulation, because variances cannot be negative, the intraclass correlation ranges from 0 to 1; it equals 0 if subjects explain none of the variance (*i.e.*, $\sigma_\pi^2 = 0$), and it equals 1 if subjects explain all of the variance (*i.e.*, $\sigma_e^2 = 0$). Thus, as described later, it can also be interpreted as the proportion of the total variance that is attributable to subjects.

Given the above assumptions, the variance covariance matrix of the repeated measures has the "compound symmetry" structure:

$$\Sigma_{\boldsymbol{y}_i} = \begin{bmatrix}
\sigma_e^2 + \sigma_\pi^2 & \sigma_\pi^2 & \sigma_\pi^2 & \cdots & \sigma_\pi^2 \\
\sigma_\pi^2 & \sigma_e^2 + \sigma_\pi^2 & \sigma_\pi^2 & \cdots & \sigma_\pi^2 \\
\cdots & \cdots & \cdots & \cdots & \cdots \\
\sigma_\pi^2 & \cdots & \sigma_\pi^2 & \sigma_e^2 + \sigma_\pi^2 & \sigma_\pi^2 \\
\sigma_\pi^2 & \sigma_\pi^2 & \cdots & \sigma_\pi^2 & \sigma_e^2 + \sigma_\pi^2
\end{bmatrix}, \tag{2.3}$$

where the variance is homogeneous across time $(\sigma_e^2 + \sigma_\pi^2)$, the covariances are homogeneous across time (σ_π^2), and the correlation is given by $\sigma_\pi^2/(\sigma_e^2 + \sigma_\pi^2)$. Unfortunately, compound symmetry is not very realistic for longitudinal data. First, variances often change over time, where subjects are generally more similar at the start of the trial than at the end of the trial where some have responded to treatment and others have not. Second, covariances close in time are usually greater than covariances that are further separated in time.

In the balanced case (*i.e.*, no missing data and all subjects measured on the same occasions), the ANOVA table for a model with random subject effects and fixed time effects is of the following form:

Source	df	SS	MS	E(MS)
Subjects	$N-1$	$\text{SS}_S = n\sum_{i=1}^{N}(\bar{y}_{i.} - \bar{y}_{..})^2$	$\frac{\text{SS}_S}{N-1}$	$\sigma_e^2 + n\sigma_\pi^2$
Time	$n-1$	$\text{SS}_T = N\sum_{j=1}^{n}(\bar{y}_{.j} - \bar{y}_{..})^2$	$\frac{\text{SS}_T}{n-1}$	$\sigma_e^2 + N\sum(\tau_j - \tau_.)^2$
Residual	$(N-1) \times(n-1)$	$\text{SS}_R = \sum_{i=1}^{N}\sum_{j=1}^{n}(y_{ij} - \bar{y}_{i.} - \bar{y}_{.j} + \bar{y}_{..})^2$	$\frac{\text{SS}_R}{(N-1)(n-1)}$	σ_e^2
Total	$Nn-1$	$\text{SS}_y = \sum_{i=1}^{N}\sum_{j=1}^{n}(y_{ij} - \bar{y}_{..})^2$		

where

SS = sum of squares

MS = mean squares

E(MS) = expected MS

$\bar{y}_{..}$ = grand mean (averaged over time and subjects)

$\bar{y}_{i.}$ = subject mean ($i = 1, \ldots, N$)

$\bar{y}_{.j}$ = timepoint mean ($j = 1, \ldots, n$)

Tests of hypothesis are constructed as follows:

$$
\begin{aligned}
&H_S: \quad \sigma_\pi^2 = 0 && F_S = \frac{\text{MS}_S}{\text{MS}_R} \overset{H_S}{\sim} F_{N-1,(N-1)(n-1)} \\
&H_T: \quad \tau_1 = \tau_2 = \ldots = \tau_n = 0 && F_T = \frac{\text{MS}_T}{\text{MS}_R} \overset{H_T}{\sim} F_{n-1,(N-1)(n-1)}
\end{aligned}
\qquad (2.4)
$$

Testing the significance of the time effect is typically the focus, because we generally usually assume that $\sigma_\pi^2 > 0$ (*i.e.*, subjects have significant influence on their longitudinal data). To quantify the degree of the subject effect, the intraclass correlation (ICC) for this design describes the relative magnitude of σ_π^2, namely

$$\text{ICC} = \frac{\hat{\sigma}_{\pi}^2}{\hat{\sigma}_{\pi}^2 + \hat{\sigma}_e^2} \,. \tag{2.5}$$

As subjects' data are highly correlated, the ICC becomes large. Here, the variance parameters are estimated as

$$\hat{\sigma}_{\pi}^2 = (\text{MS}_S - \text{MS}_R) \;/\; n \tag{2.6}$$

and

$$\hat{\sigma}_e^2 = \text{MS}_R. \tag{2.7}$$

When $\text{MS}_S \leq \text{MS}_R$, then $\hat{\sigma}_{\pi}^2 = 0$. Notice that the ICC is akin to a R^2 statistic; it represents the proportion of (unexplained) variation due to subjects. In the present model, the term "unexplained" refers to the variation that is not explained by the time effect. More generally, in models with more independent variables or covariates, "unexplained" refers to variation not explained by the set of independent variables. Thus, the value of the ICC can vary depending on model covariates. As more and more of the between-subject variability is explained by model covariates, the value of the ICC decreases.

2.1.2 Decomposing the Time Effect

As mentioned, the testing of the time effect is typically the focus in the current design. However, the overall test of the null hypothesis of no difference over time,

$$H_T : \tau_1 = \tau_2 = \ldots = \tau_n = 0 \tag{2.8}$$

is a very global test. It tests whether there is any difference in the population means across time, namely,

$$H_T : \mu_1 = \mu_2 = \ldots = \mu_n. \tag{2.9}$$

For more specific time-related comparisons, it is useful to construct a set of $n-1$ contrasts $L_{j'}$ of the timepoint means as follows:

$$L_{j'} = \sum_{j=1}^{n} c_{j'j} \, \bar{y}_{.j} \,, \qquad j' = 1, \ldots, n-1, \tag{2.10}$$

where $c_{j'j}$ represent contrast coefficients. Specific examples of sets of contrast coefficients will be given subsequently. Note, for a given contrast $L_{j'}$, there is a restriction that the sum of these contrast coefficients equals 0 across the n timepoints,

$$\sum_{j=1}^{n} c_{j'j} = 0. \tag{2.11}$$

Tests of these contrasts ($H_{j'} : L_{j'} = 0$) can be obtained using

$$\text{MS}_{j'} = \text{SS}_{j'} = \frac{NL_{j'}^2}{\sum_{j=1}^{n} c_{j'j}^2} \tag{2.12}$$

in terms of either an F or t statistic:

$$F_{j'} = \frac{\text{MS}_{j'}}{\text{MS}_R} \overset{H_{j'}}{\sim} F_{1,(N-1)(n-1)} ,$$

$$t_{j'} = \frac{L_{j'}}{\sqrt{\text{MS}_R \left[\sum_{j=1}^{n} \frac{c_{j'j}^2}{N} \right]}} \overset{H_{j'}}{\sim} t_{(N-1)(n-1)}.$$

Note that these two tests are the same, the F statistic is simply the square of the t statistic.

A set of $n-1$ contrasts partitions the variation attributable to time (*i.e.*, differences across time) in terms of specific timepoint comparisons. If the set of $n-1$ contrasts are orthogonal (*i.e.*, independent of each other), then

$$\text{SS}_T = \sum_{j'=1}^{n-1} \text{SS}_{j'}, \tag{2.13}$$

and we have an independent partitioning of the variation due to time. If the set of contrasts are not orthogonal, then adding the contrast sums of squares together does not equal SS $_T$.

Selecting a set of $n-1$ contrasts clearly depends on the set of scientific questions that are of interest in a particular study. Below, we present several sets that are commonly (and perhaps not so commonly) applied in longitudinal models, and we indicate some of the characteristics of each set. Before selecting a particular set for a given analysis, the analyst should think carefully and match the set with the scientific aims of the study. By choosing a set of contrasts, one is choosing how changes in the dependent variable over time are modeled in an analysis. It should be noted, however, that the overall test of the time effect (*i.e.*, the F-test of $H_T : \tau_1 = \tau_2 = \ldots = \tau_n = 0$) is unaffected by the choice of contrasts (as long as there are $n-1$ contrasts in a set). It is the decomposition of this overall test that changes as the set of contrasts varies.

2.1.2.1 *Trend Analysis—Orthogonal Polynomial Contrasts* One approach to testing specific time-related contrasts is to characterize the $n-1$ time effects as $n-1$ orthogonal polynomials (see Bock [1975], Draper and Smith [1981], or Fleiss [1986]). For example, the $(n-1) \times n$ contrast matrix for $n = 4$ is

$$\boldsymbol{C} = \begin{bmatrix} -3 & -1 & 1 & 3 \\ 1 & -1 & -1 & 1 \\ -1 & 3 & -3 & 1 \end{bmatrix} \begin{matrix} \div\sqrt{20} & \text{linear} \\ \div\sqrt{4} & \text{quadratic} \\ \div\sqrt{20} & \text{cubic} \end{matrix} \tag{2.14}$$

for linear, quadratic, and cubic trend components. Here, the rows of the matrix indicate the $n-1$ polynomial contrasts, and the columns indicate the n timepoints. Thus, the values in a given row are the contrast coefficient values $c_{j'j}$ for a particular contrast $L_{j'}$. The division sign to the right of the matrix indicates that the elements of each row are to be divided by the indicated square root quantity for that row. These quantities are simply the sum of squares of the row elements, and so dividing the elements of the matrix by these yields polynomial contrasts (*i.e.*, linear, quadratic, and cubic) that are on the same scale and thus can be more directly compared to each other. As the name implies, orthogonal polynomials are orthogonal (*i.e.*, independent of each other). Additionally, they can be useful for determining (a) the "degree" of change across time and (b) the relative contribution of each polynomial component of the trend. While the above matrix is appropriate if the timepoints are equally spaced, it is possible to generalize orthogonal polynomials for unequally spaced timepoints. We will describe this in more detail in Chapter 5. Also, for some problems it is common to specify fewer than $n - 1$ contrasts, especially as n gets large, because higher-order polynomials beyond cubic, or so, can be difficult to interpret and hard to justify.

2.1.2.2 Change Relative to Baseline—Reference Cell Contrasts In some cases, we are less concerned with the form of the growth curves and are more concerned about testing whether any change has occurred whatsoever. In this case, we can construct contrasts for each timepoint relative to the first timepoint, presumably baseline, as follows (again, for the case of 4 timepoints):

$$C = \begin{bmatrix} -1 & 1 & 0 & 0 \\ -1 & 0 & 1 & 0 \\ -1 & 0 & 0 & 1 \end{bmatrix}. \tag{2.15}$$

Again, here the rows indicate the contrasts and the columns the timepoints, with the values indicating the contrast coefficient values. Denoting the four timepoints as T1, T2, T3, and T4, the above three contrasts represent the timepoint differences of T2 versus T1, T3 versus T1, and T4 versus T1, respectively. These contrasts are not orthogonal, and they are sometimes called simple contrasts. Although the above matrix has the first timepoint as the reference cell, of course any of the timepoints can be treated as the reference cell (*e.g.*, the last timepoint).

2.1.2.3 Consecutive Time Comparisons—Profile Contrasts In other cases, we may have interest in determining whether each consecutive timepoint is significantly different from the immediately previous timepoint. These contrasts are sometimes referred to as "profile contrasts," and are constructed as follows:

$$C = \begin{bmatrix} -1 & 1 & 0 & 0 \\ 0 & -1 & 1 & 0 \\ 0 & 0 & -1 & 1 \end{bmatrix} \tag{2.16}$$

to test T2-T1, T3-T2, T4-T3, respectively. Profile contrasts are not orthogonal, but are useful for identifying when change begins (and ends). Use of profile contrasts in repeated measures ANOVA and MANOVA models is sometimes called "profile analysis," as described in detail by Morrison [1976].

2.1.2.4 Contrasting Each Timepoint to the Mean of Subsequent Timepoints—Helmert Contrasts When interest is in contrasting each timepoint to the mean of all subsequent timepoints, we can construct Helmert contrasts [Bock, 1975] as follows:

$$C = \begin{bmatrix} 1 & -1/3 & -1/3 & -1/3 \\ 0 & 1 & -1/2 & -1/2 \\ 0 & 0 & 1 & -1 \end{bmatrix} \tag{2.17}$$

for T1 versus the average of T2, T3, and T4; T2 versus the average of T3 and T4; and T3 versus T4. Helmert contrasts are orthogonal, and are useful for "ordered" tests. They can also be reversed to compare each timepoint to the mean of the previous timepoints:

$$C = \begin{bmatrix} -1 & 1 & 0 & 0 \\ -1/2 & -1/2 & 1 & 0 \\ -1/3 & -1/3 & -1/3 & 1 \end{bmatrix} \tag{2.18}$$

for T2 versus T1; T3 versus the average of T1 and T2; and T4 versus the average of T1, T2, and T3. In this form, these contrasts are somewhat similar to profile contrasts, except that the reference is all prior timepoints rather than simply the single previous timepoint.

2.1.2.5 Contrasting Each Timepoint to the Mean of Others—Deviation Contrasts

When there is interest in contrasting each timepoint to the mean of all other timepoints, deviation contrasts can be computed as

$$C = \begin{bmatrix} 1 & -1/3 & -1/3 & -1/3 \\ -1/3 & 1 & -1/3 & -1/3 \\ -1/3 & -1/3 & 1 & -1/3 \end{bmatrix} \tag{2.19}$$

for T1 versus the average of T2, T3, and T4; T2 versus the average of T1, T3, and T4; and T3 versus the average of T1, T2, and T4. Here, all timepoints except the last one are compared to the average of the other timepoints. Which timepoint is excluded can, of course, be changed. Deviation contrasts are not orthogonal, but are useful for situations in which there is "vague prior knowledge" about the changes over time.

2.1.2.6 Multiple Comparisons

Although $n - 1$ multiple comparisons can often be specified *a priori*, we are nevertheless faced with making multiple comparisons that will potentially lead to elevated experiment-wise Type I error rates. The most conservative solution is to use the so-called Bonferroni corrected α level, which is given as $\alpha^* = \alpha/(n-1)$. Here, each test is evaluated at the modified α^* level to ensure that the experiment-wise Type I error rate is not inflated.

A far less conservative alternative is to apply the so-called Fisher protected test logic, in which each individual test is conducted at the α level, but the individual tests are only applied when the global test $H_T : \tau_1 = \tau_2 = \ldots = \tau_n = 0$ is rejected. For the special case of orthogonal polynomials, we can start with the highest-order polynomial and eliminate each degree polynomial in a backwards manner until we encounter the first significant one. In a simulation study, Hummel and Sligo [1971] support the use of Fisher protected tests if n is not too large. Also, from Rosner [1995] (page 319):

> "If a few linear contrasts, which have been specified in advance, are to be tested, then it may not be necessary to use a multiple-comparisons procedure, since if such procedures are used, there will be less power to detect differences for linear contrasts whose means are truly different from zero. Conversely, if many contrasts are to be tested, which have not been specified before looking at the data, then multiple-comparisons procedures may be useful in protecting against declaring too many significant differences."

Furthermore, some notable statisticians even argue against using any kind of adjustment for multiple comparisons [Cook and Farewell, 1996; Rothman, 1990; Saville, 1990]. As one can appreciate, it is not always clear whether to adjust or how to adjust for multiple comparisons. Additionally, there are many multiple comparison adjustment procedures to choose from (*e.g.*, see Westfall et al. [1999]). In this book we will generally use the aforementioned Fisher protected test logic, specifying the contrasts *a priori*. The reader should realize, though, that this is a "grey area" of statistics with many varying opinions.

2.2 MULTIPLE-SAMPLE REPEATED MEASURES ANOVA

In the multiple-sample case, the ANOVA model for repeated measurements is referred to as a "split-plots" ANOVA model. This is a common design in randomized clinical trials, where subjects are randomized to different treatment groups and followed across time. Here, with $h = 1, \ldots, s$ groups, $i = 1, \ldots, N_h$ subjects in group h (with $N = \sum_{h=1}^{s} N_h$), and $j = 1, \ldots, n$ timepoints, the ANOVA model is:

$$y_{hij} = \mu + \gamma_h + \tau_j + (\gamma\tau)_{hj} + \pi_{i(h)} + e_{hij}, \tag{2.20}$$

where

μ = grand mean,

γ_h = effect of group h $(\sum_h \gamma_h = 0)$,

τ_j = effect of time j $(\sum_j \tau_j = 0)$,

$(\gamma\tau)_{hj}$ = interaction effect of time j and group h $[\sum_h \sum_j (\gamma\tau)_{hj} = 0]$,

$\pi_{i(h)}$ = individual difference component for subject i nested in group h,

and e_{hij} = error for subject i in group h at time j.

The distributional assumptions for this model are the same as the previous randomized blocks ANOVA, namely,

$$\pi_{i(h)} \sim N(0, \sigma_\pi^2) \quad \text{and} \quad e_{hij} \sim N(0, \sigma_e^2),$$

which imply the same compound symmetry structure for $V(\boldsymbol{y}_i)$ as in (2.3). Also, as in the randomized blocks ANOVA, the model is a mixed model because subjects are considered random effects and group and time are considered fixed effects. The data are assumed to be balanced in terms of n (*i.e.*, timepoints), but not necessarily in terms N_h, (*i.e.*, group sample sizes). An example data layout is

Group	Subject	Timepoint 1	2	...	n
1	1	y_{111}	y_{112}	...	y_{11n}
1	2	y_{121}	y_{122}	...	y_{12n}
1	.	.	.	...	.
1	.	.	.	...	.
1	N_1	y_{1N_11}	y_{1N_12}	...	y_{1N_1n}
.	.	.	.	...	.
.	.	.	.	...	.
.	.	.	.	...	.
.	.	.	.	...	.
s	1	y_{s11}	y_{s12}	...	y_{s1n}
s	2	y_{s21}	y_{s22}	...	y_{s2n}
s	.	.	.	...	.
s	.	.	.	...	.
s	N_s	y_{sN_s1}	y_{sN_s2}	...	y_{sN_sn}

Here, subjects are nested within groups and crossed with the time factor. Thus, each subject is observed only within a single group, and each subject is observed once at each timepoint. The ANOVA table is as follows:

Source	df	SS	MS	E(MS)
Group	$s-1$	$\text{SS}_G = n\sum_{h=1}^{s} N_h\,(\bar{y}_{h..} - \bar{y}_{...})^2$	$\frac{\text{SS}_G}{s-1}$	$\sigma_e^2 + n\sigma_\pi^2 + D_G$
Time	$n-1$	$\text{SS}_T = N\sum_{j=1}^{n}(\bar{y}_{..j} - \bar{y}_{...})^2$	$\frac{\text{SS}_T}{n-1}$	$\sigma_e^2 + D_T$
Group × Time	$(s-1)$ $\times(n-1)$	$\text{SS}_{GT} = \sum_{h=1}^{s}\sum_{j=1}^{n} N_h\,(\bar{y}_{h.j} - \bar{y}_{h..} - \bar{y}_{..j} + \bar{y}_{...})^2$	$\frac{\text{SS}_{GT}}{(s-1)(n-1)}$	$\sigma_e^2 + D_{GT}$
Subjects in Grps	$N-s$	$\text{SS}_{S(G)} = n\sum_{h=1}^{s}\sum_{i=1}^{N_h}(\bar{y}_{hi.} - \bar{y}_{h..})^2$	$\frac{\text{SS}_{S(G)}}{N-s}$	$\sigma_e^2 + n\sigma_\pi^2$
Residual	$(N-s)$ $\times(n-1)$	$\text{SS}_R = \sum_{h=1}^{s}\sum_{i=1}^{N_h}\sum_{j=1}^{n}(y_{hij} - \bar{y}_{h.j} - \bar{y}_{hi.} + \bar{y}_{h..})^2$	$\frac{\text{SS}_R}{(N-s)(n-1)}$	σ_e^2
Total	$Nn-1$	$\text{SS}_y = \sum_{h=1}^{s}\sum_{i=1}^{N_h}\sum_{j=1}^{n}(y_{hij} - \bar{y}_{...})^2$		

Here, D_G, D_T, and D_{GT} represent differences among groups, timepoints, and group by time interaction, respectively. Also, in terms of notation, the bar (*i.e.*, $\bar{y}$) indicates averaging and the dot subscript indicates the unit(s) that the averaging is over. Thus,

$\bar{y}_{...}$ = average across groups, timepoints, and subjects,
$\bar{y}_{h..}$ = average for group h across timepoints and subjects,
$\bar{y}_{..j}$ = average for timepoint j across groups and subjects,
$\bar{y}_{hi.}$ = average for subject i in group h across timepoints,
$\bar{y}_{h.j}$ = average for group h at timepoint j across subjects.

2.2.1 Testing for Group by Time Interaction

The group by time interaction, which is typically the test of primary interest, is constructed as

$$H_{GT} : D_{GT} = 0, \qquad F_{GT} = \frac{\mathrm{MS}_{GT}}{\mathrm{MS}_R} \overset{H_{GT}}{\sim} F_{(s-1)(n-1),(N-s)(n-1)}. \tag{2.21}$$

If the null hypothesis of no group by time interaction is rejected, we conclude that (a) the between-group differences are not the same across time, (b) the between-group curves across time are not parallel, and (c) group and time effects are confounded with the interaction and cannot be separately tested (or estimated). In this case, there is no "one" overall group effect, because it varies across time. Likewise, there is no single overall time effect, because it varies by groups.

If the null hypothesis of no group by time interaction cannot be rejected, then tests of the main effects of time and group are, respectively,

$$H_T : \tau_1 = \tau_2 = \ldots = \tau_n = 0, \qquad F_T = \frac{\mathrm{MS}_T}{\mathrm{MS}_R} \overset{H_T}{\sim} F_{n-1,(N-s)(n-1)}, \tag{2.22}$$

$$H_G : \gamma_1 = \gamma_2 = \ldots = \gamma_s = 0, \qquad F_G = \frac{\mathrm{MS}_G}{\mathrm{MS}_{S(G)}} \overset{H_G}{\sim} F_{s-1,N-s}. \tag{2.23}$$

In this case, the main effects of time (H_T) and group (H_G) are separately and independently testable. Note that the correct denominator for the F test of the group effect is not the usual error term MS_R, but instead the subjects within groups mean squares $\mathrm{MS}_{S(G)}$.

2.2.2 Testing for Subject Effect

To test for the significance of random subject effects, we construct the following statistic:

$$H_{S(G)} : \sigma_\pi^2 = 0, \qquad F_{S(G)} = \frac{\mathrm{MS}_{S(G)}}{\mathrm{MS}_R} \overset{H_{S(G)}}{\sim} F_{N-s,(N-s)(n-1)}. \tag{2.24}$$

As in the randomized blocks ANOVA, we generally assume that $\sigma_\pi^2 > 0$, and estimate the intraclass correlation as

$$ICC = \hat{\sigma}_\pi^2 / (\hat{\sigma}_\pi^2 + \hat{\sigma}_e^2) \tag{2.25}$$

The intraclass correlation represents the proportion of (unexplained) variation in the dependent variable that is due to subjects. Here, "unexplained" refers to the variation not explained by the fixed effects of the model: group, time, and group by time. As the intraclass correlation approaches zero, we can conclude that there is little correlation among the repeated observations over time, and a traditional fixed-effects ANOVA model will suffice. However, this is rarely the case, it is much more common for the intraclass correlation to be moderately large for most longitudinal data.

2.2.3 Contrasts for Time Effects

As in the single-group case, it is advantageous to characterize time and group by time effects using time-related contrasts. These include the same sets of contrasts described earlier for single-group designs, namely, orthogonal polynomials, profile contrasts, Helmert contrasts etc. As before, it may be necessary to consider the multiple comparisons issue and utilize Fisher protected tests or Bonferroni correction for the $(n-1)(s-1)$ group by time contrasts (and for the $n-1$ time contrasts).

2.2.3.1 Orthogonal Polynomial Partition of SS An example orthogonal polynomial decomposition of the time effect for a four-timepoint design is given by

$$C = \begin{bmatrix} -3 & -1 & 1 & 3 \\ 1 & -1 & -1 & 1 \\ -1 & 3 & -3 & 1 \end{bmatrix} \begin{matrix} \div\sqrt{20} & \mathbf{c}_1, \\ \div\sqrt{4} & \mathbf{c}_2, \\ \div\sqrt{20} & \mathbf{c}_3. \end{matrix} \tag{2.26}$$

The decomposition of the Time sum of squares is given by

Time	df	SS
Linear	1	$\mathrm{SS}_{T_1} = N\,\mathbf{c}_1\,\bar{\mathbf{y}}_{..}\,\bar{\mathbf{y}}'_{..}\,\mathbf{c}'_1$
Quadratic	1	$\mathrm{SS}_{T_2} = N\,\mathbf{c}_2\,\bar{\mathbf{y}}_{..}\,\bar{\mathbf{y}}'_{..}\,\mathbf{c}'_2$
.	.	.
$(n-1)$th	1	$\mathrm{SS}_{T_{n-1}} = N\,\mathbf{c}_{n-1}\,\bar{\mathbf{y}}_{..}\,\bar{\mathbf{y}}'_{..}\,\mathbf{c}'_{n-1}$
	$n-1$	SS_T

where $\bar{\mathbf{y}}_{..}$ is the $n \times 1$ vector of timepoint means (*i.e.*, the averages at each timepoint, averaging over groups and subjects), and $\mathbf{c}_j$ is the $1 \times n$ vector of contrasts of order j. Corresponding F-statistics for each trend component are

$$\begin{aligned} F_{T_{n-1}} &= \mathrm{SS}_{T_{n-1}} \,/\, \mathrm{MS}_R, \\ &\cdot \quad \cdot \\ F_{T_1} &= \mathrm{SS}_{T_1} \,/\, \mathrm{MS}_R. \end{aligned}$$

Typically, we determine the polynomial of least degree by working backwards. Similarly, the Group by Time sum of squares is decomposed as

G × T	df	SS
Linear	$s-1$	$\mathrm{SS}_{GT_1} = \sum_h N_h \boldsymbol{c}_1 \bar{\boldsymbol{y}}_{h.} \bar{\boldsymbol{y}}_{h.}' \boldsymbol{c}_1' - \mathrm{SS}_{T_1}$
Quadratic	$s-1$	$\mathrm{SS}_{GT_2} = \sum_h N_h \boldsymbol{c}_2 \bar{\boldsymbol{y}}_{h.} \bar{\boldsymbol{y}}_{h.}' \boldsymbol{c}_2' - \mathrm{SS}_{T_2}$
.	.	.
$(n-1)$th	$s-1$	$\mathrm{SS}_{GT_{n-1}} = \sum_h N_h \boldsymbol{c}_{n-1} \bar{\boldsymbol{y}}_{h.} \bar{\boldsymbol{y}}_{h.}' \boldsymbol{c}_{n-1}' - \mathrm{SS}_{T_{n-1}}$
	$(s-1)(n-1)$	SS_{GT}

where $\bar{\boldsymbol{y}}_{h.}$ is the $n \times 1$ vector of timepoint means for group h, and $\boldsymbol{c}_j$ is the $1 \times n$ vector of contrasts of order j. F-statistics are given by

$$\begin{aligned} F_{GT_{n-1}} &= \frac{\mathrm{SS}_{GT_{n-1}} / (s-1)}{\mathrm{MS}_R}, \\ &\quad \cdots \\ &\quad \cdots \\ F_{GT_1} &= \frac{\mathrm{SS}_{GT_1} / (s-1)}{\mathrm{MS}_R}. \end{aligned}$$

Again, the order of the polynomial of least degree for the Group by Time interaction is determined working backwards from highest degree (most complex) to lowest degree (*i.e.*, linear).

2.2.4 Compound Symmetry and Sphericity

For both univariate randomized blocks and split-plot ANOVA models, the variance of the vector of responses is given by

$$V(\boldsymbol{y}_i) = \sigma_\pi^2 \mathbf{1}_n \mathbf{1}_n' + \sigma_e^2 \boldsymbol{I}_n. \tag{2.27}$$

As mentioned, the compound symmetry (CS) structure is given by

$$\begin{aligned} V(y_{ij}) &= \sigma_\pi^2 + \sigma_e^2 && \forall\, j, \\ C(y_{ij}, y_{ij'}) &= \sigma_\pi^2 && \forall\, j \text{ and } j' \;\; (j \neq j'), \end{aligned} \tag{2.28}$$

where $\forall$ is the mathematical symbol denoting "for all." CS implies that variances are equal across time and that the covariances are all equal. Likewise, the correlation between responses across timepoints, given by the intraclass correlation,

$$Corr(y_{ij}, y_{ij'}) = \sigma_\pi^2 \,/\, (\sigma_\pi^2 + \sigma_e^2), \tag{2.29}$$

is the same across all pairs of timepoints. The CS assumption is highly restrictive, and often unrealistic (especially as n gets large). CS is a special case of the more general situation, sphericity, under which F-tests for time-related terms from ANOVA models are valid. If sphericity holds, then these F-tests are valid, otherwise, if sphericity doesn't hold, then these F-tests are generally too liberal.

2.2.4.1 Sphericity Sphericity, sometimes called circularity, can be expressed in different ways. The most general is that all variances of all pairwise differences between variables are equal

$$\begin{aligned} V(y_{ij} - y_{ij'}) &= V(y_{ij}) + V(y_{ij'}) - 2C(y_{ij}, y_{ij'}) \\ &= \text{constant } \forall\, j \text{ and } j'. \end{aligned}$$

The variance–covariance structure of compound symmetry satisfies this condition because the variances are all the same, as are the covariances. More generally, Crowder and Hand [1990] (page 50) note:

> "MS ratios derived by the univariate approach follow exact F-distributions if and only if the covariance matrix of the orthonormal contrasts has equal variances and zero covariances."

This statement is equivalent to

$$\underset{(n-1)\times n}{\boldsymbol{C}} \quad \underset{n\times n}{V(\boldsymbol{y}_i)} \quad \underset{(n-1)\times n}{\boldsymbol{C}'} \quad = \quad \text{constant} \quad \underset{(n-1)\times(n-1)}{\boldsymbol{I}_{t-1}}, \tag{2.30}$$

where $\boldsymbol{C}$ is a matrix of orthonormal polynomials (*i.e.*, orthogonal polynomials with unit variance). Testing whether sphericity holds is therefore a test of whether the matrix product above results in the form on the right-hand side of the equality. A chi-square goodness-of-fit test of this was developed by Mauchly [1940] and is implemented in many standard statistical software packages. As Mauchly noted, this sphericity test has relatively low statistical power for small sample sizes. Alternatively, for large samples, the test is likely to be significant even though the effect on the F-test may be negligible. The sphericity test is sensitive to departures from normality and to the presence of outliers. As such, it should be used as a guide and not as a strict rule.

If the sphericity assumption is rejected, or deemed implausible, one can use a multivariate repeated measures analysis (MANOVA), which allows for general $V(\boldsymbol{y}_i)$ but does not allow for any missing data across time. In the following chapter we discuss the MANOVA approach to analysis of longitudinal data. Other classical statistical alternatives include adjusted univariate F-tests as described by Greenhouse and Geisser [1959] and by Huynh and Feldt [1976], both of which are generally overly conservative.

2.3 ILLUSTRATION

Bock [1975] presents data on vocabulary growth measured in a cohort of
the University of Chicago Laboratory school. The longitudinal data consist of
measurements of the vocabulary section of the cooperative reading test [Davis, 19
Alternate forms of the test were administered in eighth through eleventh grade. Since this age range marks the period of time that physical growth begins to decelerate, he hypothesized that a similar deceleration might be observed in the acquisition of new vocabulary as well. Figure 2.1 displays a plot of the average scores versus grade and visually suggests that the rate of change is indeed decelerating.

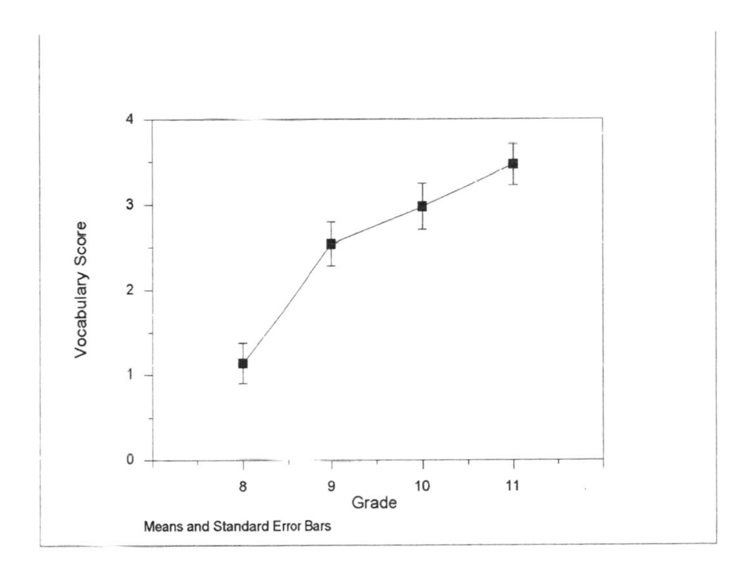

Figure 2.1. Average vocabulary scores of 64 students.

Summary statistics (means, standard deviations, and correlations) are presented in Table 2.1. Inspection of these summary statistics reveals that the variances and covariances are reasonably homogeneous over time, supporting the assumption of compound symmetry underlying the repeated measures ANOVA. Performing Mauchly's test yields a chi-square statistic of 6.32, on five degrees of freedom, which is not statistically significant. Thus, the assumption of sphericity, and therefore compound symmetry, is reasonable for these data.

Table 2.1. Means, Standard Deviations, and Correlations for the Vocabulary-Growth Data

Grade	Mean	SD	Correlations			
8	1.137	1.889	1.000			
9	2.542	2.085	.810	1.000		
10	2.988	2.169	.868	.785	1.000	
11	3.472	1.925	.785	.757	.811	1.000

This example can be used to nicely illustrate the simplest case of a one-sample repeated measures ANOVA. In the model, subject represents a random effect, and time represents a fixed effect. The ANOVA is illustrated in Table 2.2, which presents the ANOVA results.

Table 2.2. Repeated Measures ANOVA Results for the Vocabulary-Growth Data

Source	df	SS	MS	F	$p <$
Subjects	63	$SS_S = 873.60$	13.87	16.91	.0001
Grade (*i.e.*, "Time")	3	$SS_T = 194.34$	64.78	79.02	.0001
Residual	189	$SS_R = 154.94$	0.82		
Total	255	$SS_y = 1{,}222.88$			

The estimate of the error variance, which equals MS_R, is

$$\hat{\sigma}_e^2 = \frac{SS_R}{(N-1)(n-1)} = \frac{154.94}{189} = 0.82.$$

The estimate of the subject variance is gotten as

$$\hat{\sigma}_\pi^2 = \frac{MS_S - MS_R}{n} = \frac{13.87 - 0.82}{4} = 3.26,$$

and the intraclass correlation equals

$$ICC = \frac{\hat{\sigma}_\pi^2}{\hat{\sigma}_\pi^2 + \hat{\sigma}_e^2} = \frac{3.26}{3.26 + 0.82} = .80.$$

Thus, as one would expect, there is a tremendous effect of subjects on their vocabulary scores. 80% of the variation in vocabulary, that is not explained by grade, is attributable to subjects.

In terms of the grade effect, the ANOVA table reveals that we must reject the null hypothesis of no grade effect. This is clearly supported by Figure 2.1, which shows that vocabulary generally increases with grade. However, to obtain a more sensitive analysis, we can examine the significance of the individual polynomial terms based on the 4-timepoint orthogonal polynomial matrix:

$$\boldsymbol{C} = \begin{bmatrix} -.67082 & -.22361 & .22361 & .67082 \\ .5 & -.5 & -.5 & .5 \\ -.22361 & .67082 & -.67082 & .22361 \end{bmatrix}.$$

Premultiplying the 4×1 vector of column means

$$\bar{y} = \begin{bmatrix} 1.14 \\ 2.54 \\ 2.99 \\ 3.47 \end{bmatrix},$$

by C, we obtain the following orthogonal estimates:

$$\text{Linear} = 1.67, \quad \text{Quadratic} = -0.46, \quad \text{Cubic} = 0.22.$$

Note that the squares of these estimates multiplied by 64 (*i.e.*, the number of subjects) are the numerators of the F-ratios for linear, quadratic, and cubic trends shown in Table 2.3. The denominator is $\hat{\sigma}_e^2 = 0.82$ and the denominator degrees of freedom equals 189 for these F-ratios. Clearly, the positive linear trend is highly significant. However, the significant quadratic term, coupled with its negative sign, reveals that the observed deceleration in the growth rate is statistically significant. Finally, the cubic term is only marginally significant suggesting that the deceleration reverses to some extent with increasing age (see Figure 2.1). Examining the trend estimates, we see that the cubic estimate also pales in comparison with the dominant linear and moderate quadratic trend components: There is clearly diminishing importance as the order of the polynomial is increased.

Table 2.3. Orthogonal Polynomial Decomposition of the Grade Effect

Source	df	SS	F	$p <$
Grade				
Linear	1	$SS_{T_1} = 177.58$	216.63	.0001
Quadratic	1	$SS_{T_2} = 13.58$	16.56	.0001
Cubic	1	$SS_{T_3} = 3.17$	3.86	.051

Taken together, these results indicate a decelerating positive trend across grade, supporting the notion that vocabulary acquisition is slowing down as students approach maturity.

2.4 SUMMARY

In summary, the ANOVA approach to analysis of longitudinal data represents a well-understood and well-developed statistical methodology. In addition, there is considerable available computer software for ANOVA computation. The results are based on relatively simple and noniterative calculations. Unfortunately, the ANOVA model for repeated measurements assumes sphericity, which is unrealistic for most applications where variances tend to increase with time and correlation decreases with increasing intervals in time. Other limitations include limited treatment of missing data, and the requirement that all subjects are measured on the same occasions.

CHAPTER 3

MANOVA APPROACHES TO LONGITUDINAL DATA

As discussed in the previous chapter, there are two classical approaches to the analysis of longitudinal data: The first is variably called univariate mixed-model, split-plot, or repeated-measures ANOVA, and the second is based on multivariate ANOVA (MANOVA). In this chapter, we discuss the general multivariate growth curve model. The primary advantage of the MANOVA approach versus the ANOVA approach is that the MANOVA assumes a general form for the correlation of repeated measurements over time, whereas the ANOVA assumes the much more restrictive compound-symmetric form. The disadvantage of the MANOVA model is that it requires complete data. Subjects with incomplete data must be removed from the analysis, leading to potential bias, or have their missing values imputed in some way. In addition, both MANOVA and ANOVA models focus on comparison of group means and provide no information regarding subject-specific growth curves. Finally, both ANOVA and MANOVA models require that the timepoints are fixed across subjects (either evenly or unevenly spaced) and are treated as a classification variable in the ANOVA or MANOVA model. This precludes analysis of unbalanced designs in which different subjects are measured on different occasions. Finally, the MANOVA approach precludes use of time-varying covariates that are often essential to modeling dynamic relationships between predictors and outcomes.

In the following sections, we present the MANOVA model drawing upon the description in Bock [1975]. This text has more details for the interested reader. While this model is no longer recommended for routine application (if at all), it is important in that it helps to fix ideas for the development of the more modern and advanced methods that are the primary focus of this book.

3.1 DATA LAYOUT FOR ANOVA VERSUS MANOVA

A primary distinction between application of the ANOVA and MANOVA models for repeated measures concerns the data arrangement. For the ANOVA approach, each subject's data across n timepoints consists of n different observations in the dataset. For example, for a study with three timepoints, we might have the following lines for the first two subjects:

Subject	Time	y
1	1	y_{11}
1	2	y_{12}
1	3	y_{13}
2	1	y_{21}
2	2	y_{22}
2	3	y_{23}

Here, subject 1 is observed at three timepoints (coded 1, 2, and 3), and the values of the time-varying dependent variable are given on consectutive lines. There is one dependent variable y and two variables indicating subjects and time. This data layout is sometimes called the "univariate setup." Of course, there might be more variables in the datafile, but the essential point is that the repeated measures are considered to be multiple instances of the one dependent variable y.

Alternatively, in the "multivariate setup," used in MANOVA, the dependent variable across n timepoints consists of n different variables. Continuing with the same example, we would have

Subject	$y1$	$y2$	$y3$
1	$y1_1$	$y2_1$	$y3_1$
2	$y1_2$	$y2_2$	$y3_2$

Notice that the time variable is not explicitly provided, though it is implicit in the number of repeated measures (*e.g.*, here it is three for $y1$, $y2$, and $y3$). This data setup provides a clue as to why MANOVA does not include subjects with incomplete data across time. Essentially, MANOVA treats the repeated measures as one data vector (*i.e.*, one multivariate observation of the dependent variable), and the entire data vector must be complete for the subject to be included in the analysis.

Because these two models require different data setups, it is useful to be able to translate datasets between these two formats. Below is SAS code, adapted from Littell et al. [1991], for reading in a dataset in multivariate form and translating it into univariate form. This code reads in two subjects with dependent variable values of 1, 2, and 3 for the first subject, and values of 4, 5, and 6 for the second subject. All SAS-specific syntax is indicated in upper-case letters, optional user-specified names are indicated in lower case. The multivariate dataset, named `multdat`, contains three variables: `y1`, `y2`, and `y3`. The univariate dataset, named `unidat`, contains three variables `subject`, `time`, and `y`. The code below utilizes an `ARRAY` and a `DO` loop to convert the data to the univariate setup.

The array yv is designated to contain the three variables y1, y2, and y3. The subject assignment statement creates the subject variable which is incremented (by 1) for each subsequent line in the multivariate dataset. The DO loop creates the time variable, taking on values of 1, 2, and 3. The assignment statement, within the DO loop, assigns the values of the variables y1, y2, and y3 to the univariate y. Following each assignment in the loop, the OUTPUT statement creates a new observation.

```
DATA multdat;
INPUT y1 y2 y3;
DATALINES;
1 2 3
4 5 6
;
DATA unidat;
SET multdat;
ARRAY yv(3) y1 y2 y3;
subject + 1;
DO time = 1 TO 3;
    y = yv(time);
    OUTPUT;
END;
DROP y1 y2 y3;
RUN;
```

To go in the opposite direction, the SAS code below reads in the same dataset in univariate form and then translates it to multivariate form. Again, an ARRAY and a DO loop are used in the conversion. Note the different placement of the SET statement below, relative to the code above, and also the difference in the statement, within the DO loop, assigning the dependent variable values of the univariate y to the multivariate y1, y2, and y3.

```
DATA unidat;
INPUT subject time y;
DATALINES;
1 1 1
1 2 2
1 3 3
2 1 4
2 2 5
2 3 6
;
DATA multdat;
ARRAY yv(3) y1 y2 y3;
DO time = 1 TO 3;
    SET unidat;
    yv(time) = y;
END;
DROP y subject time;
RUN;
```

3.2 MANOVA FOR REPEATED MEASUREMENTS

In the MANOVA approach to analysis of repeated measurements, the n repeated measures are treated as a $n \times 1$ response vector $\boldsymbol{y}_i$. Due to the multivariate nature of the analysis, subjects with any missing y_{ij} (across time) are omitted from the analysis. This is the "Achilles heel" of the MANOVA model for repeated measurements and why it is largely of only limited use in many research fields. The one-sample MANOVA model is given by

$$\boldsymbol{y}_i = \boldsymbol{\mu} + \boldsymbol{\varepsilon}_i, \tag{3.1}$$

where $\boldsymbol{\mu} = n \times 1$ mean vector for timepoints, and $\boldsymbol{\varepsilon}_i = n \times 1$ vector of errors, distributed as $N(\mathbf{0}, \boldsymbol{\Sigma})$ in the population. Unlike the univariate repeated measures model of the last chapter, this specification indicates that the error variance–covariance matrix is allowed to be completely general. For the one-sample case, we can characterize the timepoint vector $\boldsymbol{\mu}$ and choose contrasts depending on the structure and hypotheses of interest. Below, we will present the model using orthogonal polynomial contrasts, but other sets of contrasts, like those described in the previous chapter, can be used instead.

Notice that under the univariate model assumptions, the variance–covariance matrix is $\boldsymbol{\Sigma} = \sigma_\pi^2 \mathbf{1}_n \mathbf{1}_n' + \sigma_e^2 \boldsymbol{I}_n$ and the mean vector is $\boldsymbol{\mu} = \mu + \boldsymbol{\tau}$, which are the grand mean and time effects in the univariate model notation. As a result, as we shall see, all univariate results can be extracted from the multivariate model.

3.2.1 Growth Curve Analysis—Polynomial Representation

The mean vector of the polynomial growth-curve model can be characterized as

$$\begin{bmatrix} \mu_1 \\ \mu_2 \\ \cdot \\ \cdot \\ \mu_n \end{bmatrix} = \begin{bmatrix} 1 \\ 1 \\ \cdot \\ \cdot \\ 1 \end{bmatrix} \beta_0 + \begin{bmatrix} t_1 \\ t_2 \\ \cdot \\ \cdot \\ t_n \end{bmatrix} \beta_1 + \begin{bmatrix} t_1^2 \\ t_2^2 \\ \cdot \\ \cdot \\ t_n^2 \end{bmatrix} \beta_2 + \cdots + \begin{bmatrix} t_1^{q-1} \\ t_2^{q-1} \\ \cdot \\ \cdot \\ t_n^{q-1} \end{bmatrix} \beta_{q-1},$$

$$\underset{n\times 1}{\boldsymbol{\mu}} = \underset{n\times q}{\boldsymbol{T}'} \; \underset{q\times 1}{\boldsymbol{\beta}},$$

where $t_1, t_2, \ldots, t_n$ represent timepoint values, and $q \leq n$ represents the degree of the polynomial. It is generally advantageous to orthogonalize $\boldsymbol{T}$ as $\boldsymbol{\mu} = \boldsymbol{P}'\boldsymbol{\theta}$, where $\boldsymbol{P}$ is the $q \times n$ matrix of orthogonal polynomials. The first row of $\boldsymbol{P}$ is for the constant term, and the remaining rows correspond sequentially to the linear, quadratic, etc. Note that $\boldsymbol{P} = \boldsymbol{S}^{-1}\boldsymbol{T}$ and $\boldsymbol{SS}' = (\boldsymbol{TT}')$, which is the Cholesky factorization, where $\boldsymbol{S}$ is a $q \times q$ lower triangular matrix. For equal time intervals, these orthogonal polynomial contrasts can be found in several statistics texts—for example, in Pearson and Hartley [1976].

Alternatively, the SAS PROC IML statements below can be used to produce $\boldsymbol{P}$ based on $\boldsymbol{T}$, here considering four timepoints. Note that in the code below, $\boldsymbol{T}$ is named `time` and $\boldsymbol{P}$ is named `orthpoly`.

```
TITLE 'producing orthogonal polynomial matrix';
PROC IML;
 time = {  1  1  1  1   ,
           0  1  2  3   ,
           0  1  4  9   ,
           0  1  8  27  } ;
orthpoly =T(INV(ROOT(time*T(time))))*time;
PRINT 'time matrix', time [FORMAT=8.4];
PRINT 'orthogonalized time matrix', orthpoly [FORMAT=8.4];
```

This code uses several built-in SAS matrix routines: `T` which obtains the transpose of a matrix, `INV` which performs matrix inversion, and `ROOT` which yields the transpose of the Cholesky factor (*i.e.*, `ROOT` yields the upper triangular matrix $\boldsymbol{S}'$). Additionally, matrix multiplication is performed using the $*$ operator, which is also the ordinary scalar multiplication operator.

Using either the tabled contrast values, or those obtained using the SAS code above, yields the following values for $q = n = 4$:

$$\boldsymbol{P}_4 = \begin{bmatrix} 1 & 1 & 1 & 1 \\ -3 & -1 & 1 & 3 \\ 1 & -1 & -1 & 1 \\ -1 & 3 & -3 & 1 \end{bmatrix} \begin{array}{ll} \div\sqrt{4} & \boldsymbol{p}_0, \\ \div\sqrt{20} & \boldsymbol{p}_1, \\ \div\sqrt{4} & \boldsymbol{p}_2, \\ \div\sqrt{4} & \boldsymbol{p}_3. \end{array} \tag{3.2}$$

The division sign to the right of the matrix indicates that the elements of each row are to be divided by the indicated square root quantity for that row. These quantities are simply the sum of squares of the row elements, and so dividing the elements of the matrix by these yields polynomial contrasts (*i.e.*, $\boldsymbol{p}_0$ = constant, $\boldsymbol{p}_1$ = linear, $\boldsymbol{p}_2$ = quadratic, and $\boldsymbol{p}_3$ = cubic) that are on the same scale, and so can be more directly compared to each other. The orthogonal polynomial trend model is therefore

$$\begin{aligned} \boldsymbol{P}\boldsymbol{y}_i &= \boldsymbol{P}\boldsymbol{\mu} + \boldsymbol{P}\boldsymbol{\varepsilon}_i \\ &= \boldsymbol{\theta} + \boldsymbol{\varepsilon}_i^*, \end{aligned}$$

where

- $\boldsymbol{\theta} = n \times 1$ vector of transformed population means with its least squares estimate given by the transformed sample mean vector, namely $\hat{\boldsymbol{\theta}} = \boldsymbol{P}\bar{\boldsymbol{y}}_.$ (where $\bar{\boldsymbol{y}}_.$ is the $n \times 1$ vector of timepoint means), and
- $\boldsymbol{\varepsilon}_i^* \sim N(\boldsymbol{0}, \boldsymbol{\Sigma}^* = \boldsymbol{P}\boldsymbol{\Sigma}\boldsymbol{P}')$.

Note that the univariate ANOVA of the last chapter assumes that $\boldsymbol{P}\boldsymbol{\Sigma}\boldsymbol{P}'$ is diagonal with equal values below the first element (since $\boldsymbol{P}$, as defined above, includes the zero-order term, *i.e.*, the grand mean, as the first row). Specifically, the test of sphericity examines whether the lower $(n-1) \times (n-1)$ partition of the $n \times n$ matrix $\boldsymbol{P}\boldsymbol{\Sigma}\boldsymbol{P}'$ equals a constant on the diagonal and zero for all off-diagonal elements.

The MANOVA table, with SSCP denoting a sum of squares and cross-product matrix (or more simply a cross-product matrix), is given as

Source	df	SSCP $(n \times n)$	E(SSCP)
Time	1	$\text{SST}^* = N\boldsymbol{P}\bar{\boldsymbol{y}}_{.}\bar{\boldsymbol{y}}_{.}'\boldsymbol{P}'$	$\boldsymbol{P}\left[\Sigma + N\boldsymbol{\mu}\boldsymbol{\mu}'\right]\boldsymbol{P}'$
Residual	$N-1$	$\text{SSR}^* = \boldsymbol{P}\,\text{SSR}\,\boldsymbol{P}'$ $= \boldsymbol{P}(\boldsymbol{Y}'\boldsymbol{Y} - N\bar{\boldsymbol{y}}_{.}\bar{\boldsymbol{y}}_{.}')\boldsymbol{P}'$	$(N-1)\boldsymbol{P}\Sigma\boldsymbol{P}'$
Total	N	$\text{SSY}^* = \boldsymbol{P}\boldsymbol{Y}'\boldsymbol{Y}\boldsymbol{P}'$	

In the above table, $\boldsymbol{Y}$ is the $N \times n$ matrix of all data, and $\bar{\boldsymbol{y}}_{.}$ is the $n \times 1$ vector of timepoint means. The degrees of freedom for each $n \times n$ cross-product matrix reflects the amount of between-subjects information. Based on the above formulation, SST * has as its first diagonal element $Nn\bar{y}_{..}^2$, which is a function of the grand mean, while the remaining $n-1$ diagonal elements are the orthogonal polynomial decomposition of the Time SS (*i.e.*, $N\sum_{j=1}^{n}(\bar{y}_{.j} - \bar{y}_{..})^2$). SSR* has its first diagonal element as the subjects SS (*i.e.*, $n\sum_{i=1}^{N}(\bar{y}_{i.} - \bar{y}_{..})^2$), with the other $n-1$ diagonal elements as the orthogonal polynomial decomposition of Error (*i.e.*, Subject by Time) SS.

The orthogonal polynomial partition of sum of squares and products is

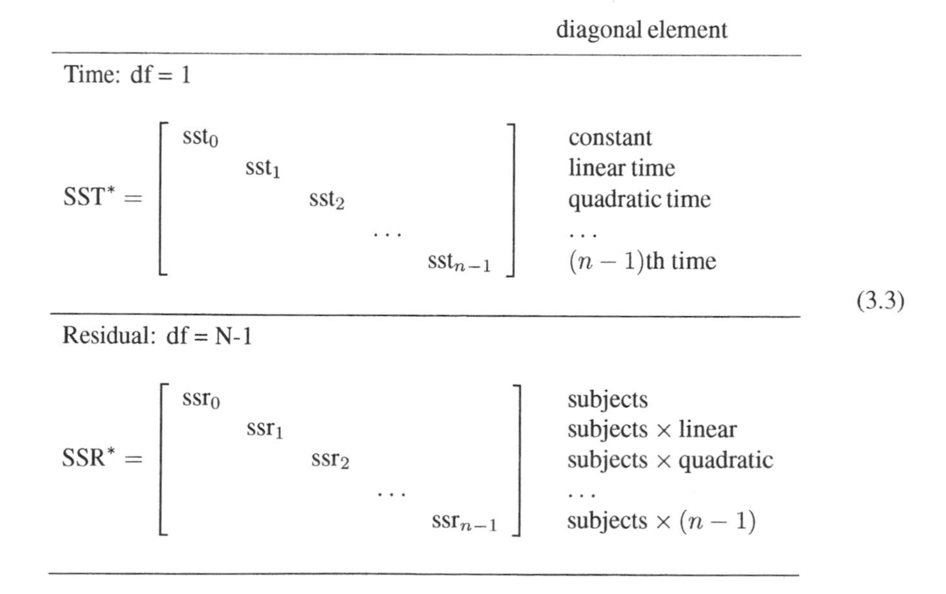

diagonal element

Time: df = 1

$$\text{SST}^* = \begin{bmatrix} \text{sst}_0 & & & & \\ & \text{sst}_1 & & & \\ & & \text{sst}_2 & & \\ & & & \cdots & \\ & & & & \text{sst}_{n-1} \end{bmatrix} \quad \begin{array}{l} \text{constant} \\ \text{linear time} \\ \text{quadratic time} \\ \cdots \\ (n-1)\text{th time} \end{array} \tag{3.3}$$

Residual: df = N-1

$$\text{SSR}^* = \begin{bmatrix} \text{ssr}_0 & & & & \\ & \text{ssr}_1 & & & \\ & & \text{ssr}_2 & & \\ & & & \cdots & \\ & & & & \text{ssr}_{n-1} \end{bmatrix} \quad \begin{array}{l} \text{subjects} \\ \text{subjects} \times \text{linear} \\ \text{subjects} \times \text{quadratic} \\ \cdots \\ \text{subjects} \times (n-1) \end{array}$$

Note that both of these are symmetric matrices, and they contain all of the information that is necessary either for extracting the univariate repeated measures ANOVA results (if sphericity is satisfied) or for the more general MANOVA model. Both approaches will now be described.

3.2.2 Extracting Univariate Repeated Measures ANOVA Results

If sphericity holds, then the univariate repeated measures results can be extracted directly from the SST* and SSR* matrices as follows:

Source	df	SS	MS	E(MS)
Subjects	$N-1$	$\text{SS}_S = n\sum_{i=1}^{N}(\bar{y}_{i.} - \bar{y}_{..})^2$	$\frac{\text{SS}_S}{N-1}$	$\sigma_e^2 + n\sigma_\pi^2$
Time	$n-1$	$\text{SS}_T = N\sum_{j=1}^{n}(\bar{y}_{.j} - \bar{y}_{..})^2$	$\frac{\text{SS}_T}{n-1}$	$\sigma_e^2 + N\sum(\tau_j - \tau_.)^2$
Residual	$(N-1) \times (n-1)$	$\text{SS}_R = \sum_{i=1}^{N}\sum_{j=1}^{n}(y_{ij} - y_{i.} - \bar{y}_{.j} + \bar{y}_{..})^2$	$\frac{\text{SS}_R}{(N-1)(n-1)}$	σ_e^2
Total	$Nn-1$	$\text{SS}_y = \sum_{i=1}^{N}\sum_{j=1}^{n}(y_{ij} - \bar{y}_{..})^2$		

where

$\text{SS}_S = \text{ssr}_0$ from the SSR* matrix,

SS_T = sum of the lower $n-1$ diagonal elements of SST*, and

SS_R = sum of the lower $n-1$ diagonal elements of SSR*.

Thus, these SS quantities are easily obtained from the SST* and SSR* matrices. For the F-tests, the MS quantities for Subjects and Time are divided by MS_R in the usual way. In terms of the residual mean square, notice that

$$\begin{aligned}\text{MS}_R &= \frac{\text{sum of the lower } n-1 \text{ diagonal elements of SSR}^*}{(N-1)(n-1)} \\ &= \frac{\text{average of the } n-1 \text{ SSR}^* \text{ diagonal elements}}{(N-1)}.\end{aligned}$$

As a result, the univariate F-test for Time, which use MS_R as the denominator, is sometimes referred to as an "averaged" test because of this averaging across the the $n-1$ elements of the SSR* matrix to yield a single error term. Notice that sphericity maintains that these $n-1$ elements are equal and uncorrelated, which justifies the averaging of these error variance elements. As a result of this averaging, the denominator degrees of freedom are greater for the univariate F-test of Time, relative to the MANOVA test which is subsequently described. This results in greater power for the univariate test, relative to the multivariate test, and is the primary reason why, if sphericity holds, there is generally an advantage to using the univariate repeated measures ANOVA model rather than the MANOVA model.

3.2.3 Multivariate Test of the Time Effect

In order to test the null hypothesis of no effect of time, *i.e.*, H_0 : $\boldsymbol{\mu}$ elements are all equal such that $H_0 : \boldsymbol{\tau} = \mathbf{0}$, we must extract and compare the lower $(n-1) \times (n-1)$ submatrices of SST* and SSR*. As indicated in the MANOVA table, these two submatrices have the same expectation if the null hypothesis is true. Thus, to the extent that there is a time effect, and the null hypothesis is not true, the time SST* submatrix will contain larger elements than the residual SSR* submatrix. Thus, the logic here is similar to a univariate F-test where we compare two mean squares (*i.e.*, MS_T / MS_R) to test the null hypothesis. Here, of course, we cannot form a simple ratio because we are dealing with matrices. Instead, to compare them, we can solve the determinantal equation:

$$| \, \text{SST}^*_{(n-1)} \; - \; \lambda \, \text{SSR}^*_{(n-1)}| \; = \; 0 \qquad (3.4)$$

which has one nonzero latent root or eigenvalue λ_1. Note that this latent root equals one if $\text{SST}^*_{(n-1)} = \text{SSR}^*_{(n-1)}$. Thus, to the extent that the null hypothesis is true, the latent root will be approximately equal to one.

The above equation is a two-matrix eigenproblem. It can be simplified to a one-matrix eigenproblem, or characteristic equation, using the Cholesky factorization, $\text{SSR}^*_{(n-1)} = \boldsymbol{E}\boldsymbol{E}'$, to yield

$$|\boldsymbol{E}^{-1} \, \text{SST}^*_{(n-1)}(\boldsymbol{E}^{-1})' \; - \; \lambda \boldsymbol{I}_{(n-1)}| \; = \; 0. \qquad (3.5)$$

Overall test statistics for the null hypothesis of no time effect include Roy's largest root statistic (the latent root or eigenvalue λ_1), and Wilk's Lambda ($\Lambda = 1/(1+\lambda_1)$). Functions of these test statistics approximately follow an F-distribution (under the null hypothesis), though sometimes interpolation is necessary, giving rise to fractional df in some cases. Other multivariate test statistics include the Hotelling–Lawley trace and the Pillai–Bartlett trace.

3.2.4 Tests of Specific Time Elements

To test specific components of the time effect, there are two options depending on whether sphericity is reasonable or not. First, if sphericity is reasonable and the univariate repeated measures ANOVA is used, then we can obtain univariate test statistics by extracting the lower $n-1$ diagonal elements of SST*, with MS_R as a common denominator for all trend components, *i.e.*,

$$\begin{aligned}
F_1 &= \frac{\text{sst}_1}{\sum_{j=1}^{n-1} \text{ssr}_j/[(n-1)(N-1)]} \quad \text{linear}, \\
F_2 &= \frac{\text{sst}_2}{\sum_{j=1}^{n-1} \text{ssr}_j/[(n-1)(N-1)]} \quad \text{quadratic}, \\
&\ldots \qquad\qquad\qquad\qquad . \\
F_{n-1} &= \frac{\text{sst}_{n-1}}{\sum_{j=1}^{n-1} \text{ssr}_j/[(n-1)(N-1)]} \quad (n-1)\text{th}.
\end{aligned} \qquad (3.6)$$

Note that the summation in the denominator is over the lower $n-1$ diagonal elements of the SSR* matrix (*i.e.*, it does not include ssr_0). Again, this denominator is akin to an averaged estimate of the error variance across time, and each of these F-statistics is evaluated with 1 numerator and $(n-1)(N-1)$ denominator degrees of freedom.

If sphericity is rejected and a MANOVA is performed, then the aforementioned multivariate test of the overall time effect, extracting and comparing the lower $n-1$ diagonal elements of SSM* and SSR*, is followed up with the following univariate F-tests of specific trend components:

$$\begin{aligned} F_1 &= \frac{\text{sst}_1}{\text{ssr}_1/(N-1)} && \text{linear,} \\ F_2 &= \frac{\text{sst}_2}{\text{ssr}_2/(N-1)} && \text{quadratic,} \\ &\cdot\ \cdot\ \cdot && \cdot \\ F_{n-1} &= \frac{\text{sst}_{n-1}}{\text{ssr}_{n-1}/(N-1)} && (n-1)\text{th.} \end{aligned} \tag{3.7}$$

Here, there is no pooling across time to obtain an averaged error term. Instead, each trend component has its own error term from the SSR * matrix, and so these F-statistics are evaluated with 1 numerator and only $(N-1)$ denominator degrees of freedom. This difference in the denominator degrees of freedom is why, in general, the univariate repeated measures model tests (*i.e.*, assuming sphericity) of the trend components in (3.6), which use the pooled error term, are more powerful than the corresponding tests under the MANOVA model in (3.7).

3.3 MANOVA OF REPEATED MEASURES—S SAMPLE CASE

In the case of multiple groups, let $h = 1, \ldots, s$ groups, $i = 1, \ldots, N_h$ subjects in group h, $j = 1, \ldots, n$ timepoints, and $N = \sum N_h$ total number of subjects. Thus, the number of subjects per group can vary, but each subject is measured at n timepoints. The model is written as

$$\boldsymbol{y}_{hi} = \boldsymbol{\mu} + \boldsymbol{\gamma}_h + \boldsymbol{\varepsilon}_{hi} \tag{3.8}$$

where

$\boldsymbol{\mu}$ is the $n \times 1$ vector of timepoint means,

$\boldsymbol{\gamma}_h$ is the $n \times 1$ vector effect for the population from which the hth group of subjects was drawn,

$\boldsymbol{\varepsilon}_{hi}$ is the $n \times 1$ vector of errors distributed as $N(\mathbf{0}, \boldsymbol{\Sigma})$ in each of the populations.

The model assumes homogeneity of variance–covariance across the s groups (*i.e.*, the same general error variance–covariance matrix $\boldsymbol{\Sigma}$ for all s groups). Again, with orthogonal transformation for the time effects, the model is written as

$$\boldsymbol{P}\boldsymbol{y}_{hi} = \boldsymbol{P}\boldsymbol{\mu} + \boldsymbol{P}\boldsymbol{\gamma}_h + \boldsymbol{P}\boldsymbol{\varepsilon}_{hi}, \tag{3.9}$$

with $\boldsymbol{\varepsilon}^*_{hi} \sim N(\mathbf{0}, \boldsymbol{\Sigma}^* = \boldsymbol{P}\boldsymbol{\Sigma}\boldsymbol{P}')$. As in the one-sample case, we can test $\boldsymbol{\Sigma}^*$ for sphericity and proceed using univariate "averaged" tests if appropriate, or the tests of the MANOVA model if sphericity is violated. The resulting MANOVA table is given by

Source	df	SSCP $(n \times n)$
Time	1	$\text{SST}^* = \boldsymbol{P}\,\text{SST}\,\boldsymbol{P}' = N\boldsymbol{P}\bar{\boldsymbol{y}}_{..}\bar{\boldsymbol{y}}'_{..}\boldsymbol{P}'$
Group	$s-1$	$\text{SSG}^* = \boldsymbol{P}\,\text{SSG}\,\boldsymbol{P}' = \boldsymbol{P}(\sum_h N_h\bar{\boldsymbol{y}}_{h.}\bar{\boldsymbol{y}}'_{h.} - \text{SST})\boldsymbol{P}'$
Residual	$N-s$	$\text{SSR}^* = \boldsymbol{P}\,\text{SSR}\,\boldsymbol{P}' = \boldsymbol{P}(\text{SSY} - \text{SSG} - \text{SST})\boldsymbol{P}'$
Total	$N = \sum N_h$	$\text{SSY}^* = \boldsymbol{P}\,\text{SSY}\,\boldsymbol{P}' = \boldsymbol{P}(\sum_h\sum_i \boldsymbol{y}_{hi}\boldsymbol{y}'_{hi})\boldsymbol{P}'$

where the degrees of freedom for each $n \times n$ cross-product matrix reflects the amount of between-subjects information. Note that the results only depend on the following summary statistics: (a) the cross-product matrix from the overall mean vector of the repeated measures $\bar{\boldsymbol{y}}_{..}\bar{\boldsymbol{y}}'_{..}$, (b) the sum of cross-product matrices from the group mean vectors of the repeated measures $\sum_h N_h\bar{\boldsymbol{y}}_{h.}\bar{\boldsymbol{y}}'_{h.}$, and (c) the sum of cross-product matrices from the subject data vectors of the repeated measures $\sum_h\sum_i \boldsymbol{y}_{hi}\boldsymbol{y}'_{hi}$.

Under the present orthogonal polynomial parameterization of the model, the information in the cross-product matrices is as follows.

$$\begin{array}{ll}
 & \text{diagonal element} \\
\hline
\text{Time: df} = 1 & \\
\text{SST}^* = \begin{bmatrix} \text{sst}_0 & & & & \\ & \text{sst}_1 & & & \\ & & \text{sst}_2 & & \\ & & & \cdots & \\ & & & & \text{sst}_{n-1} \end{bmatrix} & \begin{array}{l} \text{constant} \\ \text{linear time} \\ \text{quadratic time} \\ \cdots \\ (n-1)\text{th time} \end{array} \\
\hline
\text{Between groups: df} = \text{s-1} & \\
\text{SSG}^* = \begin{bmatrix} \text{ssg}_0 & & & & \\ & \text{ssg}_1 & & & \\ & & \text{ssg}_2 & & \\ & & & \cdots & \\ & & & & \text{ssg}_{n-1} \end{bmatrix} & \begin{array}{l} \text{groups} \\ \text{groups} \times \text{linear} \\ \text{groups} \times \text{quadratic} \\ \cdots \\ \text{groups} \times (n-1)\text{th time} \end{array} \\
\hline
\text{Subjects within groups: df} = \text{N-s} & \\
\text{SSR}^* = \begin{bmatrix} \text{ssr}_0 & & & & \\ & \text{ssr}_1 & & & \\ & & \text{ssr}_2 & & \\ & & & \cdots & \\ & & & & \text{ssr}_{n-1} \end{bmatrix} & \begin{array}{l} \text{subjects in groups} \\ \text{s(g)} \times \text{linear} \\ \text{s(g)} \times \text{quadratic} \\ \cdots \\ \text{s(g)} \times (n-1)\text{th time} \end{array} \\
\hline
\end{array} \tag{3.10}$$

All three are symmetric matrices, and they contain all of the information that is necessary either for extracting the univariate repeated measures ANOVA results (if sphericity is satisfied) or for the more general MANOVA tests.

3.3.1 Extracting Univariate Repeated Measures ANOVA Results

Given the results of the previous section, it is a simple matter to extract the univariate ANOVA results, as shown in the following.

			Source
SS Time	=	$\text{sst}_1 + \text{sst}_2 + \cdots + \text{sst}_{n-1}$	SST*
SS Group	=	ssg_0	SSG*
SS Group × Time	=	$\text{ssg}_1 + \text{ssg}_2 + \cdots + \text{ssg}_{n-1}$	SSG*
SS Subjects within Groups	=	ssr_0	SSR*
SS Residual	=	$\text{ssr}_1 + \text{ssr}_2 + \cdots + \text{ssr}_{n-1}$	SSR*

The F-tests are then obtained as

Source	df	MS	F
Time	$n-1$	$\text{MS}_T = \frac{\text{SS}_T}{(n-1)}$	MS_T / MS_R
Group	$s-1$	$\text{MS}_G = \frac{\text{SS}_G}{(n-1)}$	MS_T / $\text{MS}_{S(G)}$
Group × Time	$(s-1)(n-1)$	$\text{MS}_{GT} = \frac{\text{SS}_{GT}}{(s-1)(n-1)}$	MS_{GT} / MS_R
Subjects within Groups	$N-s$	$\text{MS}_{S(G)} = \frac{\text{SS}_{S(G)}}{(N-s)}$	$\text{MS}_{S(G)}$ / MS_R
Residual	$(N-s)(n-1)$	$\text{MS}_R = \frac{\text{SS}_R}{(N-s)(n-1)}$	

Testing of the individual trend components (*i.e.*, linear, quadratic, etc.) uses MS_R as the denominator error term as in the one-sample case. This is also the case for the individual group by time trend components (*i.e.*, group by linear, group by quadratic, etc.). Again, the tests of the time and group by time interaction, as well as the tests for the individual trend components within both, are termed averaged tests because they utilize an averaged estimate of error (*i.e.*, the common MS_R), which is appropriate only if sphericity holds.

3.3.2 Multivariate Tests

For the multivariate model, the overall group test is the same as in the univariate model. However, testing for the overall time and group by time effects both involve multivariate tests. The multivariate test of the group by time interaction is obtained by extracting the

$(n-1) \times (n-1)$ submatrices of SSG* and SSR*, and solving the determinantal equation for $\min(s-1, n-1)$ latent roots:

$$|\,\mathrm{SSG}^*_{(n-1)} \; - \; \boldsymbol{\lambda}\,\mathrm{SSR}^*_{(n-1)}| \;=\; 0. \tag{3.11}$$

Several test statistics are generally provided by standard statistical software for MANOVA: Roy's largest root statistic, Wilk's Lambda, Hotelling-Lawley Trace, and Pillai's Trace. Roy's largest root statistic is given by the first latent root or eigenvalue λ_1. Wilk's Lambda is computed as

$$\Lambda = \prod_{h=1}^{s-1} 1/(1+\lambda_h). \tag{3.12}$$

Note that these same tests can be constructed for $n-q-1$ terms if only a $q < n$ degree trend is considered (*e.g.*, only test for group by linear and quadratic trends even if $n > 3$). To do this, one extracts and compares the corresponding $n-q-1$ submatrices of the SSG* and SSR* matrices.

If the the overall group by time test is nonsignificant, we may want to pool the interaction into the residual for (multivariate) testing of the time effect. Unfortunately, despite the increase in statistical power, pooling is not easily accomplished with most MANOVA statistical software packages. Instead, the usual multivariate test of the time effect involves the same determinantal equation as in the one-sample case:

$$|\,\mathrm{SST}^*_{(n-1)} \; - \; \boldsymbol{\lambda}\,\mathrm{SSR}^*_{(n-1)}| \;=\; 0. \tag{3.13}$$

For both the group by time and time effects, testing for the significance of specific time-related components involves using a separate denominator for each, rather than the pooled MS_R that is used in the univariate model. For example, after the overall multivariate test of the group by time interaction, individual components are tested as

$$\begin{aligned}
F_{\mathrm{GT}_1} &= \frac{\mathrm{ssg}_1/(s-1)}{\mathrm{ssr}_1/(N-s)} && \text{group by linear} \\
F_{\mathrm{GT}_2} &= \frac{\mathrm{ssg}_2/(s-1)}{\mathrm{ssr}_2/(N-s)} && \text{group by quadratic} \\
&\;\;\cdot \quad \cdot \quad \cdot && \quad \cdot \\
F_{\mathrm{GT}_{n-1}} &= \frac{\mathrm{ssg}_{n-1}/(s-1)}{\mathrm{ssr}_{n-1}/(N-s)} && \text{group by } (n-1)\text{th}
\end{aligned} \tag{3.14}$$

each on $s-1$ and $N-s$ degrees of freedom.

3.4 ILLUSTRATION

To illustrate the multivariate approach, we return to the vocabulary growth data previously presented in Chapter 2 and in Bock [1975]. Summary statistics (vocabulary score means, standard deviations, and correlations between the grades) are presented in Table 2.1, and are displayed graphically in Figure 2.1. Pre-multiplying the vector mean

$$\bar{\boldsymbol{y}} = \begin{bmatrix} 1.14 \\ 2.54 \\ 2.99 \\ 3.47 \end{bmatrix},$$

by the orthogonal polynomial matrix

$$\boldsymbol{P} = \begin{bmatrix} .5 & .5 & .5 & .5 \\ -.67082 & -.22361 & .22361 & .67082 \\ .5 & -.5 & -.5 & .5 \\ .22361 & .67082 & -.67082 & .22361 \end{bmatrix},$$

yields the following orthogonal polynomial estimates.

Constant	5.0694
Linear	1.6658
Quadratic	−0.4606
Cubic	0.2224

The polynomial SST* matrix, presented in Table 3.1, is obtained by multiplying the squares and cross products of these coefficients by the number of subjects ($N = 64$). Similarly, the residual SSR* matrix in Table 3.1 is obtained by the matrix calculation:

$$\text{SSR}^* = \boldsymbol{P}(\boldsymbol{Y}'\boldsymbol{Y} - N\bar{\boldsymbol{y}}_{.}\bar{\boldsymbol{y}}_{.}')\boldsymbol{P}'.$$

As noted, these are referred to as sum of squares and cross-product (SSCP) matrices. In the present case, the SSR* matrix conforms to mixed-model assumptions in that the off-diagonal elements (of the lower three-by-three portion of this matrix) are small relative to their corresponding diagonal elements, and the last three diagonal elements are of the same magnitude. A test of this assumption (not shown) does not reject sphericity. However, for illustration, we will interpret the multivariate results for these data and compare them to the univariate results presented in Chapter 2.

To obtain the multivariate test for the grade effect, the determinantal equation

$$|\boldsymbol{E}^{-1}\,\text{SST}^*_{(n-1)}(\boldsymbol{E}^{-1})' - \lambda\boldsymbol{I}_{(n-1)}| = 0, \tag{3.15}$$

using the Cholesky factorization $\text{SSR}^*_{(n-1)} = \boldsymbol{E}\boldsymbol{E}'$, yields 4.7432 as Roy's largest root statistic (eigenvalue or latent root λ_1). Similarly, Wilks' Lambda equals $1/(1 + 4.7432) =$ .17412. It can be shown that these translate to a F-value of 96.45 (see Finn [1974]) with df = 3, 61, which is highly significant: $p < .0001$. Thus, there is clearly a grade effect on average vocabulary scores.

In terms of the individual trend components, results of the analysis are quite similar to the univariate results presented in Chapter 2, though the cubic term is even less significant here. As in the univariate analysis, both the linear and quadratic trend components are highly significant, and the above polynomial contrast estimates indicate a positive linear and negative quadratic effect. Vocabularly growth is increasing across these grades, but at a decelerating rate. As mentioned, the multivariate analysis uses separate denominators for forming these F-values for the trend components. For example, under the multivariate model the F-test for the linear trend component is calculated as

$$\text{Linear } F = 177.59/(50.42/63) = 221.88,$$

which is compared to the F-distribution with 1 numerator and 63 denominator degrees of freedom. Alternatively, under the univariate repeated measures ANOVA, this same test uses the pooled MS_R to yield

$$\text{Linear } F = 177.59/(((50.42 + 43.95 + 60.57)/3)/63) = 216.63,$$

which is compared to F with 1 numerator and 189 denominator degrees of freedom. Both are highly significant in this case, but the point is that the critical value is smaller under the univariate model assumptions because of the pooling of the error term.

The univariate sums of squares in Tables 2.2 and 2.3, may be obtained from Table 3.1 as follows: (a) The constant term is the first diagonal element of SST *, and the linear, quadratic, and cubic polynomial grade sums of squares are the remaining diagonal elements of SST*; (b) the subject sums of squares is the first diagonal element of SSR *, and (c) the residual sums of squares is the sum of the remaining three diagonal elements of SSR *.

Table 3.1. Multivariate Analysis of Variance of the Orthogonal Polynomial Transformed Vocabulary-Growth Data

Source	df	SSCP				F	p
Time SST*	1						
Constant		1644.71					
Linear		540.45	177.59			221.88	.0001
Quadratic		−149.45	−49.11	13.58		19.46	.0001
Cubic		72.16	23.71	−6.56	3.17	3.29	.075
Residual SSR*	63						
Between subjects		873.60					
Linear error		3.84	50.42				
Quadratic error		−49.82	12.05	43.95			
Cubic error		−23.76	−3.36	−4.27	60.57		

3.5 SUMMARY

In summary, ANOVA and MANOVA approaches for analysis of longitudinal data represent well-understood and well-developed statistical methodologies. In addition, there is considerable available computer software for their computation. The results are based on relatively simple and noniterative calculations. Both models, unfortunately, have features which limit their usage in longitudinal data analysis. The ANOVA model for repeated measurements assumes sphericity, which is unrealistic for many applications where variances tend to increase with time and correlation decreases with increasing intervals in time. Alternatively, while the MANOVA model allows for a general variance–covariance structure for the repeated measures, it has the disadvantage of requiring complete data for all subjects and identical measurement occasions. Unfortunately, this overly stringent requirement is violated in most cases. Furthermore, software implementations of the multivariate model often provide only limited ways of handling covariates. In the following chapters, we consider more general models that overcome the limitations of these traditional approaches.

CHAPTER 4

MIXED-EFFECTS REGRESSION MODELS FOR CONTINUOUS OUTCOMES

4.1 INTRODUCTION

The previous chapters have considered traditional analysis of variance methods for longitudinal data analysis. Unfortunately, these traditional methods are of limited use because of restrictive assumptions concerning missing data across time and the variance–covariance structure of the repeated measures. The univariate "mixed-model" analysis of variance assumes that the variances and covariances of the dependent variable across time are equal (*i.e.*, compound symmetry). Alternatively, the multivariate analysis of variance for repeated measures only includes subjects with complete data across time. Also, these procedures focus on estimation of group trends across time and provide little help in understanding about how specific individuals change across time. For these and other reasons, mixed-effects regression models (MRMs) have become popular for modeling longitudinal data.

Variants of MRMs have been developed and described under a variety of names: random-effects models [Laird and Ware, 1982], variance component models [Dempster et al., 1981], multilevel models [Goldstein, 1995], hierarchical linear models [Raudenbush and Bryk, 2002], two-stage models [Bock, 1989], random coefficient models [de Leeuw and Kreft, 1986], mixed models [Longford, 1987; Wolfinger, 1993], empirical Bayes models [Hui and Berger, 1983; Strenio et al., 1983], and random regression models [Bock, 1983a,b; Gibbons et al., 1988, 1993]. A basic characteristic of these models is the inclusion of random subject effects into regression models in order to account for the influence of subjects on their repeated observations. These random subject effects thus describe each person's trend across time, and explain the correlational structure of the longitudinal data.

Additionally, they indicate the degree of subject variation that exists in the population of subjects.

There are several features that make MRMs especially useful in longitudinal research. First, subjects are not assumed to be measured on the same number of timepoints, thus, subjects with incomplete data across time are included in the analysis. The ability to include subjects with incomplete data across time is an important advantage relative to procedures that require complete data across time because (a) by including all data, the analysis has increased statistical power, and (b) complete-case analysis may suffer from biases to the extent that subjects with complete data are not representative of the larger population of subjects. Because time is treated as a continuous variable in MRMs, subjects do not have to be measured at the same timepoints. This is useful for analysis of longitudinal studies where follow-up times are not uniform across all subjects. Both time-invariant and time-varying covariates can be included in the model. Thus, changes in the outcome variable may be due to both stable characteristics of the subject (*e.g.*, their gender or race) as well as characteristics that change across time (*e.g.*, life-events). Finally, whereas traditional approaches estimate average change (across time) in a population, MRM can also estimate change for each subject. These estimates of individual change across time can be particularly useful in longitudinal studies where a proportion of subjects exhibit change across time that deviates from the average trend.

Applications of MRMs are steadily increasing and can be found in many different fields, including studies on alcohol [Curran et al., 1997], smoking [Niaura et al., 2002], HIV/AIDS [Gallagher et al., 1997], drug abuse [Carroll et al., 1994; Halikas et al., 1997], psychiatry [Elkin et al., 1995; Serretti et al., 2000], and child development [Huttenlocher et al., 1991; Campbell and Hedeker, 2001], to name a few. Not only do these articles illustrate the wide applicability of MRMs, they also give a sense of how MRM results are typically reported in the various literatures.

This chapter will focus on describing MRMs for continuous outcomes in a very practical way. We will first illustrate how MRMs can be seen as an extension of an ordinary linear regression model. Starting with a simple linear regression model, the model will slowly be extended and described, in order to guide the reader going from familiar to less familiar territory. Following the descriptions of the statistical models, several MRM analyses will be presented using a longitudinal psychiatric dataset. These analyses will illustrate many of the key features of MRMs for longitudinal data analysis. For further illustration, readers are reminded that they can download the dataset and program files to replicate the analyses in this chapter.[1]

4.2 A SIMPLE LINEAR REGRESSION MODEL

To introduce MRMs, consider a simple linear regression model for the measurement y of individual i ($i = 1, 2, \ldots, N$ subjects) on occasion j ($j = 1, 2, \ldots, n_i$ occasions):

$$y_{ij} = \beta_0 + \beta_1 t_{ij} + \varepsilon_{ij}. \tag{4.1}$$

Ignoring subscripts, this model represents the regression of the outcome variable y on the independent variable time (denoted t). The subscripts keep track of the particulars of the data, namely whose observation it is (subscript i) and when was this observation made

[1] http://www.uic.edu/~hedeker/long.html

(the subscript j). The independent variable t gives a value to the level of time, and may represent time in weeks, months, etc. Since y and t carry both i and j subscripts, both the outcome variable and the time variable are allowed to vary by individuals and occasions.

In linear regression models, like (4.1), the errors ε_{ij} are assumed to be normally and *independently* distributed in the population with zero mean and common variance σ^2. This independence assumption makes the model given in equation (4.1) an unreasonable one for longitudinal data. This is because the outcomes y are observed repeatedly from the same individuals, and so it is much more reasonable to assume that errors within an individual are correlated to some degree. Furthermore, the above model posits that the change across time is the same for all individuals since the model parameters (β_0, the intercept or initial level, and β_1, the linear change across time) do not vary by individuals. For both of these reasons, it is useful to add individual-specific effects into the model that will account for the data dependency and describe differential time trends for different individuals. This is precisely what MRMs do. The essential point is that MRMs therefore can be viewed as augmented linear regression models.

4.3 RANDOM INTERCEPT MRM

A simple extension of the regression model given in (4.1) to allow for the influence of each individual on their repeated outcomes is provided by

$$y_{ij} = \beta_0 + \beta_1 t_{ij} + \upsilon_{0i} + \varepsilon_{ij}, \tag{4.2}$$

where υ_{0i} represents the influence of individual i on his/her repeated observations. Notice that if individuals have no influence on their repeated outcomes, then all of the υ_{0i} terms would equal 0. However, it is more likely that subjects will have positive or negative influences on their longitudinal data, and so the υ_{0i} terms will deviate from 0.

To better reflect how this model characterizes an individual's influence on their observations, it is helpful to represent the model in a hierarchical or multilevel form [Goldstein, 1995; Raudenbush and Bryk, 2002]. For this, it is partitioned into the following within-subjects (or level-1) model,

$$y_{ij} = b_{0i} + b_{1i} t_{ij} + \varepsilon_{ij}, \tag{4.3}$$

and between-subjects (or level-2) model,

$$\begin{aligned} b_{0i} &= \beta_0 + \upsilon_{0i}, \\ b_{1i} &= \beta_1. \end{aligned} \tag{4.4}$$

Here, the level-1 model indicates that individual i's response at time j is influenced by his/her initial level b_{0i} and time trend, or slope, b_{1i}. The level-2 model indicates that individual i's initial level is determined by the population initial level β_0, plus a unique contribution for that individual υ_{0i}. Thus, each individual has their own distinct initial level. Conversely, the present model indicates that each individual's slope is the same; all are equal to the population slope β_1. Another way to think about it is that each person's trend line is parallel to the population trend determined by β_0 and β_1. The difference between each individual's trend and the population trend is v_{0i}, which is constant across time.

The between-subjects, or level-2, model is sometimes referred to as a "slopes as outcomes" model [Burstein, 1980]. The hierarchical representation shows that just as within-subjects (level-1) covariates can be included in the model to explain variation in level-1 outcomes (y_{ij}), between-subjects (level-2) covariates can be included to explain variation in level-2 outcomes (the subject's intercept b_{0i} and slope b_{1i}). Note that combining the within- and between–subjects models (4.3) and (4.4) yields the previous single-equation model (4.2).

Since individuals in a sample are typically thought to be representative of a larger population of individuals, the individual-specific effects v_{0i} are treated as random effects. That is, v_{0i} are considered to be representative of a distribution of individual effects in the population. The most common form for this population distribution is the normal distribution with mean 0 and variance σ_v^2. In the model given by equation (4.2), the errors ε_{ij} are now assumed to be normally and *conditionally independently* distributed in the population with zero mean and common variance σ^2. Conditional independence here means conditional on the random individual-specific effects v_{0i}. Since the errors now have an influence due to individuals removed from them, this conditional independence assumption is much more reasonable than the ordinary independence assumption associated with (4.1).

As mentioned, individuals deviate from the regression of y on t in a parallel manner in this model (since there is only one subject effect v_{0i}). Thus, it is sometimes referred to as a random-intercept model, with each v_{0i} indicating how individual i deviates from the population trend. Figure 4.1 represents this model graphically.

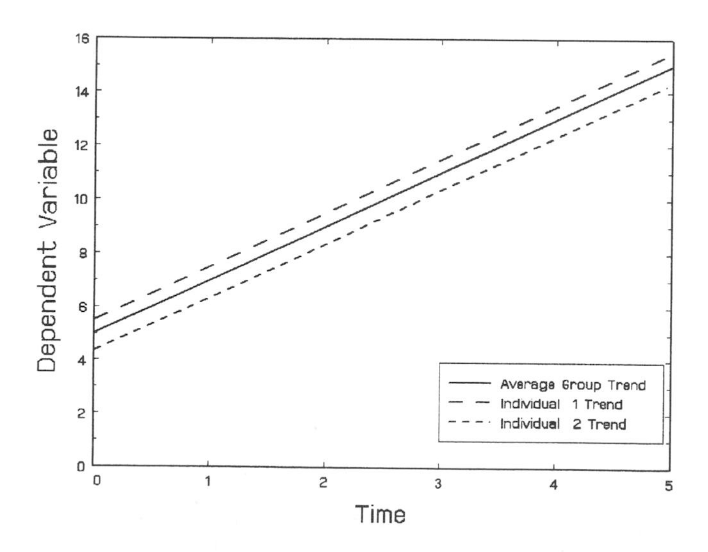

Figure 4.1. Random-intercept MRM.

In this figure, the solid line represents the population average trend, which is based on β_0 and β_1. Also depicted are two individual trends, one below and one above the population (average) trend. For a given sample there are N such lines, one for each individual. The variance term σ_v^2 represents the spread of these lines. If σ_v^2 is near-zero, then the individual lines would not deviate much from the population trend. In this case, individuals do not exhibit much heterogeneity in their change across time. Alternatively as individuals differ from the population trend, the lines move away from the population trend line and σ_v^2 increases. In this case, there is more individual heterogeneity in time trends.

4.3.1 Incomplete Data Across Time

The occasions range from $j = 1$ to n_i in the model specification, with each person being measured on n_i timepoints. Since n carries the i subscript, each subject may vary in terms of the number of measured occasions. Furthermore, there are no restrictions on the number of observations per individual, subjects who are missing at a given timepoint are not excluded from the analysis. Also, since the time variable t carries the i subscript, subjects can be measured on different occasions. It is even possible that there are no "common" timepoints or "waves" for measurements; each individual could be measured on an individualized schedule. The underlying assumption of the model is that the data that are available for a given individual are representative of how that individual deviates from the population trend across the timeframe of the study.

Chapter 14 will discuss missing data issues more thoroughly. For now, a few points are worth mentioning. As Laird [1988] points out, MRMs for longitudinal data using maximum likelihood (ML) estimation provide valid statistical tests in the presence of ignorable nonresponse. By ignorable nonresponse, it is meant that the probability of nonresponse can depend on observed covariates (*e.g.*, time) *and* observed values of the dependent variable from the subjects with missing data. The notion here is that if missingness is related to observed performance (*i.e.*, observed values of the dependent variable), in addition to other observable subject characteristics (*i.e.*, observed covariates), then MRMs provide valid statistical inferences for the model parameters. This is a very useful result, because many instances of missing data can be assumed to be related to observed performance or other subject characteristics. Thus, MRMs provide an attractive method for dealing with incomplete longitudinal data.

In considering missing data and whether they are ignorable or not, a related issue is the distinction between attrition (*i.e.*, subjects dropping out of the study and not returning) and sporadic or intermittent missing data (*i.e.*, subjects with missing data between observed timepoints). It may very well be that these arise from distinct processes and so, for example, it might be more plausible to assume ignorable missingness for intermittent missing data rather than attrition. Chapter 14 will discuss these issues in greater detail and also describe methods of analysis not requiring ignorable nonresponse.

4.3.2 Compound Symmetry and Intraclass Correlation

The random intercept model implies a compound symmetry assumption for the variances and covariances of the longitudinal data. That is, both the variances and covariances across time are assumed to be the same, namely,

$$\begin{aligned} V(y_{ij}) &= \sigma_v^2 + \sigma^2, \\ C(y_{ij}, y_{ij'}) &= \sigma_v^2, \qquad \text{where } j \neq j'. \end{aligned} \tag{4.5}$$

Expressing the covariance as a correlation yields the *intraclass correlation*, which is the ratio of the individual variance σ_v^2 to the total variance $\sigma^2 + \sigma_v^2$. This coefficient represents the degree of association of the longitudinal data within subjects, and specifically indicates the proportion of variance in the data attributable to individuals.

As an important caveat, it should be noted that a random intercept model that includes autocorrelated errors, as described in Chapter 7, provides a variance–covariance structure that is more general than the above compound symmetry structure. So, to be precise, it is a random intercept model with independent errors that implies compound symmetry.

4.3.3 Inference

Hypothesis testing for the fixed-effects parameters (*i.e.*, $\boldsymbol{\beta}$) generally involves the so-called "Wald test" [Wald, 1943], which uses the ratio of parameter estimate to its standard error to determine statistical significance. These test statistics (*i.e.*, Z = ratio of the parameter estimate to its standard error) are compared to a standard normal frequency table to test the null hypothesis that the parameter equals 0. Alternatively, these Z-statistics are sometimes squared, in which case the resulting test statistic is distributed as chi-square on one degree of freedom. In either case, the p-values are identical.

For the variance and covariance terms, there are concerns in using the standard errors in constructing Wald test statistics particularly when the population variance is thought to be near zero and the number of subjects is small [Bryk and Raudenbush, 1992]. This is because variance parameters are bounded; they cannot be less than zero and so using the standard normal for the sampling distribution is not reasonable. As a result, in this text we will not include the Wald tests for variance and covariance terms.

For nested models, the likelihood ratio test can be used to perform uni- or multi-parameter hypothesis tests. For this, one compares the model deviance values (*i.e.*, $-2 \log L$) to a chi-square distribution, where the degrees of freedom equals the number of parameters set equal to zero in the more restrictive model. It should be noted that while use of the likelihood ratio test for fixed effects is not problematic, for variance and covariance terms this test also suffers from the variance boundary problem mentioned above [Verbeke and Molenberghs, 2000]. Based on simulation studies it can be shown that the likelihood ratio test is too conservative (for testing null hypotheses about variance and covariance parameters), namely, it does not reject the null hypothesis often enough. This would then lead to accepting a more restrictive variance–covariance structure than is correct. As noted by Berkhof and Snijders [2001], this bias can largely be corrected by dividing the p-value obtained from the likelihood ratio test (of variance and covariance parameters) by two.

4.3.4 Psychiatric Dataset

Throughout this and later chapters, we will consider data from a psychiatric study described in Reisby et al. [1977]. This study focused on the longitudinal relationship between imipramine (IMI) and desipramine (DMI) plasma levels and clinical response in 66 depressed inpatients. Imipramine is the prototypic drug in the series of compounds known as tricyclic antidepressants, and is commonly prescribed for the treatment of major depression [Seiden and Dykstra, 1977]. Since imipramine biotransforms into the active metabolite desmethylimipramine (or desipramine), measurement of desipramine was also done in this study. Major depression is often classified in terms of two types. The first type, nonendogenous or reactive depression, is associated with some tragic life event such as the death of a close friend or family member, whereas the second type, endogenous depression, is not a result of any specific event and appears to occur spontaneously. It is sometimes held that antidepressant medications are more effective for endogenous depression [Willner, 1985]. In this sample, 29 patients were classified as nonendogenous and the remaining 37 patients were deemed to be endogenous.

The study design was as follows. Following a placebo period of 1 week, patients received 225 mg/day doses of imipramine for four weeks. In this study, subjects were rated with the Hamilton Depression Rating Scale (HDRS) [Hamilton, 1960] twice during the baseline placebo week (at the start and end of this week) as well as at the end of each of the four treatment weeks of the study. These HDRS scores represent the dependent variable that is measured across time. Higher scores on the HDRS represent higher levels of depression and lower scores indicate less depression. Plasma level measurements of both IMI and its metabolite DMI were made at the end of each week; these will be treated as time varying covariates. The sex and age of each patient was recorded and a diagnosis of endogenous or nonendogenous depression was made for each patient. These time-invariant (*i.e.*, individual-level) variables are all potential covariates, though our analyses will only focus on diagnosis.

Although the total number of subjects in this study was 66, the number of subjects with all measures at each of the weeks fluctuated: 61 at week 0 (start of placebo week), 63 at week 1 (end of placebo week), 65 at week 2 (end of first drug treatment week), 65 at week 3 (end of second drug treatment week), 63 at week 4 (end of third drug treatment week), and 58 at week 5 (end of fourth drug treatment week). Of the 66 subjects, only 46 had complete data at all timepoints. Thus, complete-case analysis under repeated measures MANOVA, for example, would discard approximately one-third of the dataset. MRM, alternatively, uses the data that are available from all 66 subjects.

Table 4.1 presents observed HDRS means, standard deviations, and sample sizes across the six study timepoints. Because the HDRS means are decreasing across time, there appears to be consistent improvement across time. Additionally, it is clear that the standard deviations are increasing across time. There is more spread in HDRS scores as time goes by. This is reasonable because some patients likely improved across time, to varying degrees, while others did not.

Table 4.1. Observed HDRS Means, Standard Deviations (sd), and n Across Time

	Week 0	Week 1	Week 2	Week 3	Week 4	Week 5
Mean	23.44	21.84	18.31	16.42	13.62	11.95
sd	4.53	4.70	5.49	6.42	6.97	7.22
n	61	63	65	65	63	58

Correlations, both pairwise and listwise, of the repeated HDRS outcomes are given in Table 4.2. Note that pairwise correlations are calculated based on all available data for a given pair of variables (*e.g.*, all subjects with week 0 and week 1 measurements are included in the calculation of the week 0 and week 1 correlation), whereas listwise correlations requires subjects to have complete data on all variables (*i.e.*, only subjects with data at all six timepoints are included). The correlations follow the commonly seen pattern of diminishing in value as one goes further away from the diagonal. This is true for both the pairwise and listwise correlations, which are similar. Also, within a given time lag, it appears that the association may increase across time. For example, for the time lag of one week, the pairwise correlation equals .49 at the beginning (*i.e.*, week 0 and week 1 correlation) and increases to .65 at the end (*i.e.*, week 4 and week 5 correlation). Taken together, these data do not appear to satisfy a compound symmetry assumption of equal variances and covariances across time.

Table 4.2. Observed HDRS Correlations: **Listwise (n = 46)** and *Pairwise* $(46 \leq n \leq 66)$

	Week 0	Week 1	Week 2	Week 3	Week 4	Week 5
Week 0	1.0	*.49*	*.41*	*.33*	*.23*	*.18*
Week 1	**.49**	1.0	*.49*	*.41*	*.31*	*.22*
Week 2	**.42**	**.49**	1.0	*.74*	*.67*	*.46*
Week 3	**.44**	**.51**	**.73**	1.0	*.82*	*.57*
Week 4	**.30**	**.35**	**.68**	**.78**	1.0	*.65*
Week 5	**.22**	**.23**	**.53**	**.62**	**.72**	1.0

Figure 4.2 presents the so-called "spaghetti plot" of the data. This plot is obtained by constructing a scatterplot of the HDRS scores by time, and then connecting the dots of each individual's data across time. This plot is useful for assessing overall aspects of the data. For example, the plot above suggests that there is a general linear decline in the HDRS scores across time, though clearly there is considerable individual heterogeneity in this. Also, the plot clearly shows the increasing variance in HDRS scores across time.

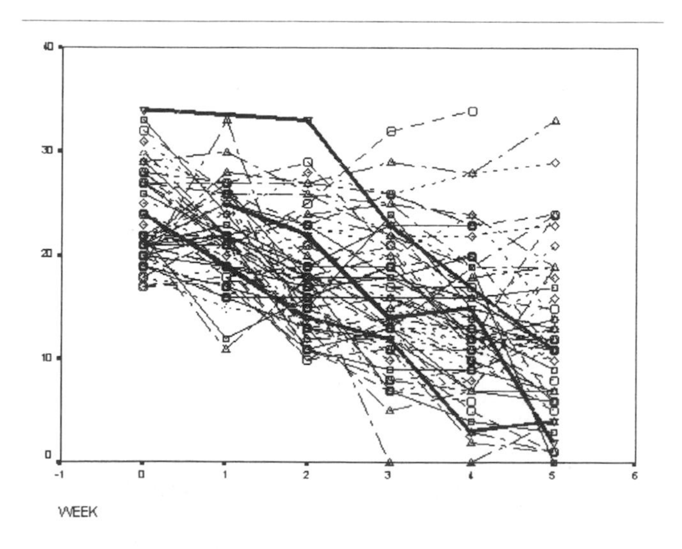

Figure 4.2. Reisby data: Spaghetti plot of observed data.

4.3.5 Random Intercept Model Example

The first model fit to these data corresponds to within-subjects model (4.3) and between-subjects model (4.4). This is the random-intercept model with only time as a regressor, where time is treated using incremental values from 0 to 5. Though the descriptive statistics above indicate that the compound symmetry assumption of this model is very dubious, we will fit this simple model here as a starting point in our examination of these data. Table 4.3 presents the results from this analysis using maximum likelihood (ML) estimation.

Table 4.3. MRM Results for Level-1 Model (4.3) and Level-2 Model (4.4)

Parameter	Estimate	SE	Z	$p <$
β_0	23.55	0.64	36.80	.0001
β_1	-2.38	0.14	-17.00	.0001
$\sigma^2_{\upsilon_0}$	16.15	3.41		
σ^2	19.04	1.53		

Note. $-2 \log L = 2285.19$. SE = standard error.

Focusing first on the estimated regression parameters, this model indicates that patients start off, on average, with a HDRS score of 23.55 and change by -2.38 points each week. Lower scores on the HDRS reflect less depression, so patients are improving across time by a little over 2 points per week. Both the intercept and slope are statistically significant ($p < .0001$) in this analysis. The intercept being significant is not particularly meaningful; it just indicates that HDRS scores are different than zero at baseline. However, because the slope is significant, the rate of improvement is significantly different from zero based on this analysis. On average, patients are improving across time.

The model estimates can be used to generate estimated values of the mean HDRS scores across time. Specifically, $\hat{y} = 23.552 - 2.376$ Week. These are displayed in Table 4.4 along with the observed means.

Table 4.4. Observed and Estimated Means

	Week					
	0	1	2	3	4	5
Observed	23.44	21.84	18.31	16.42	13.62	11.95
Estimated	23.55	21.18	18.80	16.42	14.05	11.67

Comparing the estimated to observed means in Table 4.4 indicates excellent model fit of these marginal means. Thus, overall it appears that the change across time in HDRS scores is linear. In their report, Reisby et al. [1977] classified patients into three groups based on their final HDRS scores: responders had scores below 8, partial responders were between 8 and 15, and nonresponders had final HDRS scores above 15. By this criteria, the estimated average trend is in the partial response range at the final timepoint (*i.e.*, = 11.67). For a more quantitative assessment of model fit, the interested reader is referred to Kaplan and

George [1998], which describes use of econometric forecasting statistics to assess various forms of fit between observed and estimated means.

The model fit of the variances and covariances can also be examined. Here, the estimated variance, which is assumed to be constant over time, is 16.15 + 19.04 = 35.19, or expressed as a standard deviation it yields 5.93. Since the observed standard deviations displayed in Table 4.1 clearly increase across time, this estimate of constant variance is an oversimplification. Turning to the correlations of the repeated measures, the intraclass correlation here equals $r = 16.15/(16.15 + 19.04) = .46$, which indicates that 46% of the unexplained variance in HDRS scores (*i.e.*, that part of the HDRS scores not explained by the linear effect of week) is at the individual level. Thus, subjects display considerable heterogeneity in depression levels. Comparing this value of .46 to the correlation matrix in Table 4.2 again suggests that this model is an oversimplification; while the average of the correlations might be approximately .46, there is considerable variation in these correlations and so assuming that they are all the same does not appear to be reasonable.

As a final point of comparison, note that performing an OLS simple linear regression on these data, as described in Section 4.2 ignoring the clustering of observations within subjects, yields $\hat{\beta}_0 = 23.60$ (SE = .55), $\hat{\beta}_1 = -2.41$ (SE = .18), and $\hat{\sigma}^2 = 35.40$ (SE = 2.59) [a similar analysis using ML estimation yields the same regression results and $\hat{\sigma}^2 = 35.21$ with SE = 2.57]. Thus, our results from the random-intercept model are in very close agreement with these, though of course the standard errors are considerably different. What is very interesting to note is that what the ordinary regression model lumped together into error variance (35.40 or 35.21), the random-intercept model separates into within-subjects and between-subjects variances (19.04 and 16.15, respectively). This illustrates a golden rule of statistics: One statistician's error term is another's career!

4.4 RANDOM INTERCEPT AND TREND MRM

For longitudinal data, the random intercept model is often too simplistic for a number of reasons. First, it is unlikely that the rate of change across time is the same for all individuals. It is more likely that individuals differ in their time trends; not everyone changes at the same rate. Furthermore, the compound symmetry assumption of the random intercept model is usually untenable for most longitudinal data. In general, measurements at points close in time tend to be more highly correlated than measurements further separated in time. Also, in many studies, subjects are more similar at baseline, and they grow at different rates across time. Thus, it is natural to expect that variability will increase over time.

For these reasons, a more realistic MRM allows both the intercept and time trend to vary by individuals. For this, the level-1 model is as before in (4.3), but the level-2 model is augmented as

$$\begin{aligned} b_{0i} &= \beta_0 + \upsilon_{0i}, \\ b_{1i} &= \beta_1 + \upsilon_{1i}. \end{aligned} \tag{4.6}$$

In this model, β_0 is the overall population intercept, β_1 is the overall population slope, υ_{0i} is the intercept deviation for subject i, and υ_{1i} is the slope deviation for subject i. As before, ε_{ij} is an independent error term distributed normally with mean 0 and variance σ^2. The assumption regarding the independence of the errors is one of conditional independence, that is, they are independent conditional on υ_{0i} and υ_{1i}. With two random individual-specific effects, the population distribution of intercept and slope deviations is assumed to be bivariate normal $\mathcal{N}(0, \mathbf{\Sigma}_\upsilon)$, with the random-effects variance–covariance matrix as

$$\mathbf{\Sigma}_\upsilon = \begin{bmatrix} \sigma^2_{\upsilon_0} & \sigma_{\upsilon_0 \upsilon_1} \\ \sigma_{\upsilon_0 \upsilon_1} & \sigma^2_{\upsilon_1} \end{bmatrix}.$$

This model can be thought of as a personal trend or change model since it represents the measurements of y as a function of time, both at the individual (υ_{0i} and υ_{1i}) and population (β_0 and β_1) levels. The intercept parameters indicate the starting point, and the slope parameters indicate the degree of change over time. The population intercept and slope parameters represent the overall (population) trend, while the individual parameters express how subjects deviate from the population trend.

Figure 4.3 represents this model graphically. Again, the figure represents the population trend with the solid line and the trends from two individuals, who now deviate both in terms of the intercept and slope. Because the slope varies for individuals, this model allows the possibility that some individuals do not change across time, while others can exhibit dramatic change. The population trend is the average across the individuals and the variance terms indicate how much heterogeneity there is in the population. Specifically, the variance term $\sigma^2_{\upsilon_0}$ indicates how much spread there is around the population intercept, and $\sigma^2_{\upsilon_1}$ represents the spread in slopes. To the degree that each individual's deviation from the population trend is only due to random error, these variance terms will approach zero. Alternatively, as each individual's deviation from the population trend is nonrandom, but characterized by the individual trend parameters υ_{0i} and υ_{1i} as being nonzero, these variance terms will increase from zero. Additionally, the covariance term, $\sigma_{\upsilon_0 \upsilon_1}$, represents the degree to which the individual intercept and slope parameters covary. For example, a positive covariance term would suggest that individuals with higher initial values have greater positive slopes, while a negative covariance would suggest the opposite.

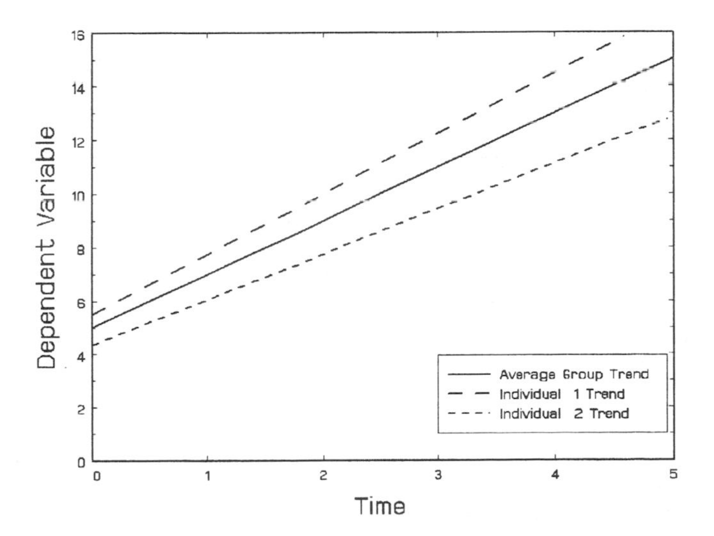

Figure 4.3. Random intercept and trend MRM.

4.4.1 Random Intercept and Trend Example

Continuing with our psychiatric example, we will fit the within-subjects model (4.3) and between-subjects model (4.6). As before, time is treated using incremental values from 0 to 5. The ML results are presented in Table 4.5.

Table 4.5. MRM Results for Level-1 Model (4.3) and Level-2 Model (4.6)

Parameter	Estimate	SE	Z	$p <$
β_0	23.58	0.55	43.22	.0001
β_1	−2.38	0.21	−11.39	.0001
$\sigma^2_{\upsilon_0}$	12.63	3.53		
$\sigma_{\upsilon_0 \upsilon_1}$	−1.42	1.04		
$\sigma^2_{\upsilon_1}$	2.08	0.52		
σ^2	12.22	1.12		

Note. $-2 \log L = 2219.04$. SE = standard error.

The results for the regression coefficients are very similar to the previous random-intercept analysis of these data. This model indicates that patients start off, on average, with a HDRS score of 23.58 and change by −2.38 points each week. As before, the effect of time is significant, and so we can conclude that the rate of improvement is significantly different from zero in this study.

For the variance and covariance terms, statistical significance is not indicated in the table because of the problem with use of the Wald test for these parameters (discussed in Section 4.3.3). However, the magnitude of the estimates does reveal the degree of individual heterogeneity in both the intercepts and slopes. For example, while the average intercept in the population is estimated to be 23.58, the estimated population standard deviation for the intercept is 3.55 ($= \sqrt{12.63}$). Similarly, the average population slope is −2.38, but the estimated population standard deviation for the slope equals 1.44 ($= \sqrt{2.08}$), and so approximately 95% of subjects in the population are expected to have slopes in the interval $-2.38 \pm (1.96 \times 1.44) = -5.20$ to .44. That the interval includes positive slopes reflects the fact that not all subjects improve across time. Thus, there is considerable heterogeneity in terms of patients' initial level of depression and in their change across time. Finally, the covariance between the intercept and linear trend is negative; expressed as a correlation it equals −.28, which is moderate in size. This suggests that patients who are initially more depressed (*i.e.*, greater intercepts) improve at a greater rate (*i.e.*, more pronounced negative slopes). An alternative explanation, though, is that of a floor effect due to the HDRS rating scale. Simply put, patients with less depressed initial scores have a more limited range of lower scores than those with higher initial scores.

An interesting question, at this point, is whether the between-subjects model in equation (4.6) is necessary over that in equation (4.4). In other words, is the assumption of compound symmetry rejected or not. Because these are nested models, they can be compared using a likelihood ratio test, albeit with the caveat that because the testing involves variance terms the p-value obtained from the likelihood ratio test should be divided by two. In the present case, the difference in model deviance values equals $2285.19 - 2219.04 = 66.15$ on 2 degrees of freedom (the 2 degrees of freedom are for the addition of the slope variance and the slope-intercept covariance), and so the p-value is less than .001 regardless. Thus, there is clear evidence that the assumption of compound symmetry is rejected.

Finally, empirical Bayes estimates of the individual random effects, $\hat{b}_{0i}$ and $\hat{b}_{1i}$, are often of interest. The derivation of these estimates will be described later. For now, these are plotted in Figure 4.4. The dashed lines indicate the estimated population intercepts and slopes. Thus, $\hat{v}_{0i}$ is represented by the vertical distance between a point and the horizontal line, while $\hat{v}_{1i}$ is represented by the horizontal distance between a point and the vertical line. This scatterplot reveals the wide range of observed intercepts and slopes in this sample. In particular, there are some patients who are very depressed initially but who improve to a great degree (upper left-hand corner). Similarly, there are some patients who show little or no improvement over time (towards the right side).

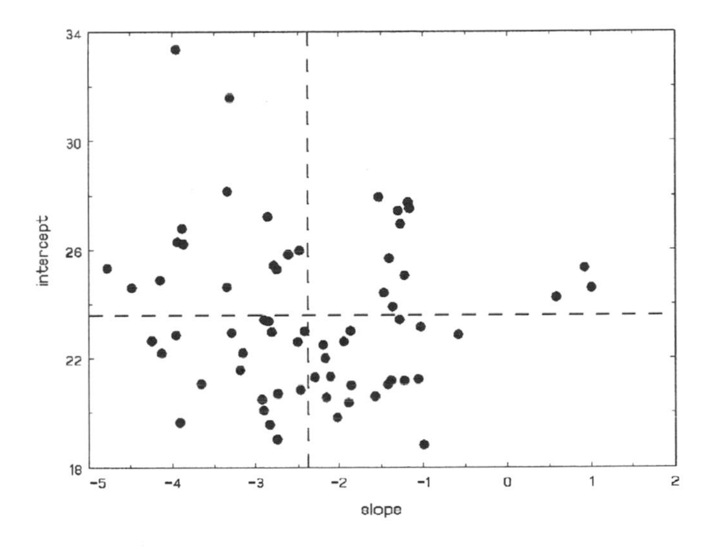

Figure 4.4. Reisby data: Estimated random effects.

4.4.2 Coding of Time

The coding of the time variable t has implications for the interpretation of the model parameters. For example, as in our Reisby analysis, t can start with the value 0 for baseline and be incremented according to the measurement timeline (*e.g.*, 1, 2, 3, 4, and 5 for the weekly follow-ups). In this formulation, the intercept parameters (β_0, v_{0i}, and $\sigma^2_{v_0}$) characterize aspects of the baseline timepoint. Alternatively, t can be expressed in centered form, where the average of time is subtracted from each time value (*e.g.*, -2.5, -1.5, $-.5$, .5, 1.5, 2.5). In this case, the meaning of the intercept parameters changes to reflect aspects about the midpoint of time, and not the baseline timepoint.

Figure 4.5 represents how the "intercept" variance $\sigma^2_{v_0}$ can change dramatically between baseline and centered codings of time. In the former, $\sigma^2_{v_0}$ represents the degree of individual heterogeneity at time 0, whereas in the latter it would represent heterogeneity at week 2.5 (the center of time). The figure portrays the average trend across time (the solid line) and two individual trends (the dot-dashed lines); the latter are meant to reflect the range of trends in the population of individuals (for an actual dataset there would be N such individual trends). The normal distributions on the figure represent the degree of individual heterogeneity at time equal to 0 and 2.5. Notice that the spread in the normal distribution, and thus $\sigma^2_{\upsilon_0}$, is much greater when time equals its midpoint relative to its baseline value. Thus, analysis of the same dataset would yield very different estimates of $\sigma^2_{\upsilon_0}$ if baseline-incremented versus centered coding of time was used.

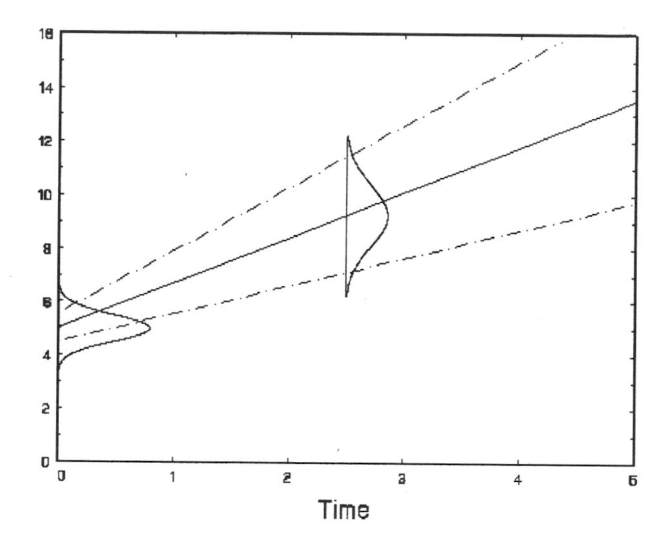

Figure 4.5. Intercept variance changes with coding of time.

As yet another coding choice, sometimes substantive interest focuses on the end of the measurement timeline. Here, time could be coded as -5, -4, -3, -2, -1, and 0 (in this case with six timepoints), so that the intercept parameters reflect aspects of the final timepoint. The choice of which representation to use often depends on ease of interpretation and the hypotheses of interest. It is important, though, that careful consideration is given to what is treated as "zero time," because all of the intercept parameters are effected by this choice. In particular, the zero time value should, in general, be one that is within the range of the observed data. Otherwise the intercept parameters represent extrapolations (in time) of the data. For example, a common mistake is to code time sequentially as 1, 2, 3, ..., n. In such a model, with time included, the intercept parameters represent a timepoint that is one unit *before* the first study timepoint. Clearly, one is typically not interested in trying to estimate what happened before the study began. For the interested reader, Biesanz et al. [2004], Horwitz et al. [1990], and Mehta and West [2000] describe this issue of zero time, and related issues, in great detail.

4.4.2.1 Example To illustrate the effect that the coding of time has, we reran the random intercept and trend model using a centered version of week, namely weekc = week − 2.5. The ML results for this analysis are presented in Table 4.6. Comparing these results to those in Table 4.5 illustrates the effect that the coding of time has. First, notice that the deviance values and slope estimates are identical. This includes both the estimate of average slope $\hat{\beta}_1$ and the heterogeneity in slopes $\hat{\sigma}^2_{\upsilon_1}$. The slope estimates haven't changed because the scale of the time variable has not changed. However, the parameters involving the location of the time variable have changed because the origin of this variable has shifted. Specifically, the estimate of the intercept is now 17.63, which corresponds to the average HDRS level when time equals 2.5 weeks (at the center of time). This intercept estimate is less than its counterpart in Table 4.5 because individuals are, on average, improving across time. Likewise, the intercept variance estimate of 18.52 reflects the degree of individual heterogeneity at 2.5 weeks. This value is greater than its counterpart in Table 4.5, indicating that the phenomenon illustrated in Figure 4.5 is occurring for these data. Namely, subjects are more alike at the beginning of the study than at the middle of the study.

Table 4.6. MRM Results for Level-1 Model (4.3) and Level-2 Model (4.6) with Centered Week

Parameter	Estimate	SE	Z	$p <$
β_0	17.63	0.56	31.47	.0001
β_1	−2.38	0.21	−11.39	.0001
$\sigma^2_{\upsilon_0}$	18.52	3.62		
$\sigma_{\upsilon_0\upsilon_1}$	3.78	1.08		
$\sigma^2_{\upsilon_1}$	2.08	0.52		
σ^2	12.22	1.12		

Note. $-2 \log L = 2219.04$. SE = standard error.

Finally, the covariance has not only changed values, but signs as well. How could this happen? Easier than one might at first imagine. Notice that from Table 4.5, the interpretation of the negative covariance would be that subjects with higher initial HDRS values have more negative slopes across time. Thus, subjects who are very depressed initially improve at a greater rate than those who are not so depressed to begin with. Turning to the centered results in Table 4.6, the positive covariance suggests that subjects with higher mid-study HDRS values have less negative (or more positive) slopes across time. In other words, subjects who are more depressed at mid-study are those that have improved less than subjects with lower mid-study depression levels. Clearly both of these interpretations are reasonable and consistent with each other; Figure 4.6 illustrates the spaghetti plot of two subjects following these patterns. The two dashed horizontal lines in the figure indicate the mean HDRS at week 0 (23.6) and week 2.5 (17.6). As the figure shows, the subject with the above average HDRS value at week 0 has the more negative slope, while the subject with the above average HDRS value at week 2.5 has the less negative (or more positive) slope. Thus, this example has highlighted the fact that correct model interpretation depends on an understanding of the coding of the variables in the analysis.

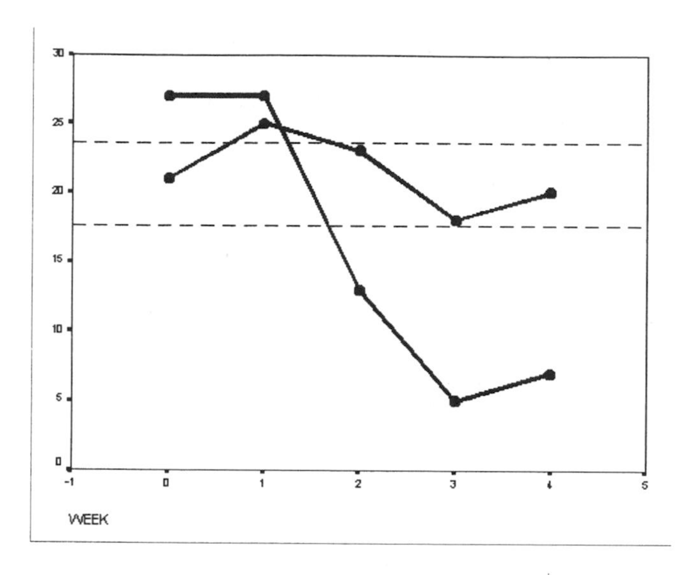

Figure 4.6. Spaghetti plot of two subjects supporting differential interpretation of intercept slope covariance.

4.4.3 Effect of Diagnosis on Time Trends

At this point, it may be interesting to examine whether we can explain some of the heterogeneity in intercepts and slopes, depicted earlier in Figure 4.4, in terms of particular subject characteristics. For example, in this study it may be that a subject's diagnosis (endogenous versus nonendogenous depression) is related to their initial depression level and change across time. Preparing for this analysis, note the observed HDRS means across time stratified by diagnostic group in Table 4.7.

Table 4.7. Observed HDRS Means and n Across Time Stratified by Group

	Week 0	Week 1	Week 2	Week 3	Week 4	Week 5
Endogenous	24.0	23.0	19.3	17.3	14.5	12.6
n	33	34	37	36	34	31
Nonendogenous	22.8	20.5	17.0	15.3	12.6	11.2
n	28	29	28	29	29	27

As the means indicate, both groups are clearly improving across time, though the endogenous group is consistently higher than the nonendogenous group. To explore this, we will augment the level-2 model to include a covariate DX which equals 0 if the patient's diagnosis is nonendogenous (NE) and equals 1 if the patient is endogenous (E). This variable enters the level-2 model rather than the level-1 model because it varies only with subjects (i) and not with time (j).

$$\begin{aligned} b_{0i} &= \beta_0 + \beta_2 DX_i + \upsilon_{0i} \\ b_{1i} &= \beta_1 + \beta_3 DX_i + \upsilon_{1i} \, . \end{aligned} \tag{4.7}$$

Here, β_0 represents the average week 0 HDRS level for NE patients, and β_1 the average HDRS weekly improvement for NE patients. Similarly, β_2 represents the average week 0 HDRS difference for E patients (relative to NE patients) and β_3 the average difference in HDRS weekly improvement rates for E patients (relative to NE patients). Thus, β_3 represents the diagnosis by time interaction, indicating the degree to which the time trends vary by diagnostic group. In this augmented model, υ_{0i} is the individual's deviation from their diagnostic group intercept and υ_{1i} is the individual's deviation from their diagnostic group slope. To the degree that the variable DX is useful in explaining intercept and slope variation, these individual deviations and their corresponding variances, ($\sigma^2_{\upsilon_0}$ and $\sigma^2_{\upsilon_1}$), will be reduced. Results for this model are listed in Table 4.8.

A likelihood ratio test comparing this model to the previous one can be used to test the null hypothesis that the diagnosis-related effects (*i.e.*, β_2 and β_3) are zero. This yields $X^2_2 = 2219.04 - 2214.94 = 4.1$, which is not statistically significant. Inspection of the estimates in Table 4.8 reveals a marginally significant difference in terms of their initial scores, with endogenous patients about 2 points higher, and absolutely no difference in their trends across time. This is also borne out if one compares the variance estimates from Tables 4.5 and 4.8. Notice that the intercept variance has diminished slightly from to 12.63 to 11.64, as a result of the marginally significant intercept difference, whereas the slope variance is the same. Taken together, there is no real evidence that the two diagnostic groups differ in terms of their HDRS scores across time.

Table 4.8. MRM Results for Level-1 Model (4.3) and Level-2 Model (4.7)

Parameter	Estimate	SE	Z	$p <$
NE intercept β_0	22.48	0.79	28.30	.0001
NE slope β_1	−2.37	0.31	−7.59	.0001
E intercept difference β_2	1.99	1.07	1.86	.063
E slope difference β_3	−0.03	0.42	−0.06	.95
$\sigma^2_{\upsilon_0}$	11.64	3.53		
$\sigma_{\upsilon_0 \upsilon_1}$	−1.40	1.00		
$\sigma^2_{\upsilon_1}$	2.08	0.50		
σ^2	12.22	1.11		

Note. $-2 \log L = 2214.94$.

Figure 4.7 illustrates the observed and estimated trends for these two groups. For the latter, these are simply computed as $22.48 - 2.37\, Week$ for the nonendogenous group, and $(22.48 + 1.99) + (-2.37 - 0.03)\, Week$ for the endogenous group. The figure helps to illustrate the conclusions of the analysis. As can be seen, there is only a marginal difference between the two groups that is consistent across time. Also, the observed and estimated means are in close agreement for both groups.

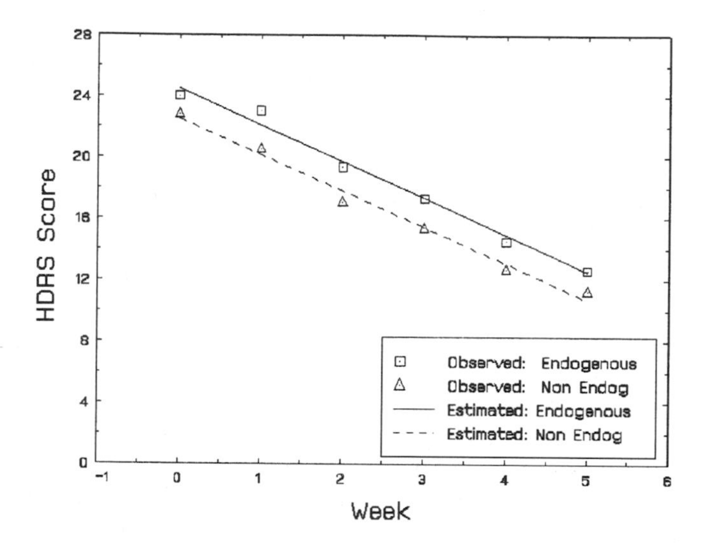

Figure 4.7. Reisby data: Estimated and observed means across time and diagnostic groups.

4.5 MATRIX FORMULATION

A more compact representation of the model is afforded using matrices and vectors. This formulation is particularly useful in summarizing statistical aspects of the model. For this, the MRM for the $n_i \times 1$ response vector $\boldsymbol{y}$ for individual i can be written as

$$\underset{n_i \times 1}{\boldsymbol{y}_i} = \underset{n_i \times p}{\boldsymbol{X}_i} \underset{p \times 1}{\boldsymbol{\beta}} + \underset{n_i \times r}{\boldsymbol{Z}_i} \underset{r \times 1}{\boldsymbol{v}_i} + \underset{n_i \times 1}{\boldsymbol{\varepsilon}_i} \tag{4.8}$$

with $i = 1 \ldots N$ individuals and $j = 1 \ldots n_i$ observations for individual i. Here, $\boldsymbol{y}_i$ is the $n_i \times 1$ dependent variable vector for individual i, $\boldsymbol{X}_i$ is the $n_i \times p$ covariate matrix for individual i, $\boldsymbol{\beta}$ is the $p \times 1$ vector of fixed regression parameters, $\boldsymbol{Z}_i$ is the $n_i \times r$ design matrix for the random effects, $\boldsymbol{v}_i$ is the $r \times 1$ vector of random individual effects, and $\boldsymbol{\varepsilon}_i$ is the $n_i \times 1$ error vector.

For example, in the random intercept and slope MRM just considered, we would have

$$\boldsymbol{y}_i = \begin{bmatrix} y_{i1} \\ y_{i2} \\ \cdots \\ \cdots \\ y_{in_i} \end{bmatrix} \quad \text{and} \quad \boldsymbol{X}_i = \boldsymbol{Z}_i = \begin{bmatrix} 1 & t_{i1} \\ 1 & t_{i2} \\ \cdots & \cdots \\ \cdots & \cdots \\ 1 & t_{in_i} \end{bmatrix}$$

for the data matrices, and

$$\boldsymbol{\beta} = \begin{bmatrix} \beta_0 \\ \beta_1 \end{bmatrix} \quad \text{and} \quad \boldsymbol{v}_i = \begin{bmatrix} v_{0i} \\ v_{1i} \end{bmatrix}$$

for the population and individual trend parameter vectors, respectively. For the model including diagnosis and the diagnosis by time interaction, the data matrix for $\boldsymbol{X}$ would be changed to

$$\boldsymbol{X}_i = \begin{bmatrix} 1 & t_{i1} & DX_i & DX_i \times t_{i1} \\ 1 & t_{i2} & DX_i & DX_i \times t_{i2} \\ \dots & \dots & \dots & \dots \\ \dots & \dots & \dots & \dots \\ 1 & t_{in_i} & DX_i & DX_i \times t_{in_i} \end{bmatrix},$$

while the $\boldsymbol{Z}$ matrix would be the same. The matrix representation does not really distinguish between person-varying (level-2) and time-varying (level-1) covariates, it's all $\boldsymbol{X}$. Thus, again, the multilevel or hierarchical representation of the model into the level-1 and level-2 submodels might give the impression that several models are simultaneously being estimated, but actually there is only one model (that is broken apart to aid in our interpretation of that model).

The distributional assumptions about the random effects and errors are

$$\begin{aligned} \boldsymbol{\varepsilon}_i &\sim \mathcal{N}(\boldsymbol{0}, \sigma^2 \boldsymbol{I}_{n_i}), \\ \boldsymbol{v}_i &\sim \mathcal{N}(\boldsymbol{0}, \boldsymbol{\Sigma}_v). \end{aligned}$$

As a result, it can be shown that the observations $\boldsymbol{y}_i$ and random effects $\boldsymbol{v}_i$ have the joint multivariate normal distribution:

$$\begin{bmatrix} \boldsymbol{y}_i \\ \boldsymbol{v}_i \end{bmatrix} \sim \mathcal{N}\left(\begin{bmatrix} \boldsymbol{X}_i\boldsymbol{\beta} \\ \boldsymbol{0} \end{bmatrix}, \begin{bmatrix} \boldsymbol{Z}_i\boldsymbol{\Sigma}_v\boldsymbol{Z}_i' + \sigma^2\boldsymbol{I}_{n_i} & \boldsymbol{Z}_i\boldsymbol{\Sigma}_v \\ \boldsymbol{\Sigma}_v\boldsymbol{Z}_i' & \boldsymbol{\Sigma}_v \end{bmatrix} \right). \tag{4.9}$$

Using results from multivariate statistics, it can be further shown that the mean of the posterior distribution of $\boldsymbol{v}_i$, given $\boldsymbol{y}_i$, yields the formula for the empirical Bayes estimates of the random effects,

$$\hat{\boldsymbol{v}}_i = \left[\boldsymbol{Z}_i'(\sigma^2\boldsymbol{I}_{n_i})^{-1}\boldsymbol{Z}_i + \boldsymbol{\Sigma}_v^{-1}\right]^{-1} \boldsymbol{Z}_i'(\sigma^2\boldsymbol{I}_{n_i})^{-1}(\boldsymbol{y}_i - \boldsymbol{X}_i\boldsymbol{\beta}). \tag{4.10}$$

Similarly, the corresponding posterior covariance matrix is given by

$$\boldsymbol{\Sigma}_{v|y_i} = \left[\boldsymbol{Z}_i'(\sigma^2\boldsymbol{I}_{n_i})^{-1}\boldsymbol{Z}_i + \boldsymbol{\Sigma}_v^{-1}\right]^{-1}. \tag{4.11}$$

Note that the variance–covariance matrix of the repeated measures $\boldsymbol{y}$ is of the form

$$V(\boldsymbol{y}_i) = \boldsymbol{Z}_i\boldsymbol{\Sigma}_v\boldsymbol{Z}_i' + \sigma^2\boldsymbol{I}_{n_i}. \tag{4.12}$$

As an example of what this equation implies, consider a model including a random intercept and time trend, $r = 2$, and a subject with three timepoints, $n = 3$. Then, their random-effects design matrix might be

$$\boldsymbol{Z}_i = \begin{bmatrix} 1 & 0 \\ 1 & 1 \\ 1 & 2 \end{bmatrix},$$

and so the variance–covariance matrix equals

$$\sigma^2 \boldsymbol{I}_{n_i} + \begin{bmatrix} \sigma^2_{\upsilon_0} & \sigma^2_{\upsilon_0} + \sigma_{\upsilon_0\upsilon_1} & \sigma^2_{\upsilon_0} + 2\sigma_{\upsilon_0\upsilon_1} \\ \sigma^2_{\upsilon_0} + \sigma_{\upsilon_0\upsilon_1} & \sigma^2_{\upsilon_0} + 2\sigma_{\upsilon_0\upsilon_1} + \sigma^2_{\upsilon_1} & \sigma^2_{\upsilon_0} + 3\sigma_{\upsilon_0\upsilon_1} + 2\sigma^2_{\upsilon_1} \\ \sigma^2_{\upsilon_0} + 2\sigma_{\upsilon_0\upsilon_1} & \sigma^2_{\upsilon_0} + 3\sigma_{\upsilon_0\upsilon_1} + 2\sigma^2_{\upsilon_1} & \sigma^2_{\upsilon_0} + 4\sigma_{\upsilon_0\upsilon_1} + 4\sigma^2_{\upsilon_1} \end{bmatrix}.$$

Notice that this structure, which is based on a model with a random intercept and time trend, allows the variances and covariances to change across time. For example, if $\sigma_{\upsilon_0\upsilon_1}$ is positive, then clearly the variance increases across time. Diminishing variance across time is also possible if, for example, $-2\sigma_{\upsilon_0\upsilon_1} > \sigma^2_{\upsilon_1}$. Other patterns are possible depending on the values of these variance and covariance parameters.

Models with more than random intercepts and linear trends are also possible, as are models that allow autocorrelated errors, that is $\varepsilon_i \sim \mathcal{N}(0, \sigma^2 \boldsymbol{\Omega}_i)$; these will be described in subsequent chapters. For now, note that by including both multiple random effects, and possibly autocorrelated errors, a wide range of variance–covariance structures for the repeated measures is possible. This flexibility is in sharp contrast to the traditional ANOVA models which assume either a compound symmetry structure (univariate ANOVA) or a totally general structure (MANOVA). Typically, compound symmetry is too restrictive and a general structure is not parsimonious. MRMs, alternatively, provide these two and everything in between, and so allow efficient modeling of the variance–covariance structure of the repeated measures. More discussion about this is included in Chapters 6 and 7.

4.5.1 Fit of Variance–Covariance Matrix

For the Reisby dataset considered in this chapter, it is of interest to consider model fit of the variances and covariances associated with the repeated outcomes. Below is the observed variance–covariance matrix for the six study timepoints. These are calculated based on the pairwise data for the covariances and the available data for each of the variances.

$$V(\boldsymbol{y}) = \begin{bmatrix} 20.55 & & & & & \\ 10.50 & 22.07 & & & & \\ 10.20 & 12.74 & 30.09 & & & \\ 9.69 & 12.43 & 25.96 & 41.15 & & \\ 7.17 & 10.10 & 25.56 & 36.54 & 48.59 & \\ 6.02 & 7.39 & 18.25 & 26.31 & 32.93 & 52.12 \end{bmatrix}.$$

As noted above, the mixed model formulates that the variance–covariance matrix of the repeated measures follow the equation $V(\boldsymbol{y}_i) = \boldsymbol{Z}_i \boldsymbol{\Sigma}_\upsilon \boldsymbol{Z}'_i + \sigma^2 \boldsymbol{I}_{n_i}$. This is specifically the variance–covariance matrix of the repeated measures given the model covariates $\boldsymbol{X}$. If the only covariates in $\boldsymbol{X}$ are time trends (*e.g.*, week and week squared) and if the model fits the observed marginal timepoint means well, then we can compare the estimated variance–covariance matrix given by the model to the observed variance–covariance matrix. For the Reisby dataset, the random intercept and trend model did fit the observed timepoint means well, and so using the estimates in Table 4.5, namely $\hat{\sigma}^2 = 12.22$ and

$$\hat{\boldsymbol{\Sigma}}_\upsilon = \begin{bmatrix} 12.63 & -1.42 \\ -1.42 & 2.08 \end{bmatrix},$$

with the design matrix of the random effects

$$\boldsymbol{Z}' = \begin{bmatrix} 1 & 1 & 1 & 1 & 1 & 1 \\ 0 & 1 & 2 & 3 & 4 & 5 \end{bmatrix},$$

yields

$$\hat{V}(\boldsymbol{y}) = \boldsymbol{Z}\hat{\boldsymbol{\Sigma}}_{\upsilon}\boldsymbol{Z}' + \hat{\sigma}^2\boldsymbol{I} = \begin{bmatrix} 24.85 & & & & & \\ 11.21 & 24.08 & & & & \\ 9.79 & 12.52 & 27.48 & & & \\ 8.37 & 13.18 & 18.00 & 35.03 & & \\ 6.95 & 13.84 & 20.73 & 27.63 & 46.74 & \\ 5.53 & 14.50 & 23.47 & 32.44 & 41.41 & 62.60 \end{bmatrix}$$

as the estimated variance–covariance matrix. Given that this variance–covariance matrix of 21 elements is represented by only 4 parameter estimates, the fit appears reasonably good. The model is clearly picking up on the increasing variance across time and the diminishing covariance away from the diagonal.

Grady and Helms [1995] describe graphical techniques to aid in examining model fit of the variance–covariance structure. These authors suggest plots of the covariances or correlations as a function of the "lag" (*i.e.*, the time between measures). For example, in the covariance plot, lag 0 would correspond to the variance of the repeated measures at the different study timepoints, lag 1 would correspond to the covariances of the repeated measures one unit of time apart (*e.g.*, week 0 and week 1 covariance, week 1 and week 2 covariance, . . ., week 4 and 5 covariance), lag 2 would correspond to covariances 2 units of time apart (*e.g.*, week 0 and week 2 covariance, week 1 and week 3 covariance, . . ., week 3 and 5 covariance), etc. Similarly for the correlation plot, with the exception that there is no lag 0 in the plot because these would all be correlations of 1.

Figures 4.8 and 4.9 show the observed covariance and correlation plots. Note the plot of the variances at lag = 0 in Figure 4.8; these increase steadily from week 0 to week 5. Similarly, examining the covariances or correlations show that, for a given lag, the level of association generally increases across timepoints. These plots make very clear that a compound symmetric structure (*i.e.*, a random intercept model) of equal variances and covariances would not fit well.

Instead, it is of interest to see the graphs for the random intercept and trend model of Table 4.5. These are provided in Figures 4.10 and 4.11. These plots suggest reasonable model fit of the variances and covariances, though as one would expect the patterns are more systematic than the actual data. The model is emphasizing the increasing (co)variance values within a lag to a greater extent than the observed data. While in Chapters 6 and 7 we will describe more statistical tools that can be used for model selection regarding the variance–covariance structure, these plots can be quite useful in getting a "feel" for this.

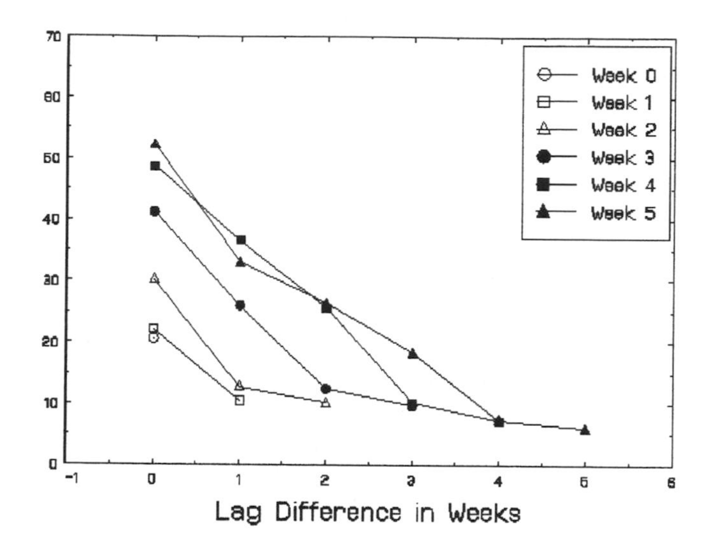

Figure 4.8. Reisby data: Observed covariance plot.

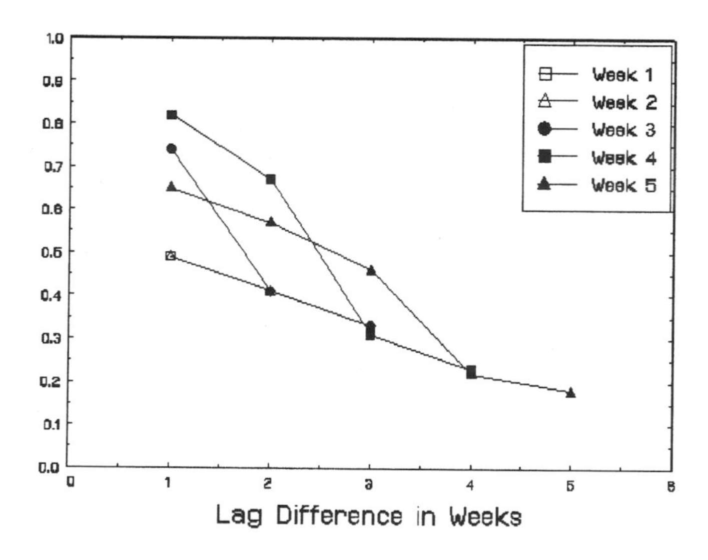

Figure 4.9. Reisby data: Observed correlation plot.

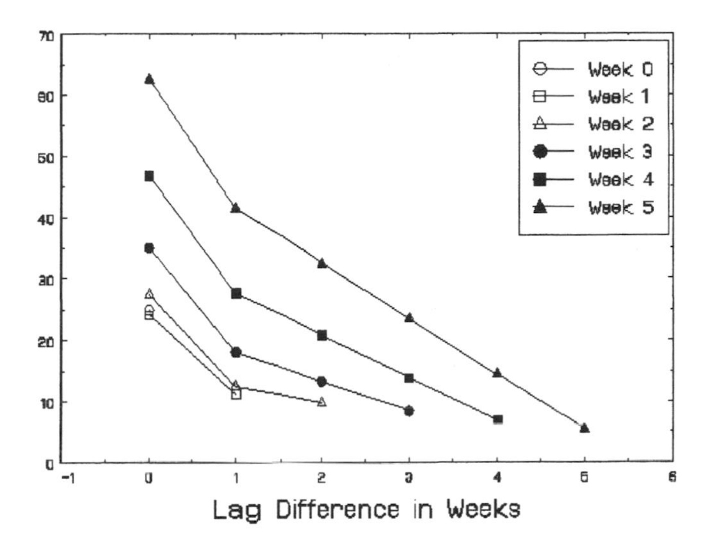

Figure 4.10. Reisby data: Random intercept and trend model estimated covariance plot.

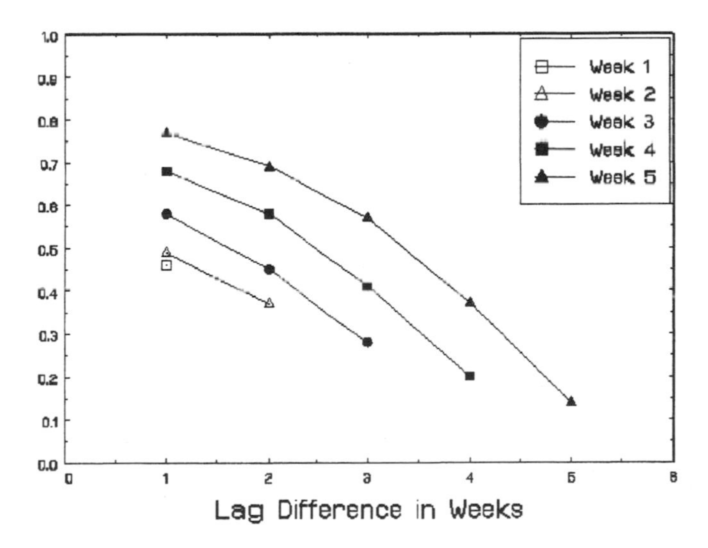

Figure 4.11. Reisby data: Random intercept and trend model estimated correlation plot.

4.5.2 Model with Time-Varying Covariates

In this section, we examine the effects of the time-varying drug plasma levels IMI and DMI. Since an inspection of the data indicated that the magnitude of these measurements varied greatly between individuals (from 4 to 312 μg/L for IMI and from 1 to 740 μg/L for DMI), a log transformation is used for these covariates. This helps to ensure that the estimated regression coefficients are not unduly influenced by extreme values on these covariates.

Also, these variables, ln IMI and ln DMI, are expressed in grand-mean centered form so that the model intercept represents HDRS scores for patients with average drug levels. To obtain the grand-mean centered versions of these variables, the variable's sample mean is subtracted from each observation. For notational simplicity in the model equations, I_{ij} and D_{ij} will represent the grand-mean centered versions of ln IMI and ln DMI, respectively, in what follows. Also, whereas the previous models considered HDRS outcomes from weeks 0 to 5, the models of this section only include HDRS outcome data from weeks 2 to 5. This is because the drug plasma levels are not available at the first two timepoints of the study (*i.e.*, week 0, or baseline, and week 1, or the end of the drug-washout period). While MRM does allow incomplete data across time, data must be complete within a given timepoint (in terms of both the dependent variable and covariates) for that timepoint to be included in the analysis. Thus, the analyses that follow are for the four week period following the drug-washout period with t_{ij} coded as 0, 1, 2, and 3 for these four respective timepoints. As a result, the intercept represents HDRS scores for week 2 of the study (*i.e.*, when $t_{ij} = 0$).

The first level-1 model is given by

$$y_{ij} = b_{0i} + b_{1i}t_{ij} + b_{2i}I_{ij} + b_{3i}D_{ij} + \varepsilon_{ij}, \tag{4.13}$$

where, b_{0i} is the week 2 HDRS level for patient i under average levels of both ln IMI and ln DMI, b_{1i} is the weekly change in HDRS for patient i, b_{2i} is the patient's change in HDRS due to ln IMI, and b_{3i} is the change in HDRS due to ln DMI. The between-subjects model is given as

$$\begin{aligned} b_{0i} &= \beta_0 + \upsilon_{0i}, \\ b_{1i} &= \beta_1 + \upsilon_{1i}, \\ b_{2i} &= \beta_2, \\ b_{3i} &= \beta_3, \end{aligned} \tag{4.14}$$

where β_0 is the average week 2 HDRS level for patients with average ln IMI and ln DMI values, β_1 is the average HDRS weekly change, β_2 is the average HDRS difference for a unit change in ln IMI, and β_3 is the average HDRS difference for a unit change in ln DMI. Also, υ_{0i} is the individual intercept deviation, and υ_{1i} is the individual slope deviation. Notice that the level-2 model indicates that the drug effects could also be treated as random. This would be accomplished by adding υ_{2i} and υ_{3i} to the model, and would allow individual variation in terms of the drug level effect on HDRS scores. Given that antidepressants like IMI and DMI are not effective for all individuals, it is plausible that the drug levels are more strongly related to changes in depression for some individuals, whereas for others they are less so. Similarly, one could add individual-level covariates (*e.g.*, endogenous/nonendogenous group) into the models for b_{2i} and b_{3i} to examine whether the drug effects vary with individual-level covariates. Again, it is feasible that the drug effects on outcome are stronger for endogenous than nonendogenous patients. An example of an MRM allowing such individual variation in relationships is described by Hedeker et al. [1996]. Fitting the present model yields the results given in Table 4.9.

It is interesting to note that neither of the drug levels seems to be significantly related to the depression scores across time. However, note that the model given in (4.13) specifies that a person's drug level is related to their depression score at that same timepoint. It might be more plausible to instead posit that a person's drug level is related to their *change* in

Table 4.9. MRM Results for Level-1 Model (4.13) and Level-2 Model (4.14)

Parameter	Estimate	SE	Z	$p <$
intercept β_0	18.17	0.71	25.70	.0001
time slope β_1	−2.03	0.28	−7.15	.0001
ln IMI β_2	0.60	0.85	0.71	.48
ln DMI β_3	−1.20	0.63	−1.90	.06
$\sigma^2_{\upsilon_0}$	24.83	5.79		
$\sigma_{\upsilon_0 \upsilon_1}$	−0.72	1.74		
$\sigma^2_{\upsilon_1}$	2.73	0.95		
σ^2	10.46	1.37		

Note. $-2 \log L = 1502.5$.

depression score, or improvement, at that same timepoint. For this, the following alternative level-1 model is considered:

$$(y_{ij} - y_{i0}) = b_{0i} + b_{1i}t_{ij} + b_{2i}I_{ij} + b_{3i}D_{ij} + \varepsilon_{ij}, \tag{4.15}$$

where y_{i0} is the individual's HDRS score at baseline (or at week 1 for those few subjects with a missing baseline score). This yields the results presented in Table 4.10.

Table 4.10. MRM Results for Level-1 Model (4.15) and Level-2 Model (4.14)

Parameter	Estimate	SE	Z	$p <$
intercept β_0	−5.18	0.66	−7.87	.0001
slope β_1	−1.97	0.29	−6.90	.0001
ln IMI β_2	0.63	0.82	0.77	ns
ln DMI β_3	−1.97	0.60	−3.26	.0014
$\sigma^2_{\upsilon_0}$	20.50	5.13		
$\sigma_{\upsilon_0 \upsilon_1}$	0.84	1.61		
$\sigma^2_{\upsilon_1}$	2.78	0.97		
σ^2	10.53	1.39		

Note. $-2 \log L = 1498.8$.

Interestingly, now the effect of DMI, the metabolite of IMI, is highly significant and negative. Thus, greater DMI values are associated with greater improvement (*i.e.*, more negative HDRS change scores). However, the parent drug IMI is not significantly related to HDRS change scores and in fact its coefficient is positive. It's important to remember that the model estimates the IMI effect controlling for the DMI effect, and vice versa. These two drug levels are moderately correlated with each other (r = .18, .23, .22, and .18 for the four respective timepoints) and so the results above are not necessarily indicative of the marginal relationships of each drug with depression scores.

Correlations of the drug plasma levels with the HDRS scores, both raw and expressed as change scores, are given in Table 4.11. These bear out the fact that the drug levels are much more associated with the HDRS change scores than the actual scores. These correlations also show the greater association between HDRS change scores and DMI, rather than IMI, drug levels.

Table 4.11. Correlation Between HDRS Scores and Plasma Levels (Natural Log Units)

Drug	Week 2	Week 3	Week 4	Week 5
		HDRS total score		
ln IMI	−0.034	−0.038	−0.003	−0.189
ln DMI	−0.177	−0.075	−0.246	−0.293
		HDRS change from baseline		
ln IMI	−0.049	−0.106	−0.046	−0.240
ln DMI	−0.366	−0.281	−0.363	−0.361

4.5.2.1 Within and Between-Subjects Effects for Time-Varying Covariates

When time-varying covariates are included in a MRM, as in the manner of the last analysis, an assumption is made that the between and within-subjects effects of these variables are equal. To see this, express the time-varying covariates I_{ij} and D_{ij} as

$$\begin{aligned} I_{ij} &= \overline{I}_i + (I_{ij} - \overline{I}_i) \\ D_{ij} &= \overline{D}_i + (D_{ij} - \overline{D}_i) \end{aligned}$$

where $\overline{I}_i$ and $\overline{D}_i$ are the means of these two time-varying covariates computed for each individual. Thus, the first term following the equality represents the individual's mean on the time-varying covariate (*i.e.*, a between-subjects variable) and the second term represents the individual's deviation around their mean (*i.e.*, a within-subjects variable). Figure 4.12 shows a plot, considering the exact same data points, illustrating the difference between (a) a purely between-subjects effect versus (b) a purely within-subjects effect.

Focusing first on (a), notice that for these three subjects there is no within-subjects effect of the covariate, since the value of the dependent variable is constant within subjects. There is, however, a large between-subjects effect indicating that y increases as the subject average on x increases. Turning attention to (b), one can see a large within-subjects effect of the covariate for the two subjects. For a given subject, y increases as the value of x

increases. However, there is no between-subjects effect of x in (b) since the mean of y is identical for these two subjects.

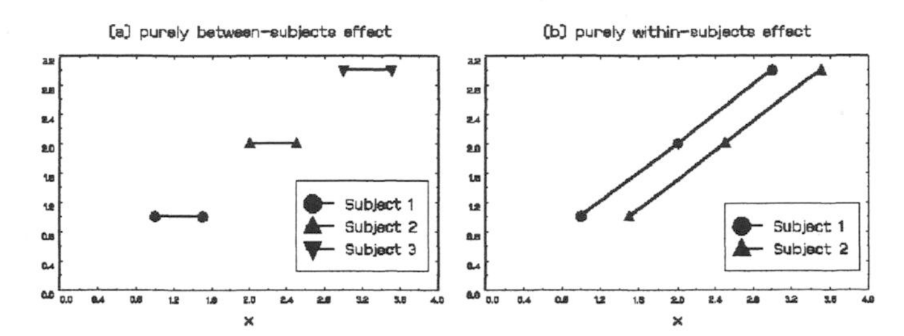

Figure 4.12. Time-varying covariate effects: (a) Purely between-subjects and (b) purely within-subjects.

To separate the within- and between-subjects effects of time-varying covariates one can include both the subject's average $\overline{x}_i$ and the subject's time-varying deviation $x_{ij} - \overline{x}_i$ into the model. In the present example including both of these terms into the MRM yields

$$(y_{ij} - y_{i0}) = b_{0i} + b_{1i}t_{ij} + b_{2i}(I_{ij} - \overline{I}_i) + b_{3i}(D_{ij} - \overline{D}_i) + \varepsilon_{ij}, \tag{4.16}$$

and

$$\begin{aligned} b_{0i} &= \beta_0 + \beta_4 \overline{I}_i + \beta_5 \overline{D}_i + v_{0i}, \\ b_{1i} &= \beta_1 + v_{1i}, \\ b_{2i} &= \beta_2, \\ b_{3i} &= \beta_3 \end{aligned} \tag{4.17}$$

for the level-1 and level-2 models. Thus, the total effect of IMI, for example,

$$\beta_2(I_{ij} - \overline{I}_i) + \beta_4 \overline{I}_i$$

is partitioned into its within- and between-subjects effects (*i.e.*, β_2 and β_4, respectively). The between-subjects part indicates the degree to which the individual's average drug level is related to their average depression level, averaging across time. In other words, it may be that subjects with consistently high drug levels have consistently low depression scores. Alternatively, the within-subjects component represents the degree to which variation in an individual's drug level is associated with a change in their depression scores (*i.e.*, a within-subject change). Thus, it may be that a higher relative drug level for an individual is associated with a lower relative depression score for that individual at a particular timepoint. If these two are equal ($\beta_2 = \beta_4$), then the IMI effect is

$$\beta_2(I_{ij} - \overline{I}_i) + \beta_2 \overline{I}_i = \beta_2 I_{ij},$$

which is exactly what was used in the last analysis. Thus, we implicitly assumed that the within- and between-subjects effects of these two drug levels were the same in the previous analysis. This assumption can be tested by comparing the previous model with the more general model of (4.16) and (4.17). Table 4.12 includes the results of this latter analysis.

Table 4.12. MRM Results for Level-1 Model (4.16) and Level-2 Model (4.17)

Parameter	Estimate	SE	Z	$p <$
intercept β_0	−5.09	0.66	−7.71	.0001
slope β_1	−2.02	0.29	−6.94	.0001
within ln IMI β_2	2.44	1.46	1.68	.10
within ln DMI β_3	−1.80	1.00	−1.80	.075
between ln IMI β_4	−0.31	1.00	−0.31	ns
between ln DMI β_5	−2.37	0.80	−2.97	.004
$\sigma^2_{\upsilon_0}$	20.32	5.10		
$\sigma_{\upsilon_0 \upsilon_1}$	0.50	1.64		
$\sigma^2_{\upsilon_1}$	2.83	0.97		
σ^2	10.38	1.36		

Note. $-2 \log L = 1495.8$.

Comparing the two models yields a likelihood ratio test statistic of $X^2_2 = 3.0$, which is not statistically significant; the assumption of homogeneity of the between- and within-subjects regressions is not rejected. Inspecting the estimated coefficients for DMI supports this: −1.8 and −2.4 for the within- and between-subjects effects, respectively. Conversely, the estimates for IMI are very different, and even of opposite sign. However, neither is statistically significant and the standard errors for these two IMI estimates are quite large. In conclusion, for these data, there is not sufficient evidence to reject the assumption of equality in the within- and between-subjects effects for these two drug levels.

4.5.2.2 Time Interactions with Time-Varying Covariates In some cases, it can be of substantive interest to examine whether there are interactions between a time-varying covariate and time. For example, one might posit that the relationship between the time-varying covariate and the outcome either increases or decreases across time. This is clearly plausible in the present example since the effectiveness of antidepressants is not thought to be immediate, but instead to develop over time [Reisby et al., 1977]. In other words, the drug plasma levels might be minimally related or unrelated to the depression outcome initially, with effects emerging across time. Thus, it is of interest to examine the degree to which the effects of the time-varying drug plasma levels on the change in depression scores vary across time. To explore this possibility, the level-1 model can be augmented to include the time interactions, namely,

$$(y_{ij} - y_{i0}) = b_{0i} + b_{1i}t_{ij} + b_{2i}I_{ij} + b_{3i}D_{ij} + b_{4i}(I_{ij} \times t_{ij}) + b_{5i}(D_{ij} \times t_{ij}) + \varepsilon_{ij}, \quad (4.18)$$

with the accompanying level-2 model,

$$b_{0i} = \beta_0 + \upsilon_{0i},$$

$$\begin{aligned} b_{1i} &= \beta_1 + \upsilon_{1i}, \\ b_{2i} &= \beta_2, \\ b_{3i} &= \beta_3, \\ b_{4i} &= \beta_4, \\ b_{5i} &= \beta_5. \end{aligned} \tag{4.19}$$

To correctly interpret the model parameters it is important to remember that the drug levels have been grand-mean centered, that the week variable equals 0 for the second week of the study, and that interpretation of the "main effects" is altered when interactions are present (*i.e.*, they represent the effect of the variable when the interacting variable equals 0). Thus, in this model, β_0 represents the average week 2 HDRS change score for patients with average drug levels, β_1 is the average weekly change in HDRS change scores for patients with average drug levels, β_2 is the HDRS change-score difference for a unit change of ln IMI at week 2, and β_3 represents the HDRS change-score difference per unit change of ln DMI at week 2. One can think of β_2 as the regression slope corresponding to the plot of HDRS change scores versus ln IMI levels considering week 2 data only (with the caveat that this regression slope is really a partial regression slope adjusting for the other drug level). Similar comments apply for interpreting β_3 in terms of ln DMI. Turning to the interactions β_4 and β_5, these indicate the per-week change in the drug effects on the HDRS change scores. In terms of the plot analogy, these interactions correspond to the change in (partial) regression slopes associated with separate weekly plots of HDRS change scores versus drug levels as one goes across the weeks. In other words, how does the slope for a given drug vary across time. Finally, υ_{0i} represents the individual intercept deviation and υ_{1i} is the individual slope (*i.e.*, time) deviation. Table 4.13 lists the results of this analysis.

Table 4.13. MRM Results for Level-1 Model (4.18) and Level-2 Model (4.19)

Parameter	Estimate	SE	Z	$p <$
intercept β_0	−5.12	0.65	−7.82	.0001
slope (*i.e.*, time) β_1	−1.94	0.28	−7.04	.0001
ln IMI β_2	0.40	0.87	0.46	ns
ln DMI β_3	−1.51	0.62	−2.43	.017
ln IMI by time β_4	0.16	0.41	0.39	ns
ln DMI by time β_5	−0.90	0.34	−2.65	.01
$\sigma^2_{\upsilon_0}$	20.24	5.05		
$\sigma_{\upsilon_0 \upsilon_1}$	0.99	1.52		
$\sigma^2_{\upsilon_1}$	2.50	0.91		
σ^2	10.35	1.36		

Note. $-2 \log L = 1492.0$.

Comparing this model to the one without drug by time interaction (*i.e.*, from Table 4.10) yields a likelihood ratio test statistic of $X^2_2 = 6.8$, which is statistically significant at the .05 level. Thus, there is evidence that the drug effects on depression do vary across time. Inspecting the estimates and their test statistics in Table 4.13 reveals that it is ln DMI, and not ln IMI, that is interacting significantly with time. Specifically, ln DMI has an initial week 2 effect that is significant ($p < .017$), indicating that higher levels of ln

DMI are associated with greater improvement on the HDRS scale at this timepoint, and this beneficial effect of ln DMI gets more pronounced across time ($p < .01$). Concretely, for a one-unit change in ln DMI at week 2 (*i.e.*, when time is coded 0) the estimate is a 1.51 point reduction on the HDRS change score, whereas by the last timepoint (*i.e.*, when time is coded 3) it is a 4.21 point reduction ($1.51 + 3 \times .9$).

At first glance, it might seem a bit unusual that the ln DMI by time interaction is so highly significant given the reported correlations in Table 4.11. To better understand this, consider the simple linear regression slopes that are obtained from regressing HDRS change scores on ln DMI values at each of the four timepoints separately: these are -2.081, -2.195, -3.370, and -3.3765, respectively. These regression slopes provide clearer evidence of the ln DMI by time interaction, as they increase (in absolute value) more dramatically across time than the analogous correlations in Table 4.11. Why do these two sets of descriptive statistics suggest different conclusions? Remembering that the correlation is essentially a scale-free representation of the slope (*i.e.*, $r = \hat{\beta}\, s_x/s_y$), it is clear that the scales of the dependent and independent variable play a role here. Interestingly, the scale of these two go in opposite directions across time; the standard deviations of the HDRS change scores increase (5.38, 6.51, 7.35, and 7.88 across the four timepoints), whereas the standard deviations of the ln DMI values decrease (.95, .84, .79, and .76 across these same four timepoints). Thus, the metric for the slopes across time is very different (*i.e.*, the ratio of standard deviations s_x/s_y equals .18, .13, .11, and .10, respectively) which explains why the simple slopes and correlations are not in such close agreement, and why the significant ln DMI by time interaction of the MRM is a bit at odds with the apparent consistent pattern of the correlations across time. As this final MRM and the descriptive statistics make clear, it is the scale-dependent slope of ln DMI (*i.e.*, how much change in depression is associated with a unit change in the natural log of this blood level) that is increasing across time, and not the scale-free association.

4.6 ESTIMATION

Estimation of MRMs generally uses a combination of two complementary methods [Laird and Ware, 1982; Bock, 1989]. For the random individual effects $\boldsymbol{v}_i$, empirical Bayes (EB) methods are used, while maximum (marginal) likelihood (ML) or restricted (or residual) maximum likelihood (REML [McCulloch and Searle, 2001]) methods are used for estimation of variance parameters, σ^2 and $\boldsymbol{\Sigma}_v$, and regression coefficients $\boldsymbol{\beta}$. For general MRMs, the solution is fairly complex and requires iterative algorithms like Newton–Raphson or Fisher-scoring procedures. Essentially, these algorithms continue iterating through the data until all parameter estimates are changing by a very small degree, at which point the estimation process is said to have converged. In the following, we will present the formulas in the case of a random-intercept MRM to illustrate the essential ideas and show connections with estimation in ordinary (*i.e.*, fixed-effects) regression models. More complete details for specific models are presented in the subsequent chapters.

EB estimates of individual effects are sometimes termed EAP ("Expected *a Posteriori*") estimates, since they are derived as the mean of the posterior distribution of v, given $\boldsymbol{y}_i$. Denoting the EB estimate of v_i as $\tilde{v}_i$ to distinguish it from subsequent ML estimates, and given the model assumptions, we get the following EB estimator of individual parameters:

$$\tilde{v}_i = \rho_{n_i n_i} \frac{1}{n_i} \mathbf{1}_i'(\boldsymbol{y}_i - \boldsymbol{X}_i \boldsymbol{\beta}) = \rho_{n_i n_i} \frac{1}{n_i} \sum_{j=1}^{n_i} \left(y_{ij} - \boldsymbol{x}_{ij}' \boldsymbol{\beta} \right), \tag{4.20}$$

where $\boldsymbol{x}_{ij}$ is the vector of regressors for a individual i at time j, and $\rho_{n_in_i}$ is equivalent to the Spearman–Brown reliability formula [Guilford, 1954], given as $\rho_{nn} = nr/\left[1 + (n-1)r\right]$ with r as the intraclass correlation.

A property of EB estimation is that $\tilde{v}_i$ is a function of both the individual's data and the empirical prior distribution specified for v_i. As information about an individual increases (*i.e.*, the reliability $\rho_{n_in_i}$ increases toward 1), by either increasing data interdependency within the subject (increasing r) and/or increasing sample size (n_i), the EB estimate approaches the average deviation (across time) for that individual, $(\sum_{j=1}^{n_i} y_{ij} - \boldsymbol{x}'_{ij}\boldsymbol{\beta})/n_i$. Note that this latter formula yields the OLS estimator of the individual effect. Alternatively, as information about an individual decreases (*i.e.*, $\rho_{n_in_i}$ decreases toward 0), by decreasing data interdependency within the subject and/or decreasing sample size, the EB estimate approaches the posited mean of the empirical prior distribution of v_i, namely 0. Thus, given r, if a subject has few measurements, then the EB estimate will be smaller (in absolute value) than the corresponding OLS estimate. Alternatively, if the subject has many measurements across time, then the EB and OLS estimates would be very similar. Because of this, the EB estimates are said to be *shrunken to the mean*, where the mean of the random effects equals zero in the population. The degree of shrinkage depends on the number of measurements an individual has. An important advantage of EB estimates relative to OLS estimates is that they are not as prone to the undue influence of outliers.

Figure 4.13 shows a plot of the EB estimates versus their OLS counterparts for a random-intercept model of the Reisby data. Notice that the spread of the OLS estimates is greater than the spread of the EB estimates, and that the points do not fall exactly on the 45-degree line. If the EB and OLS estimates were exactly the same all of the points would be on the 45-degree line. Instead, the points suggest a line that is slightly more horizontal (*i.e.*, toward zero), which results from the shrinkage of the EB estimates toward zero.

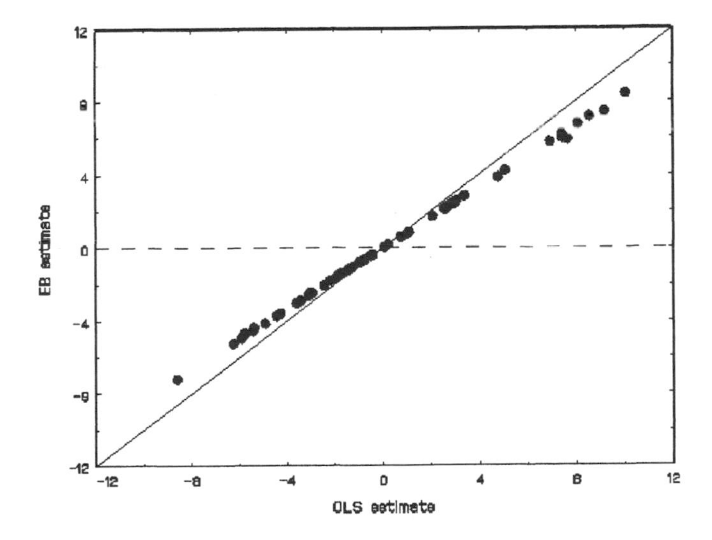

Figure 4.13. Reisby data: EB versus OLS estimates of subject effects.

In addition to the EB estimate of the posterior mean, the variance of the posterior distribution of υ is given as

$$\sigma^2_{\upsilon|y_i} = \sigma^2_\upsilon(1 - \rho_{n_i n_i}). \tag{4.21}$$

Again, the form reveals the nature of this EB estimator of the posterior variance: As information about the individual increases, the posterior variance becomes a fraction of the empirical prior variance (σ^2_υ), while as information about the individual decreases, this variance approaches the empirical prior variance.

To estimate covariate effects $\boldsymbol{\beta}$ and variance parameters σ^2_υ and σ^2, ML estimation can be used. The ML estimation procedure is more fully presented in the appendix and is described using two numerical algorithms: the EM algorithm solution [Dempster et al., 1981] and the Fisher scoring solution [Longford, 1987]. From the EM algorithm solution, we can see how the random-intercept MRM can be viewed as a generalization of the ordinary multiple linear regression model. Namely, the following equations are used in the iterative EM algorithm solution,

$$\hat{\boldsymbol{\beta}} = \left[\sum_{i=1}^N \boldsymbol{X}_i'\boldsymbol{X}_i\right]^{-1}\left[\sum_{i=1}^N \boldsymbol{X}_i'(\boldsymbol{y}_i - \mathbf{1}_i\tilde{\upsilon}_i)\right], \tag{4.22}$$

$$\hat{\sigma}^2_\upsilon = \frac{1}{N}\sum_i^N \tilde{\upsilon}_i^2 + \sigma^2_{\upsilon|y_i}, \tag{4.23}$$

$$\hat{\sigma}^2 = \frac{1}{N}\sum_i^N (\boldsymbol{y}_i - \boldsymbol{X}_i\hat{\boldsymbol{\beta}} - \mathbf{1}_i\tilde{\upsilon}_i)'(\boldsymbol{y}_i - \boldsymbol{X}_i\hat{\boldsymbol{\beta}} - \mathbf{1}_i\tilde{\upsilon}_i) + n_i\sigma^2_{\upsilon|y_i} \tag{4.24}$$

with the solution proceeding by iterating between EB equations (4.20) and (4.21) and ML equations (4.22)–(4.24) until convergence. Note that as estimates of the individual effects $\tilde{\upsilon}_i$ and variances $\sigma^2_{\upsilon|y_i}$ approach zero, the subject variance estimate (σ^2_υ) approaches zero, and the equations for regression coefficients $\boldsymbol{\beta}$ and error variance σ^2 approach the maximum likelihood solution of these parameters in the usual fixed-effects regression model, namely,

$$\hat{\boldsymbol{\beta}} = \left[\sum_{i=1}^N \boldsymbol{X}_i'\boldsymbol{X}_i\right]^{-1}\sum_{i=1}^N \boldsymbol{X}_i'\boldsymbol{y}_i \quad \text{and} \quad \hat{\sigma}^2 = \frac{1}{N}\sum_{i=1}^N(\boldsymbol{y}_i - \boldsymbol{X}_i\hat{\boldsymbol{\beta}})'(\boldsymbol{y}_i - \boldsymbol{X}_i\hat{\boldsymbol{\beta}}).$$

Thus, as the dependency of the data within individuals decreases, the solution approaches the (ML) solution for an ordinary multiple linear regression model.

The formula for the random effect variance (4.23) reveals an interesting connection between this population variance and the sample variance of the EB estimates. Notice that this equation can be written as

$$\hat{\sigma}^2_\upsilon = \frac{1}{N}\sum_i^N \tilde{\upsilon}_i^2 + \frac{1}{N}\sum_i^N \sigma^2_{\upsilon|y_i}. \tag{4.25}$$

Because the mean of the random effects is approximately zero in the sample, the first term after the equality is essentially the (ML) estimate of the sample variance of the EB estimates, while the second term is the average of the posterior variances. Thus, the estimate of the population variance of the random effects $\hat{\sigma}_v^2$ will always exceed the sample variance of the EB random effects (except for the trivial case where all of the posterior variances equal 0).

4.6.1 ML Bias in Estimation of Variance Parameters

It can be shown that the ML estimates of variance parameters in MRMs are biased downwards (*i.e.*, they are too small). This is also true of the ML estimate of the error variance in ordinary multiple regression, and the equations for this parameter illustrate the point well. Note that the ML estimate of the error variance equals

$$\sigma^2 = \frac{SSE}{N}, \tag{4.26}$$

where SSE is the sum of squared errors and N is the sample size. The unbiased ordinary least square (OLS) estimate of the same parameter is

$$\sigma^2 = \frac{SSE}{N - p - 1}, \tag{4.27}$$

where p is the number of regressors. These equations make clear that this bias is negligible if $N - p$ is relatively large (say, over 100), but can be of concern when $N - p$ is not large. While ML estimates of the fixed effects are not greatly affected by this, their standard errors can be different if the variance parameters are downwardly biased. Again, the standard errors will be too small under ML estimation, though the difference is negligible if $N - p$ is relatively large.

To correct for this bias, the MRM parameters can be estimated by REML [Patterson and Thompson, 1971]). Clearly, REML estimates are preferred over ML estimates, however there is one important consideration to keep in mind. Because REML adjusts the likelihood for the number of covariates in a model, one cannot use REML likelihood ratio tests for comparing models with different covariates. ML likelihood ratio tests do not have this limitation. Because of this, and because the sample size for the datasets in this text are relatively large, we will present ML estimates unless otherwise noted.

4.7 SUMMARY

As this chapter has demonstrated, MRMs are useful for analyzing longitudinal data. MRMs allow for the presence of missing data, irregularly-spaced measurements across time, time-varying and invariant covariates, accommodation of individual-specific deviations from the average time trend, and estimation of the population variance associated with these individual effects. Perhaps the most popular feature of MRMs is their treatment of missing data. As has been illustrated, subjects are not assumed to be measured at the same number of timepoints. Since there are no restrictions on the number of observations per individual, subjects who are missing at a given interview wave are not excluded from the analysis. The assumption of the model is that the available data for a given subject are representative of that subject's deviation from the average trends across time (which are estimated based on the whole sample). It is important to note that the missing data are not imputed in estimation of the MRM; rather the model parameters are estimated using all available data. Further treatment of missing data in longitudinal studies is described in Chapter 14.

Statistical software to perform MRM analysis has proliferated, especially for continuous outcomes. Most of the major statistical packages now include procedures for estimating continuous MRMs. Additionally, there are several independent programs that were specifically designed for MRM analysis; these include HLM [Raudenbush et al., 2000a], MLwiN [Goldstein et al., 1998], and MIXREG [Hedeker and Gibbons, 1996b], to mention a few. Review articles comparing some of these software programs include van der Leeden et al. [1996] and de Leeuw and Kreft [2001]. Rather than present software examples here, the website for this book (`http://www.uic.edu/~hedeker/long.html`) includes syntax files for the analyses presented in this chapter for some of these software programs.

This chapter has focused on the modeling aspects of MRM without a great deal of discussion on parameter estimation. Because these models are more complex than ordinary fixed-effects regression models, it is sometimes the case that the iterative procedures used for estimation of an MRM do not converge to a solution. If this occurs, it is often because the model is overly complex, relative to the data being used to estimate it, and so model simplification is necessary. Although it is not always apparent why a particular model does not converge, building models in a sequential piecewise manner can help to isolate where troubles occur.

In the example, repeated observations were observed nested within individuals. In the terminology of multilevel analysis [Goldstein, 1995] and hierarchical linear models [Raudenbush and Bryk, 2002], this is termed a two-level data structure with individuals representing level-2 and the nested repeated observations level-1. The models that we have presented are thus referred to as two-level models. Individuals themselves, though, are often observed clustered within some higher-level unit, for example, a classroom, clinic, or worksite. Cross-sectional clustered data can also be considered as two-level data, with the clusters representing level-2 and the clustered subjects level-1. Modeling of such clustered data is described in detail in several texts [Goldstein, 1995; Hox, 2002; Kreft and de Leeuw, 1998; Longford, 1993; Raudenbush and Bryk, 2002; Snijders and Bosker, 1999]. In some studies, subjects are clustered and also repeatedly measured, resulting in three-levels of data: the cluster (level-3), individual (level-2), and repeated observation (level-1). Analysis of three-level data is also described in some of the aforementioned texts, and will be treated in Chapter 13.

In this chapter, we have considered MRMs for a single longitudinal response process in which individual members of the population can systematically deviate from the overall process; however, there is a single overall process from which individuals may deviate. In some cases, this assumption may be unreasonable. In such cases, it may be more plausible to assume that there are two or more different processes in the population, and individual subjects deviate from a particular response process. From the statistical point of view, in such cases, the population may be better represented by a mixture of temporal response processes rather than a single process. Furthermore, within each component of the temporal response distribution or "latent class," the effects of treatment or other fixed-effects in the model may vary. If the data confirm the existence of multiple latent classes, each with a different temporal response process and possibly different treatment related effects, we may then be able to target specific treatments to specific subjects, based on pre-treatment characteristics that are predictive of latent class membership. Several authors have described MRMs where the random effects are drawn from either a latent class or a mixture distribution in the population [Verbeke and Lesaffre, 1996; Muth´en and Shedden, 1999; McCulloch et al., 2002; Xu and Hedeker, 2002]. These articles make clear that these extensions of MRMs can be quite useful in some contexts.

CHAPTER 5

MIXED-EFFECTS POLYNOMIAL REGRESSION MODELS

5.1 INTRODUCTION

In many situations, it is too simplistic to assume that the change across time is linear. For example, it may be that the outcome changes across time in a curvilinear manner. A curvilinear trend would allow a leveling off or accelerating of the change across time. This is clearly plausible in many situations, but especially for rating scale data, like that considered in the last chapter, where ceiling and floor effects can easily occur. In this chapter we will explore the use of polynomial trend models, illustrating how they can be used to model particular kinds of nonlinear relationships across time both at the individual and population levels.

5.2 CURVILINEAR TREND MODEL

To begin, consider the following curvilinear trend model that is obtained by adding a quadratic, or squared, term to the level-1 model:

$$y_{ij} = b_{0i} + b_{1i}t_{ij} + b_{2i}t_{ij}^2 + \varepsilon_{ij}. \tag{5.1}$$

Here, b_{0i} is the intercept for subject i, b_{1i} is the linear trend component for subject i, and b_{2i} is the quadratic trend component for subject i. Notice that this model can also be written as

$$y_{ij} = b_{0i} + (b_{1i} + b_{2i}t_{ij})t_{ij} + \varepsilon_{ij}$$

to point out that the "time effect" varies as a function of time. The level-2, between-subjects, model is now

$$\begin{aligned} b_{0i} &= \beta_0 + v_{0i}, \\ b_{1i} &= \beta_1 + v_{1i}, \\ b_{2i} &= \beta_2 + v_{2i}, \end{aligned} \tag{5.2}$$

where β_0 is the intercept, β_1 is the average linear trend component, and β_2 is the average quadratic trend component. Similarly, v_{0i} is the individual deviation from average intercept, and v_{1i} and v_{2i} represent the individual deviation the from average linear and quadratic trend components, respectively. Thus, this model allows curvilinearity at both the population (β_2) and individual (v_{2i}) levels. In some situations it may be reasonable to restrict the curvilinear part to the level-1 model only, in which case the final component of the level-2 model would simply be $b_{2i} = \beta_2$.

Figure 5.1 gives a sense of some of the possible time trends that can be fit using curvilinear trend models.

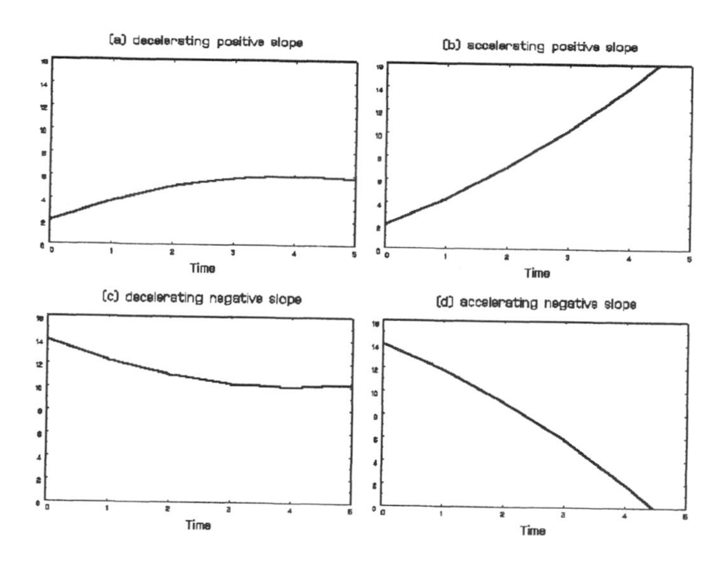

Figure 5.1. Various curvilinear models: (a) Decelerating positive slope; (b) accelerating positive slope; (c) decelerating negative slope; (d) accelerating negative slope.

In the top two graphs, positive trends are either decelerating (a) or accelerating (b). The linear component β_1 equals 2 for both, while the quadratic component β_2 is $-.25$ and $.25$, respectively. Similarly for the negative trends on the bottom, the linear component equals -2 for both, while the quadratic component is either .25 (decelerating negative) or $-.25$ (accelerating negative).

Notice that the decelerating positive trend in (a) is flat as time increases. Similarly, the negative trend in (c) flattens out. One can easily calculate the point at which a curvilinear trend "flattens out." For this, note that the derivative is

$$\frac{\partial y}{\partial t} = \beta_1 + 2\beta_2 t, \tag{5.3}$$

which equals 0 for $t = -\beta_1/(2\beta_2)$. So for our decelerating positive trend in (a) the flattening out point is when time equals 4. Similarly, the decelerating negative trend flattens out when time is 4. For interpretation of curvilinear models, it is helpful to figure out where the flattening out point occurs, since beyond that point the trend reverses sign. That is, for (a) beyond week 4 the trend goes from being a positive one to a negative one.

This changing of signs can be useful in areas were the relationship across time is thought to be J- or U-shaped. Figure 5.2 presents some examples of curves of these types. As can be seen, a wide variety of relationships are possible simply by adding in the squared term for time in the model.

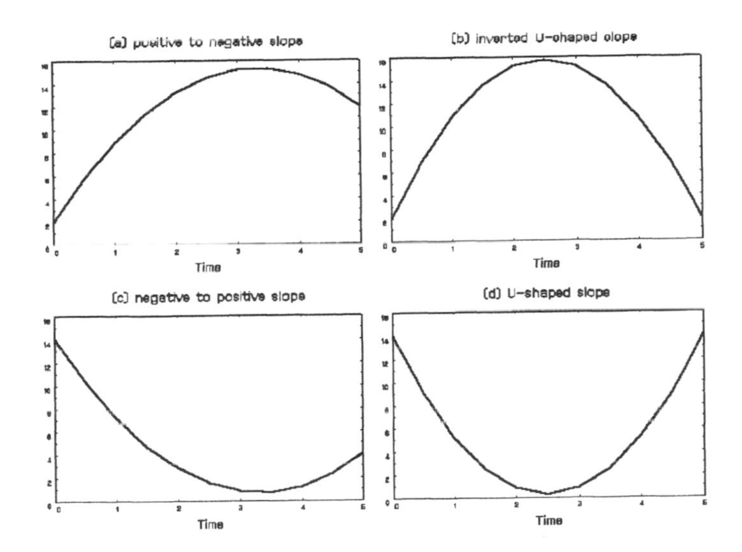

Figure 5.2. More curvilinear models: (a) Positive to negative slope ($\beta_0 = 2, \beta_1 = 8, \beta_2 = -1.2$); (b) inverted U-shaped slope ($\beta_0 = 2, \beta_1 = 11, \beta_2 = -2.2$); (c) negative to positive slope ($\beta_0 = 14, \beta_1 = -8, \beta_2 = 1.2$); (d) U-shaped slope ($\beta_0 = 14, \beta_1 = -11, \beta_2 = 2.2$).

5.2.1 Curvilinear Trend Example

Turning again to the Reisby dataset, we will add a quadratic time trend to both the population and individual trends. Fitting this model yields the results given in Table 5.1. Comparing this model to one without these quadratic terms (*i.e.*, the model listed in Table 5.5 of the previous chapter with $\beta_2 = \sigma^2_{\upsilon_2} = \sigma_{\upsilon_0\upsilon_2} = \sigma_{\upsilon_1\upsilon_2} = 0$) yields a deviance of 11.4, which is statistically significant on 4 degrees of freedom (even without an adjustment to the p-value for testing of variance terms). This is interesting given that the Wald test for β_2 is clearly nonsignificant. In fact, comparing the above model to one with $\sigma^2_{\upsilon_2} = \sigma_{\upsilon_0\upsilon_2} = \sigma_{\upsilon_1\upsilon_2} = 0$ (not shown) yields a deviance of 11.0. Nearly all of the improvement in model fit is through the inclusion of the quadratic term as a random effect, and not as a fixed effect. This suggests that although the trend across time is essentially linear at the population level, it is curvilinear at the individual level. How is this possible? Consider Figure 5.3, which plots a hypothetical example with an average linear trend and curvilinear trends for two individuals.

Table 5.1. MRM Results for Level-1 Model (5.1) and Level-2 Model (5.2)

Parameter	Estimate	SE	Z	$p <$
β_0	23.76	0.55	43.04	.0001
β_1	−2.63	0.48	−5.50	.0001
β_2	0.05	0.09	0.58	.56
$\sigma^2_{\upsilon_0}$	10.44	3.59		
$\sigma_{\upsilon_0\upsilon_1}$	−0.92	2.41		
$\sigma^2_{\upsilon_1}$	6.64	2.76		
$\sigma_{\upsilon_0\upsilon_2}$	−0.11	0.42		
$\sigma_{\upsilon_1\upsilon_2}$	−0.94	0.49		
$\sigma^2_{\upsilon_2}$	0.19	0.09		
σ^2	10.52	1.11		

Note. $-2 \log L = 2207.64$.

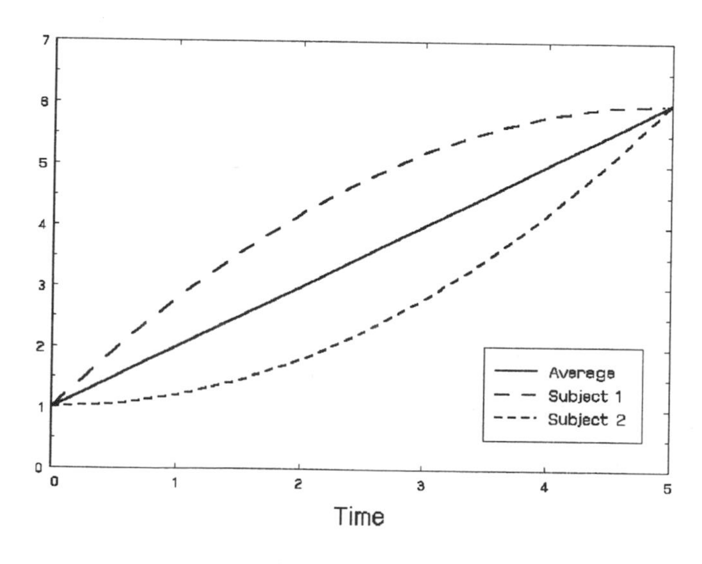

Figure 5.3. Average linear and individual quadratic trends.

Here, one individual has an accelerating positive trend (subject 2) and the other has a decelerating positive trend (subject 1). These two "cancel out" to yield an average positive trend that is strictly linear.

For the Reisby data, though the quadratic term is nonsignificant at the population level, it is instructive to calculate the point at which the decelerating negative trend flattens out:

$$\hat{t} = -\hat{\beta}_1/(2\hat{\beta}_2) = 1/2(2.63/0.05) = 26.3,$$

which is well beyond the time frame of the study. Thus, the curvilinear component is clearly very slight (at the population level).

Figure 5.4 contains a plot of the individual trend estimates from this model. These are obtained by calculating $\hat{y}_{ij} = \hat{b}_{0i} + \hat{b}_{1i}t_{ij} + \hat{b}_{2i}t_{ij}^2$ for $t = 0, 1, \ldots, 5$ and then connecting the timepoint estimates for each individual. The plot makes apparent the wide heterogeneity in trends across time, as well as the increasing variance in HDRS scores across time. Some individuals have accelerating downward trends, suggesting a delay in the drug effect. Alternatively, others have decelerating downward trends, which are consistent with a leveling off of the drug effect. Some individuals even have positive trends indicating a worsening of their depressive symptoms across time. This is not too surprising given that antidepressants, like imipramine, are known to be ineffective for some patients. The figure is also interesting in showing that many of the individual trend lines are approximately linear. Thus, the improvement that the curvilinear model provides in describing change across time is perhaps modest.

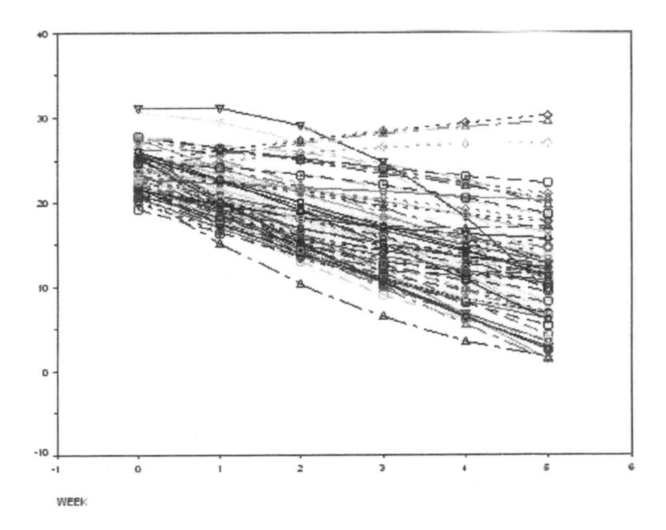

Figure 5.4. Riesby data: Estimated curvilinear trends.

The model fit of the observed variance–covariance matrix of the repeated measures is obtained as

$$\begin{aligned}\hat{V}(\boldsymbol{y}) &= \boldsymbol{Z}\hat{\boldsymbol{\Sigma}}_{\upsilon}\boldsymbol{Z}' + \hat{\sigma}^2\boldsymbol{I} \\ &= \begin{bmatrix} 20.96 & & & & & \\ 9.41 & 23.86 & & & & \\ 8.16 & 15.57 & 31.07 & & & \\ 6.68 & 16.08 & 23.11 & 38.31 & & \\ 4.98 & 14.88 & 23.26 & 30.12 & 45.98 & \\ 3.06 & 11.97 & 20.98 & 30.09 & 39.29 & 59.11 \end{bmatrix},\end{aligned}$$

where

$$\boldsymbol{Z}' = \begin{bmatrix} 1 & 1 & 1 & 1 & 1 & 1 \\ 0 & 1 & 2 & 3 & 4 & 5 \\ 0 & 1 & 4 & 9 & 16 & 25 \end{bmatrix}, \quad \hat{\boldsymbol{\Sigma}}_{\upsilon} = \begin{bmatrix} 10.44 & -0.92 & -0.11 \\ -0.92 & 6.64 & -0.94 \\ -0.11 & -0.94 & 0.19 \end{bmatrix},$$

and $\hat{\sigma}^2 = 10.52$. By comparing this matrix with the observed variance–covariance matrix, presented in the previous chapter, one can see that the estimated variances are close to the observed, and the model is clearly picking up the pattern of diminishing covariance away from the diagonal and at the earlier timepoints. We will return to this issue regarding the the relative model fit of the observed (co)variances later in this chapter.

5.3 ORTHOGONAL POLYNOMIALS

For trend models, it is often beneficial to represent the polynomials in orthogonal form [Bock, 1975]. One advantage is that it avoids collinearity problems that can result from using multiples of t (t^2, t^3, etc.) as regressors. To see this, consider a curvilinear trend model with three timepoints. Then $t = 0$, 1, and 2, while $t^2 = 0$, 1, and 4; these two variables are nearly perfectly correlated. To counter this, time is sometimes expressed in centered form, for example $(t - \bar{t}) = -1$, 0, and 1, and $(t - \bar{t})^2 = 1$, 0, and 1. If there are the same number of observations at the three timepoints, this centering removes the correlation between the linear and quadratic trend components entirely. In the more usual situation of nonequal numbers of observations across time, this greatly diminishes the correlation between the polynomials. Another aspect of centering time is that the meaning of the model intercept changes. In the previous raw form of time, the intercept represented differences at the first timepoint (*i.e.*, when time = 0). Alternatively, in centered form, the model intercept represents differences at the midpoint of time. More generally, in models with more than a linear trend, orthogonal polynomials yield an intercept term that represents an average across time.

An additional advantage of using orthogonal polynomials, over simply centering time, is that the polynomials are put on the same scale. Thus, their estimated coefficients can be compared in terms of their magnitude in the same way as standardized beta coefficients in ordinary regression analysis. This provides a way of assessing the relative contribution of each of the polynomials. Also, in the original scale it gets increasingly difficult to estimate the regression coefficients of higher-degree polynomial terms because the coefficients (and their standard errors) get smaller and smaller. This computational problem is removed by putting the polynomials on the same scale.

For equal time intervals, tables of orthogonal polynomials can be found in several statistics texts, for example in Pearson and Hartley [1976]. For example, for a study with six equally spaced timepoints the orthogonal polynomials for the constant, linear, and quadratic trend components are given as

$$\begin{bmatrix} 1 & 1 & 1 & 1 & 1 & 1 \\ -5 & -3 & -1 & 1 & 3 & 5 \\ 5 & -1 & -4 & -4 & -1 & 5 \end{bmatrix} \begin{matrix} /\sqrt{6}, \\ /\sqrt{70}, \\ /\sqrt{84}. \end{matrix}$$

The rows here represent the three orthogonal polynomials (constant, linear, and quadratic) and the columns are the six timepoints. These row vectors are independent of each other, since the sum of inner products equals 0 in all cases (*e.g.*, $-5\times5+-3\times-1+\cdots+5\times5=0$). Also, by dividing the values by the square root of the quantities on the right, which are simply the sum of squared values in a row, these polynomials have the same scale. Thus, these terms are simultaneously made independent of each other and standardized to the same (unit) scale. This holds exactly when the number of observations at each timepoint are equal, and approximately so in the more usual situation when they are unequal.

To use orthogonal polynomials in a MRM, variables representing the trend components must be created. This can be done using a series of `IF` statements in any statistical software program. For example, below is SAS code that can be used to create the orthogonal trend component variables for the first two timepoints, weeks 0 and 1, of this example with 6 timepoints (assuming a variable named `WEEK` that takes on values 0 to 5 in the dataset).

```
IF WEEK = 0 THEN DO;
       CONS = 1 / SQRT(6);
       LIN = -5 / SQRT(70);
       QUAD = 5 / SQRT(84);
END;
IF WEEK = 1 THEN DO;
       CONS = 1 / SQRT(6);
       LIN = -3 / SQRT(70);
       QUAD = -1 / SQRT(84);
END;
```

These statements assign the orthogonal polynomial values that for the first two columns of the matrix. Similar `IF` statements would be used to produce the orthogonal polynomial values for the remaining study timepoints (*i.e.*, the rightmost four columns of the matrix). The new variables `CONS`, `LIN`, and `QUAD` would then be entered into MRM analysis as regressors and random effects, with the caveat that an intercept should not be included since the `CONS` term has taken its place.

While looking up the values of orthogonal polynomials in statistical tables can be interesting (and possibly cure insomnia!), these can also be calculated directly. Bock [1975] describes a method for doing this utilizing the Cholesky factorization (*i.e.*, matrix square root) of a symmetric matrix. For this, the symmetric matrix $\boldsymbol{A}$ is factored as $\boldsymbol{A} = \boldsymbol{S}\boldsymbol{S}'$, where $\boldsymbol{S}$ is the lower triangular Cholesky factor, and $\boldsymbol{S}'$ is its upper triangular counterpart (*i.e.*, the transpose of the lower triangular matrix, since $'$ denotes the transpose of a matrix). Continuing with our quest of directly obtaining orthogonal polynomials, denote the time matrix, including the intercept, as $\boldsymbol{T}$. For example, in our case of 6 timepoints and up to quadratic trend, we have

$$\boldsymbol{T}' = \begin{bmatrix} 1 & 1 & 1 & 1 & 1 & 1 \\ 0 & 1 & 2 & 3 & 4 & 5 \\ 0 & 1 & 4 & 9 & 16 & 25 \end{bmatrix}.$$

Then, the steps one takes to obtain the orthogonal polynomial matrix are as follows:

1. Compute $\boldsymbol{T}'\boldsymbol{T}$, which yields a symmetric matrix.

$$\boldsymbol{T}'\boldsymbol{T} = \begin{bmatrix} 6 & 15 & 55 \\ 15 & 55 & 225 \\ 55 & 225 & 979 \end{bmatrix}.$$

2. Obtain the Cholesky factor $\boldsymbol{S}$ of $\boldsymbol{T}'\boldsymbol{T}$, and express it in transpose form.

$$\boldsymbol{S}' = \begin{bmatrix} 2.4495 & 6.1237 & 22.4537 \\ 0 & 4.1833 & 20.9165 \\ 0 & 0 & 6.1101 \end{bmatrix}.$$

3. Obtain the inverse $(\boldsymbol{S}')^{-1}$.

$$(\boldsymbol{S}')^{-1} = \begin{bmatrix} 0.4082 & -0.5976 & 0.5455 \\ 0 & 0.2390 & -0.8183 \\ 0 & 0 & 0.1637 \end{bmatrix}.$$

4. Multiply $\boldsymbol{T}$ by this inverse $(\boldsymbol{S}')^{-1}$.

$$\boldsymbol{T}(\boldsymbol{S}')^{-1} = \begin{bmatrix} 0.4082 & -0.5976 & 0.5455 \\ 0.4082 & -0.3586 & -0.1091 \\ 0.4082 & -0.1195 & -0.4364 \\ 0.4082 & 0.1195 & -0.4364 \\ 0.4082 & 0.3586 & -0.1091 \\ 0.4082 & 0.5976 & 0.5455 \end{bmatrix}$$

$$= \begin{bmatrix} 1/\sqrt{6} & -5/\sqrt{70} & 5/\sqrt{84} \\ 1/\sqrt{6} & -3/\sqrt{70} & -1/\sqrt{84} \\ 1/\sqrt{6} & -1/\sqrt{70} & -4/\sqrt{84} \\ 1/\sqrt{6} & 1/\sqrt{70} & -4/\sqrt{84} \\ 1/\sqrt{6} & 3/\sqrt{70} & -1/\sqrt{84} \\ 1/\sqrt{6} & 5/\sqrt{70} & 5/\sqrt{84} \end{bmatrix},$$

which yields the same orthogonal polynomial values as before.

Besides replicating the results found in statistical tables, this method has an important advantage in that it can be used to create orthogonal polynomials for cases of unequal time intervals simply by modifying the original $\boldsymbol{T}$ matrix. Notice also that to add a cubic term, a row containing the t^3 values would simply be added to the $\boldsymbol{T}'$ matrix. While it might seem difficult to implement, use of matrix algebra routines in statistical packages makes this procedure relatively easy. For instance, the SAS PROC IML statements below can be used to perform the above calculations.

```
TITLE 'producing orthogonal polynomial matrix';
PROC IML;
 time = {  1  0   0   ,
           1  1   1   ,
           1  2   4   ,
           1  3   9   ,
           1  4  16   ,
           1  5  25   } ;
orthpoly = time*INV(ROOT(T(time)*time));
PRINT 'time matrix', time [FORMAT=8.4];
PRINT 'orthogonalized time matrix', orthpoly [FORMAT=8.4];
```

This code uses several built-in SAS matrix routines: `INV`, which performs matrix inversion; `ROOT`, which yields the transpose of the Cholesky factor (*i.e.*, the upper triangular matrix $\boldsymbol{S}'$); and `T`, which yields the transpose of a matrix. Additionally, matrix multiplication is performed using the $*$ operator, which is also the ordinary scalar multiplication operator. As can be seen, SAS allows these operations to be performed all on one line, so that there is no need to save intermediate results.

5.3.1 Model Representations

To distinguish between the two representations of the MRM, the original time metric $\boldsymbol{T}$ and the orthogonal polynomial time metric $\boldsymbol{T}(\boldsymbol{S}')^{-1}$, it is helpful to modify the notation for the latter. Consider first the matrix representation of the MRM:

$$\boldsymbol{y}_i = \boldsymbol{X}_i\boldsymbol{\beta} + \boldsymbol{Z}_i\boldsymbol{v}_i + \boldsymbol{\varepsilon}_i, \tag{5.4}$$

where time is expressed (in $\boldsymbol{X}$ and $\boldsymbol{Z}$) in its original metric. The mean of the normally distributed random effects $\boldsymbol{v}_i$ is $\boldsymbol{0}$ and the variance–covariance matrix is $\boldsymbol{\Sigma}_v$. Also, the errors $\boldsymbol{\varepsilon}_i$ are normally and independently distributed with $\boldsymbol{0}$ mean and variance $\sigma^2\boldsymbol{I}_{n_i}$. Thus, $\boldsymbol{y}_i$ has mean $\boldsymbol{X}_i\boldsymbol{\beta}$ and variance–covariance matrix $\boldsymbol{Z}_i\boldsymbol{\Sigma}_v\boldsymbol{Z}_i' + \sigma^2\boldsymbol{I}_{n_i}$.

Now for the model with orthogonal polynomials for time (in both $\boldsymbol{X}$ and $\boldsymbol{Z}$), this requires a simple replacement of $\boldsymbol{X}$ with $\boldsymbol{X}(\boldsymbol{S}')^{-1}$ and $\boldsymbol{Z}$ with $\boldsymbol{Z}(\boldsymbol{S}')^{-1}$. Denote the parameters in the orthogonal polynomial metric as $\boldsymbol{\gamma}$ and $\boldsymbol{\theta}_i$ for the fixed and random effects parameters, respectively. Then, the reparameterized model is given as

$$\boldsymbol{y}_i = \boldsymbol{X}_i(\boldsymbol{S}')^{-1}\boldsymbol{\gamma} + \boldsymbol{Z}_i(\boldsymbol{S}')^{-1}\boldsymbol{\theta}_i + \boldsymbol{\varepsilon}_i. \tag{5.5}$$

The random effects $\boldsymbol{\theta}_i$ are distributed normally with mean vector $\mathbf{0}$ and variance–covariance matrix Σ_θ. As a result, the dependent variable vector $\boldsymbol{y}_i$ has its mean given as $\boldsymbol{X}_i(\boldsymbol{S}')^{-1}\gamma$ and variance–covariance as

$$\begin{aligned} V(\boldsymbol{y}_i) &= (\boldsymbol{Z}_i(\boldsymbol{S}')^{-1})\Sigma_\gamma(\boldsymbol{Z}_i(\boldsymbol{S}')^{-1})' + \sigma^2 \boldsymbol{I}_{n_i} \\ &= \boldsymbol{Z}_i\left[(\boldsymbol{S}')^{-1}\Sigma_\gamma \boldsymbol{S}^{-1}\right]\boldsymbol{Z}_i' + \sigma^2 \boldsymbol{I}_{n_i} \end{aligned} \tag{5.6}$$

in the transformed metric.

5.3.2 Orthogonal Polynomial Trend Example

Here we refit the model presented in Table 5.1, except that orthogonal polynomials are used. The results of this analysis are given in Table 5.2.

Table 5.2. MRM Results for Orthogonal Polynomial Version of Level-1 Model (5.1) and Level-2 Model (5.2)

Parameter	Estimate	SE	Z	$p <$
γ_0	43.24	1.37	31.61	.0001
γ_1	−9.94	0.86	−11.50	.0001
γ_2	0.31	0.54	0.58	.56
$\sigma^2_{\theta_0}$	111.91	21.60		
$\sigma_{\theta_0\theta_1}$	37.99	10.92		
$\sigma^2_{\theta_1}$	37.04	8.90		
$\sigma_{\theta_0\theta_2}$	−10.14	6.19		
$\sigma_{\theta_1\theta_2}$	−0.82	3.80		
$\sigma^2_{\theta_2}$	7.23	3.50		
σ^2	10.52	1.11		

Note. $-2\log L = 2207.64$.

First, notice that the log-likelihood value is identical in Tables 5.1 and 5.2. Thus, the two solutions are equivalent, one is simply a reexpressed version of the other. Comparing the regression coefficients, as before, we see that only the constant and linear terms are significant. These terms also dominate in terms of magnitude; not only is the quadratic term nonsignificant, it is negligible. Thus, at the population average level, the trend is essentially linear.

Turning to the variance estimates, we see that the estimated constant variance ($\hat{\sigma}^2_{\theta_0}$) is much larger than the estimated linear trend component ($\hat{\sigma}^2_{\theta_1}$), which is much larger than the estimated quadratic trend component ($\hat{\sigma}^2_{\theta_2}$). In terms of relative percentages, these three represent 71.7, 23.7, and 4.6, respectively, of the sum of the estimated individual variance terms. Thus, at the individual level there is heterogeneity in terms of all three components, but with diminishing return as the order of the polynomial increases. This analysis then quantifies what the plot of the empirical Bayes trends in Figure 5.4 depicts.

Inspection of the covariance terms reveals a strong positive association between the constant and linear terms ($\hat{\sigma}^2_{\theta_1\theta_0} = 37.99$, expressed as a correlation = .59). This seems to be in contrast with the results for this term from the previous analysis in Table 5.1, where there was a slight negative association between the intercept and linear terms ($\hat{\sigma}^2_{\upsilon_1 \upsilon_0} = -.92$, expressed as a correlation = $-.11$). The reason for this apparent discrepancy is that in Table 5.1, the intercept represents the first timepoint, whereas the constant term in Table 5.2 represents the average across time. Thus, an individual's linear trend is both negatively associated with their baseline depression level and positively associated with their average depression level. Subjects with higher initial depression levels have more negative linear slopes, and as a result, lower average depression values.

5.3.3 Translating Parameters

The parameters from a model using orthogonal polynomials can be directly related to those in the corresponding model that uses the original metric for time. One is simply a reexpressed or translated version of the other, with the matrix $\boldsymbol{S}'$ or $(\boldsymbol{S}')^{-1}$ serving as the translator, depending on the direction of the translation. To see the connection between the two, consider again the matrix representation of the MRM:

$$\boldsymbol{y}_i = \boldsymbol{X}_i\boldsymbol{\beta} + \boldsymbol{Z}_i\boldsymbol{\upsilon}_i + \boldsymbol{\varepsilon}_i \tag{5.7}$$

in the original metric, and consider

$$\boldsymbol{y}_i = \boldsymbol{X}_i(\boldsymbol{S}')^{-1}\boldsymbol{\gamma} + \boldsymbol{Z}_i(\boldsymbol{S}')^{-1}\boldsymbol{\theta}_i + \boldsymbol{\varepsilon}_i \tag{5.8}$$

in the orthogonal metric. Thus, the two sets of model parameters are related according to

$$\boldsymbol{\beta} = (\boldsymbol{S}')^{-1}\boldsymbol{\gamma} \tag{5.9}$$

$$\boldsymbol{\upsilon}_i = (\boldsymbol{S}')^{-1}\boldsymbol{\theta}_i \tag{5.10}$$

The two representations of the variance–covariance matrix of the random effects are related in a similar way. From before, we have that

$$V(\boldsymbol{y}_i) = \boldsymbol{Z}_i\boldsymbol{\Sigma}_\upsilon\boldsymbol{Z}_i' + \sigma^2\boldsymbol{I}_{n_i} \tag{5.11}$$

in the original MRM, and

$$V(\boldsymbol{y}_i) = \boldsymbol{Z}_i\left[(\boldsymbol{S}')^{-1}\boldsymbol{\Sigma}_\gamma\boldsymbol{S}^{-1}\right]\boldsymbol{Z}_i' + \sigma^2\boldsymbol{I}_{n_i} \tag{5.12}$$

in the transformed metric. The relationship between the two versions of the variance–covariance matrix of the random effects is therefore

$$\boldsymbol{\Sigma}_\upsilon = (\boldsymbol{S}')^{-1}\boldsymbol{\Sigma}_\gamma\boldsymbol{S}^{-1}. \tag{5.13}$$

As can be seen, $(\boldsymbol{S}')^{-1}$ is the matrix translating the orthogonal polynomial parameters to those in the original metric of time, and similarly $\boldsymbol{S}'$ translates the parameters from the original metric to the orthogonal polynomial metric, namely,

$$\boldsymbol{\gamma} = \boldsymbol{S}'\boldsymbol{\beta}, \tag{5.14}$$

$$\boldsymbol{\theta}_i = \boldsymbol{S}'\boldsymbol{\upsilon}_i, \tag{5.15}$$

$$\boldsymbol{\Sigma}_\gamma = \boldsymbol{S}'\boldsymbol{\Sigma}_\upsilon\boldsymbol{S}. \tag{5.16}$$

These equations can be used to translate the estimates from one version of the model to the other simply by substituting the parameter estimates that are obtained from an analysis. For example, one can derive the results from Table 5.1 (original time) based on those from Table 5.2 (orthogonal time) according to

$$\begin{aligned}\hat{\boldsymbol{\beta}} &= (\boldsymbol{S}')^{-1}\hat{\boldsymbol{\gamma}} \\ &= \begin{bmatrix} 0.4082 & -0.5976 & 0.5455 \\ 0 & 0.2390 & -0.8183 \\ 0 & 0 & 0.1637 \end{bmatrix} \begin{bmatrix} 43.24 \\ -9.94 \\ 0.31 \end{bmatrix} = \begin{bmatrix} 23.76 \\ -2.63 \\ 0.05 \end{bmatrix}. \end{aligned} \tag{5.17}$$

To get standard errors of the parameter estimates, a bit more work is required. For this, one must obtain the variance–covariance matrix of the parameter estimates. Note that the standard errors for a set of parameter estimates are equal to the square root of the diagonal entries of this matrix. For the regression coefficients, denote these as $V(\hat{\boldsymbol{\beta}})$ and $V(\hat{\boldsymbol{\gamma}})$ for our two versions of the model. Both of these are symmetric $p \times p$ matrices, where p is the number of polynomial trend components (including the intercept). Then, the following relationships hold:

$$V(\hat{\boldsymbol{\beta}}) = (\boldsymbol{S}')^{-1}\, V(\hat{\boldsymbol{\gamma}})\, \boldsymbol{S}^{-1}, \tag{5.18}$$

$$V(\hat{\boldsymbol{\gamma}}) = \boldsymbol{S}'\, V(\hat{\boldsymbol{\beta}})\, \boldsymbol{S}. \tag{5.19}$$

Taking the square root of the diagonal elements of these matrices then provide the standard errors for these two versions of the regression coefficients.

From our orthogonal analysis in Table 5.2, we have

$$V(\hat{\boldsymbol{\gamma}}) = \begin{bmatrix} 1.8708 & 0.5823 & -0.1402 \\ 0.5823 & 0.7470 & 0.0214 \\ -0.1402 & 0.0214 & 0.2914 \end{bmatrix}$$

and so,

$$\begin{aligned} V(\hat{\boldsymbol{\beta}}) &= \begin{bmatrix} 0.4082 & -0.5976 & 0.5455 \\ 0 & 0.2390 & -0.8183 \\ 0 & 0 & 0.1637 \end{bmatrix} \begin{bmatrix} 1.8708 & 0.5823 & -0.1402 \\ 0.5823 & 0.7470 & 0.0214 \\ -0.1402 & 0.0214 & 0.2914 \end{bmatrix} \\ &\times \begin{bmatrix} 0.4082 & 0 & 0 \\ -0.5976 & 0.2390 & 0 \\ 0.5455 & -0.8183 & 0.1637 \end{bmatrix} \\ &= \begin{bmatrix} 0.3048 & -0.1199 & 0.0146 \\ -0.1199 & 0.2294 & -0.0382 \\ 0.0146 & -0.0382 & 0.0078 \end{bmatrix}. \end{aligned}$$

Taking the square root of the diagonal entries of this matrix yields 0.5521, 0.4790, and 0.0883, which agree with the reported standard errors for the fixed effects in Table 5.1.

To obtain the standard errors for the variance–covariance parameter estimates, it is somewhat more difficult though the logic is the same. Notice that the regression coefficients are $p \times 1$ parameter vectors, whereas $\mathbf{\Sigma}_\upsilon$ and $\mathbf{\Sigma}_\theta$ are $r \times r$ symmetric matrices of parameters (where r is the number of random effects). The first step is then to vectorize these parameter matrices. This can be done using results from McCulloch [1982]. Specifically, we can vectorize the relationship $\mathbf{\Sigma}_\theta = \boldsymbol{S}'\mathbf{\Sigma}_\upsilon\boldsymbol{S}$ as

$$\begin{aligned}\text{vec}\mathbf{\Sigma}_\theta &= \text{vec}(\boldsymbol{S}'\mathbf{\Sigma}_\upsilon\boldsymbol{S}) \\ &= (\boldsymbol{S}' \otimes \boldsymbol{S}')\text{vec}\mathbf{\Sigma}_\upsilon,\end{aligned}$$

where $\text{vec}\mathbf{\Sigma}_\theta$ is the $r^2 \times 1$ vector that is obtained by stacking the r column vectors of the matrix $\mathbf{\Sigma}_\theta$ on top of each other (similarly for vec $\mathbf{\Sigma}_\upsilon$), and $\otimes$ represents the Kronecker product. For example, for a 3×3 matrix $\mathbf{\Sigma}$ the vec would be (using σ to represent either a variance or covariance parameter):

$$\text{vec}\mathbf{\Sigma} = (\sigma_{11}\,\sigma_{21}\,\sigma_{31}\,\sigma_{12}\,\sigma_{22}\,\sigma_{32}\,\sigma_{13}\,\sigma_{23}\,\sigma_{33})'.$$

Since $\mathbf{\Sigma}_\upsilon$ and $\mathbf{\Sigma}_\theta$ are symmetric matrices, not all of the r^2 elements in $\text{vec}\mathbf{\Sigma}_\theta$ or $\text{vec}\mathbf{\Sigma}_\upsilon$ are unique; instead, each have $r \times (r+1)/2$ unique parameters. For this, McCulloch [1982] defines the "vech" as the $r^* \times 1$ vector containing only the unique elements of a symmetric matrix, where $r^* = r \times (r+1)/2$. The vech stacks the on or above diagonal elements of a square matrix on top of each other to form one column vector. Thus, for the same 3×3 matrix $\mathbf{\Sigma}$ considered above, the vech would be

$$\text{vech}\mathbf{\Sigma} = (\sigma_{11}\,\sigma_{12}\,\sigma_{22}\,\sigma_{13}\,\sigma_{23}\,\sigma_{33})'.$$

Furthermore, $\boldsymbol{G}$ is the $r^2 \times r^*$ transformation matrix that carries the vech into the vec, *i.e.*, $\text{vec}\mathbf{\Sigma} = \boldsymbol{G}\text{vech}\mathbf{\Sigma}$. For example, with $r = 3$, the form of the $\boldsymbol{G}$ matrix is

$$\boldsymbol{G} = \begin{bmatrix} 1 & 0 & 0 & 0 & 0 & 0 \\ 0 & 1 & 0 & 0 & 0 & 0 \\ 0 & 0 & 0 & 1 & 0 & 0 \\ 0 & 1 & 0 & 0 & 0 & 0 \\ 0 & 0 & 1 & 0 & 0 & 0 \\ 0 & 0 & 0 & 0 & 1 & 0 \\ 0 & 0 & 0 & 1 & 0 & 0 \\ 0 & 0 & 0 & 0 & 1 & 0 \\ 0 & 0 & 0 & 0 & 0 & 1 \end{bmatrix}. \tag{5.20}$$

As a result, we can write

$$\boldsymbol{G}\text{vech}\mathbf{\Sigma}_\theta = (\boldsymbol{S}' \otimes \boldsymbol{S}')\boldsymbol{G}\text{vech}\mathbf{\Sigma}_\upsilon$$

so that

$$\text{vech}\mathbf{\Sigma}_\theta = \boldsymbol{G}^+(\boldsymbol{S}' \otimes \boldsymbol{S}')\boldsymbol{G}\text{vech}\mathbf{\Sigma}_\upsilon, \tag{5.21}$$

where $\boldsymbol{G}^+$ is the Moore–Penrose inverse of $\boldsymbol{G}$ with the property that $\text{vech}\,\mathbf{\Sigma} = \boldsymbol{G}^+\text{vec}\mathbf{\Sigma}$ [Magnus, 1988]. For $r = 3$, the $\boldsymbol{G}^+$ matrix is

$$G^{+} = \begin{bmatrix} 1 & 0 & 0 & 0 & 0 & 0 & 0 & 0 & 0 \\ 0 & \frac{1}{2} & 0 & \frac{1}{2} & 0 & 0 & 0 & 0 & 0 \\ 0 & 0 & 0 & 0 & 1 & 0 & 0 & 0 & 0 \\ 0 & 0 & \frac{1}{2} & 0 & 0 & 0 & \frac{1}{2} & 0 & 0 \\ 0 & 0 & 0 & 0 & 0 & \frac{1}{2} & 0 & \frac{1}{2} & 0 \\ 0 & 0 & 0 & 0 & 0 & 0 & 0 & 0 & 1 \end{bmatrix}. \tag{5.22}$$

One can verify that pre-multiplying the vec by G^+ does yield the vech, and that that pre-multiplying the vech by G yields the vec for this 3×3 case.

Now we have the vector of r^* unique parameters in Σ_θ expressed as a function of the r^* unique parameters in Σ_υ, pre-multiplied by the matrix product $G^+(S' \otimes S')G$. While this is a more complicated expression than that relating the regression coefficients (*i.e.*, $\gamma = S'\beta$), it has the same basic form. As a result, to obtain the variance–covariance matrix of the r^* estimates in $\hat{\Sigma}_\theta$, we get

$$V(\text{vech}\hat{\Sigma}_\theta) = G^+(S' \otimes S')G\; V(\text{vech}\hat{\Sigma}_\upsilon)\; G'(S \otimes S)G^{+'}. \tag{5.23}$$

Notice that $V(\text{vech}\hat{\Sigma}_\theta)$ is a $r^* \times r^*$ matrix, and that taking the square root of the diagonal elements of this matrix yields the standard errors for the r^* unique elements of $\hat{\Sigma}_\theta$. In a similar way, we can obtain

$$V(\text{vech}\hat{\Sigma}_\upsilon) = G^+((S')^{-1} \otimes (S')^{-1})G\; V(\text{vech}\hat{\Sigma}_\theta)\; G'(S^{-1} \otimes S^{-1})G^{+'}, \tag{5.24}$$

which again is a $r^* \times r^*$ matrix, which yields the standard errors for the r^* unique elements of $\hat{\Sigma}_\upsilon$ by taking the square root of the diagonal elements of this matrix.

5.3.4 Higher-Order Polynomial Models

Thus far we have only considered curvilinear or quadratic trend models. Adding higher-order polynomials is sometimes of interest. In general if there are n timepoints, then one can consider $n - 1$ polynomials in terms of the mean structure, where the first-order polynomial is the linear, the second-order is the quadratic, etc. For example, for the Reisby dataset with six timepoints, a model with up to the 5th-order polynomial could be fit in terms of the fixed effects. With equal numbers of observations per timepoint, such a model would fit the marginal timepoint means exactly because as many parameters (intercept plus $n - 1$ polynomial terms) are estimated as timepoints (n). In the unbalanced case, the fit would be approximately equal.

How many polynomials can be fit in the random-effects part depends on the variation in the data and its correlational structure. The maximum is $n - 1$ polynomials, though in this case it is not possible to also estimate the error variance. This is because the variance–covariance matrix of y is $n \times n$, and so a maximum of $n(n-1)/2$ variance–covariance parameters are estimable. Thus, a model with a random intercept plus $n - 1$ random polynomial terms would represent a saturated model for the variance–covariance structure.

Often, interpretation of higher-order polynomials beyond say quadratic and cubic trends is complicated and there may be insufficient rationale to expect such trends anyway. Of course this depends on the problem, but often models with linear, quadratic, and possibly cubic polynomial trends do a reasonable job of balancing model fit and interpretation.

To augment our quadratic trend model to include a cubic trend, the t^3 (original or orthogonal metric) is added to the within-subjects (level-1) model, namely,

$$y_{ij} = b_{0i} + b_{1i}t_{ij} + b_{2i}t_{ij}^2 + b_{3i}t_{ij}^3 + \varepsilon_{ij}. \tag{5.25}$$

Here, the additional term b_{3i} represents the subject's cubic trend component. The level-2, between-subjects, model is now

$$\begin{aligned} b_{0i} &= \beta_0 + \upsilon_{0i}, \\ b_{1i} &= \beta_1 + \upsilon_{1i}, \\ b_{2i} &= \beta_2 + \upsilon_{2i}, \\ b_{3i} &= \beta_3 + \upsilon_{2i}, \end{aligned} \tag{5.26}$$

where β_3 represents the average cubic change and υ_{3i} is the individual deviation from average cubic change. As in the quadratic trend model, in some cases it may be reasonable to restrict the cubic part to the level-1 model only (in which case the final component of the level-2 model would simply be $b_{3i} = \beta_3$).

Figure 5.5 illustrates some cubic curves, where the general trend is positive across time. As can be seen, a wide variety of curves are possible. One shape that is particularly useful is the S-shaped curve akin to Figure 5.5c. In this type of curve the response is first relatively flat, then rises, and finally levels off. Though such a nonlinear relationship across time may be fitted best using a logistic response curve, the cubic model does a reasonable job of approximating this type of trend in some cases.

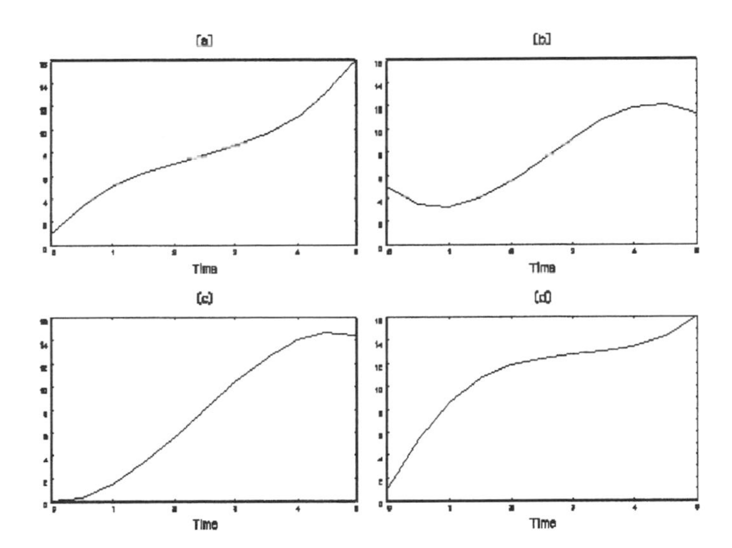

Figure 5.5. Cubic models of generally positive change across time: (a) $y = 1 + 5.5t - 1.75t^2 + 0.25t^3$; (b) $y = 5 - 4.5t + 3.15t^2 - 0.40t^3$; (c) $y = 0 - 0.5t + 2.30t^2 - 0.325t^3$; (d) $y = 1 + 10.5t - 3.25t^2 + 0.35t^3$.

One must realize that polynomial curves can yield nonsensical values of y if extrapolated in the x direction. Figure 5.6 illustrates these same cubic curves, except that now two extra weeks are added on to each side for time, so that now the range is -2 to 7 (the dashed lines in these figures represent the original timeframe of 0 to 5). Notice that these cubic trends don't level off at weeks 0 or 5, but instead keep going and going, yielding values of y that are (eventually) "off the map." Thus, it is a good idea to plot the estimated curve(s) from an analysis using higher-order polynomials to ensure that the interpretation is reasonable. Alternatively, though we won't consider them here, more general polynomial models with asymptotes are possible and have been described in an MRM context by Jones [1996].

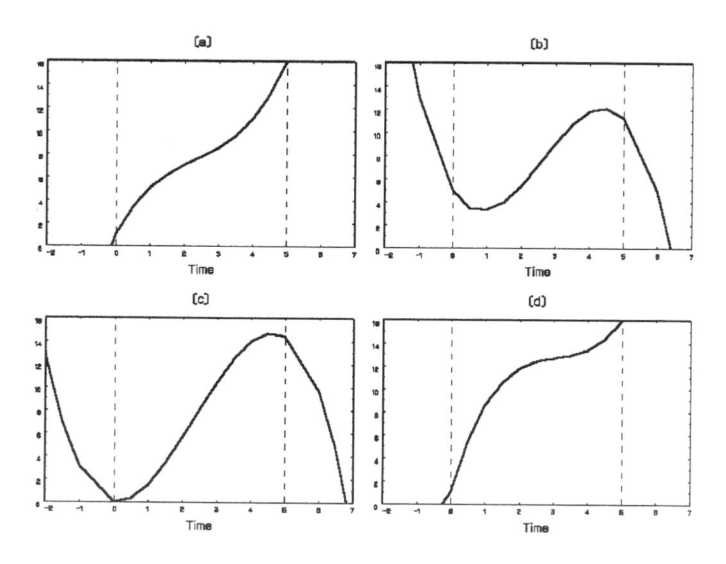

Figure 5.6. Extrapolation of cubic models across time: (a) $y = 1+5.5t-1.75t^2+0.25t^3$; (b) $y = 5-4.5t+3.15t^2-0.40t^3$; (c) $y = 0-0.5t+2.30t^2-0.325t^3$; (d) $y = 1+10.5t-3.25t^2+0.35t^3$.

5.3.5 Cubic Trend Example

The Reisby dataset is fit using a orthogonal cubic trend model and the results of this analysis are given in Table 5.3. Comparing the deviance from this model (2196.44) with that from the quadratic trend model (2207.64) yields a difference of $X^2 = 11.2$, on five degrees of freedom, which is significant at $p < .05$ even without an adjustment of the p-value for the inclusion of variance terms in this global test of

$$H_0 : \gamma_3 = \sigma^2_{\theta_3} = \sigma_{\theta_0\theta_3} = \sigma_{\theta_1\theta_3} = \sigma_{\theta_2\theta_3} = 0.$$

The estimates of the fixed effects reaffirm the linearity of the response across time at the average level. Neither the quadratic nor the cubic terms are statistically significant. Relative to the model with only quadratic trend in Table 5.2, notice how little the estimates for the intercept, linear trend, and quadratic trend have changed. Again, if the data were complete across time, these would not change at all, but because the data are not complete

Table 5.3. MRM Results for Level-1 Model (5.25) and Level-2 Model (5.26)

Parameter	Estimate	SE	Z	$p <$
Intercept γ_0	43.24	1.35	31.95	.0001
Linear trend γ_1	−9.88	0.84	−11.70	.0001
Quadratic trend γ_2	0.33	0.55	0.60	.55
Cubic trend γ_3	0.62	0.48	1.29	.20
$\sigma^2_{\theta_0}$	110.91	21.08		
$\sigma_{\theta_0\theta_1}$	34.92	10.25		
$\sigma^2_{\theta_1}$	36.25	8.25		
$\sigma_{\theta_0\theta_2}$	−11.14	6.12		
$\sigma_{\theta_1\theta_2}$	0.26	3.69		
$\sigma^2_{\theta_2}$	9.24	3.58		
$\sigma_{\theta_0\theta_3}$	7.25	5.31		
$\sigma_{\theta_1\theta_3}$	−4.33	3.27		
$\sigma_{\theta_2\theta_3}$	4.03	2.16		
$\sigma^2_{\theta_3}$	5.04	2.85		
σ^2	8.92	1.14		

Note. $-2 \log L = 2196.44$.

across time, these trend components are not exactly independent of each other and so there are slight changes when an additional polynomial is added to the model.

Turning to the variance estimates, we see that these clearly diminish as the order of the polynomial is increased. In terms of relative percentages, the four trend components represent 68.7, 22.5, 5.7, and 3.1, respectively, of the sum of the estimated individual variance terms. Thus, over 90% of the individual heterogeneity is in terms of the constant (*i.e.*, average across time) and the linear trend component. Though statistically significant, the quadratic and cubic terms do not account for a great deal of the individual differences in trends across time.

This can also be seen from Figure 5.7 which presents the observed means, estimated means, and estimated individual trends based on the cubic model. As has been seen before with this dataset, the observed means clearly descend linearly across time. Though the plot is a bit busy, one can discern a few individuals displaying nonlinear trend across time, though the extent of the nonlinearity does not appear to be that great.

To zoom in on the nonlinear trends a little more, Figure 5.8 presents the data for the ten subjects with largest estimates (in absolute value) of the cubic trend component $\hat{\gamma}_3 + \hat{\theta}_{i3}$. This plot does show a backwards S-shaped curve for at least a subset of individuals. For these subjects, the decrease in depression scores is initially not strong, then grows in the middle timepoints, and finally tapers off or increases slightly at the end. One subject even has increased depression scores following an elongated S-shaped curve. Thus, the cubic trend component does play a role in explaining some of the individual heterogeneity in trends across time, albeit a relatively modest role.

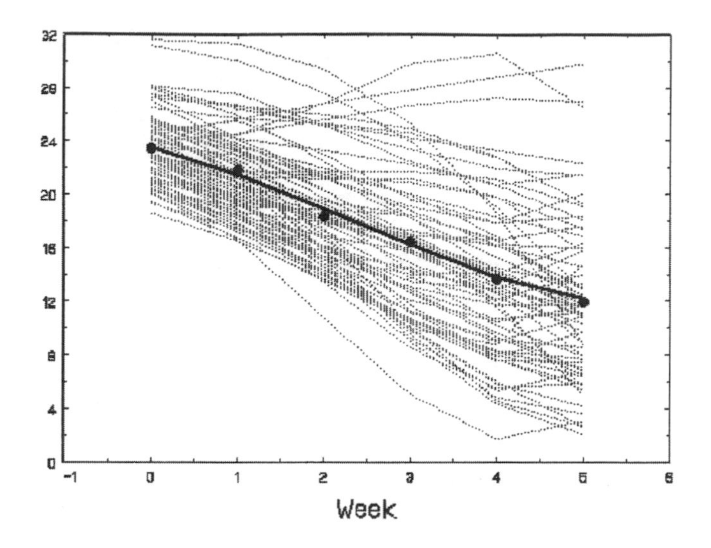

Figure 5.7. Reisby data: Observed means (solid circles), estimated means (solid line), and estimated individual trends (dotted).

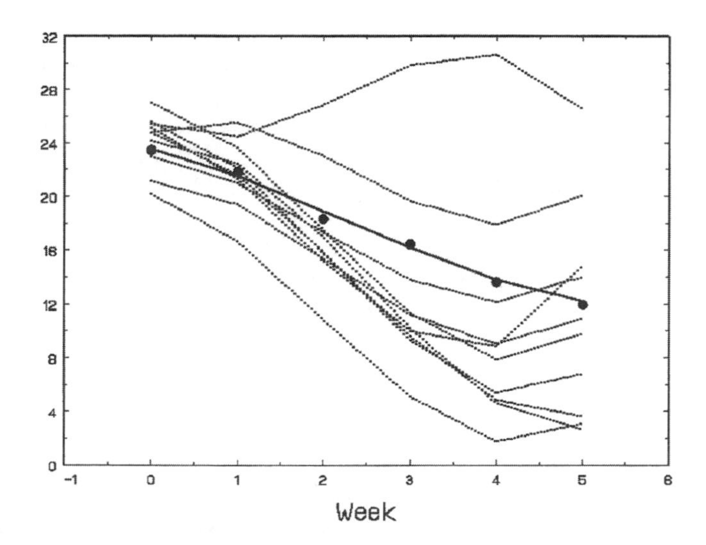

Figure 5.8. Reisby data: Observed means (solid circles), estimated means (solid line), and estimated individual trends (dotted) for 10 subjects with largest cubic trend components.

Table 5.4. Observed and Estimated Standard Deviations Across Time

	Week 0	Week 1	Week 2	Week 3	Week 4	Week 5
Observed	4.53	4.70	5.49	6.41	6.97	7.22
Model-Based Estimates						
Random intercept	5.93	5.93	5.93	5.93	5.93	5.93
Random linear trend	4.98	4.91	5.24	5.92	6.84	7.91
Random quadratic trend	4.58	4.88	5.57	6.19	6.78	7.69
Random cubic trend	4.50	4.73	5.31	6.37	7.06	7.32

Finally, Table 5.4 presents the observed standard deviations across time, along with the model-based estimates for the various MRMs fit to these data. One can see that the fit of the observed standard deviations is incrementally better as the order of the polynomial is increased. In particular, comparing the random quadratic trend to the random cubic trend, one gets a feel for why the latter is preferred by the likelihood ratio test; the cubic trend model does an excellent job of fitting the variation in the observed HDRS scores across time. It should be noted that each of these models includes all lower-order polynomials as random effects as well. Thus, for example, the random quadratic model includes random intercept and linear trend components too.

The improved fit of the cubic model for the variance–covariance structure can also be seen in the (co)variance plot in Figure 5.9. By comparing the different model estimates to the observed variances and covariances, one gets a sense of the improvement in model fit provided as the order of the polynomial is increased.

5.4 SUMMARY

Polynomial trend models are considered in this chapter, in particular models with quadratic and cubic trend components. Since trends across time can be nonlinear, the use of polynomials is a useful tool for longitudinal data analysis. It is important to remember that the model is still a linear one, since it is linear in terms of the regression coefficients. The polynomial regressors are what allow the linear model to fit some types of nonlinear relationships. In some cases, a nonlinear regression part is required, for example, for modeling of human stature across time [Bock and du Toit, 2004]. For interested readers, nonlinear mixed-effects regression models are considered extensively in Davidian and Giltinan [1995] and Vonesh and Chinchilli [1997].

Since the orthogonal polynomial representation greatly reduces any collinearity and scale differences in the regressors, it is computationally easier to obtain. For this reason, in cases where numerical difficulties are occurring with analyses using raw time values, investigators might consider using orthogonal polynomials instead. In this chapter we have considered the use of orthogonal polynomials in MRMs in detail, though they can be applied more generally in longitudinal data analysis models. For example, in later chapters we will present MRMs and GEE models for categorical outcomes, and orthogonal polynomials can be effectively applied within these types of models as well.

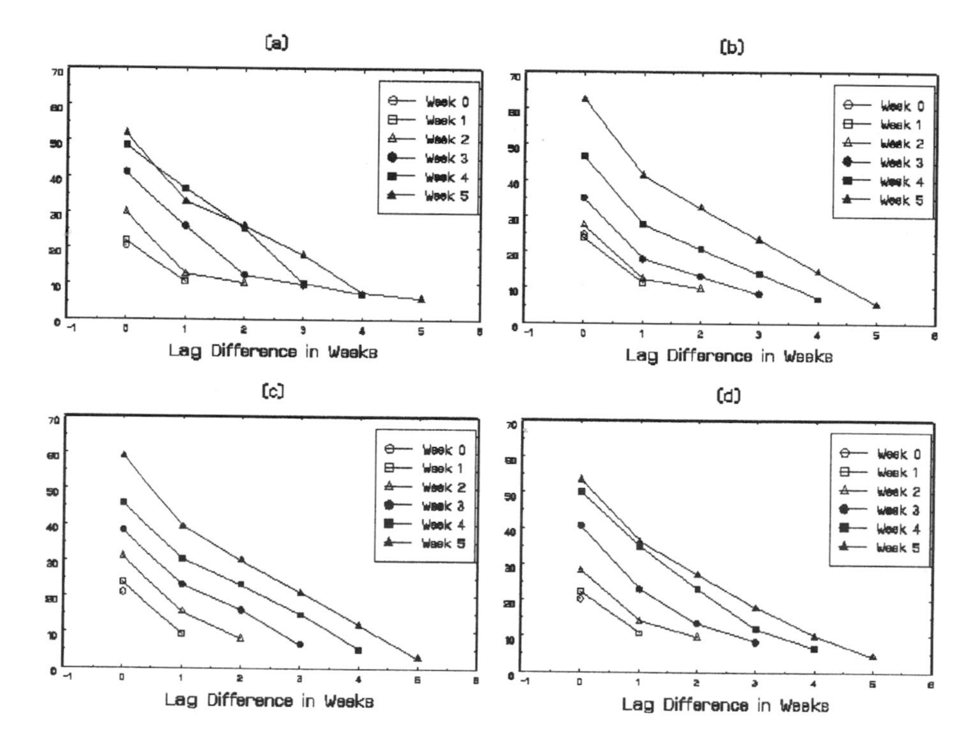

Figure 5.9. Reisby data: (a) Observed (co)variances; (b) random linear trend model (co)variance estimates; (c) random quadratic trend model (co)variance estimates; (d) random cubic trend model (co)variance estimates.

CHAPTER 6

COVARIANCE PATTERN MODELS

6.1 INTRODUCTION

The previous two chapters have focused on mixed-effects regression models (MRMs). The class of MRMs can be thought of as an extension of the univariate repeated measures ANOVA model in the sense that, like the ANOVA model, random subject effects are included to account for the clustering of the repeated observations, and their inherent correlation, within subjects. Among other things, MRMs extend the univariate repeated measures model because they can include more than one random subject effect and because they readily allow unbalanced data (*i.e.*, different numbers of observations per subject).

In a similar way, the class of models described in this chapter, covariance pattern models (CPMs), can be thought of as an extension of the multivariate analysis of variance (MANOVA) model for repeated measures. Like MANOVA, and unlike ANOVA and MRMs, CPMs do not distinguish between-subjects and within-subjects variance. Instead, the variance–covariance matrix of the repeated measures is assumed to be of a particular form and not the result of including random subject effects. Also, like ANOVA and MANOVA, CPMs treat time in a categorical sense in determining the variance–covariance matrix of the repeated measures. That is, the timing of the repeated measures is fixed or the same across all subjects. And like the MANOVA model, CPMs can allow for a general unstructured form for this variance–covariance matrix. However, unlike the MANOVA model, CPMs allow subjects to have incomplete data across the fixed number of timepoints and also allow for a variety of possible variance–covariance structures for the repeated measures. Additionally, the model is written as a regression model for greater flexibility.

6.2 COVARIANCE PATTERN MODELS

In their seminal paper Jennrich and Schluchter [1986] describe this class of models, which were then implemented in the BMDP 5V software program. Schluchter [1988] further develops and describes this approach. As mentioned, it is assumed that the timing of the measurements is fixed, in the sense that subjects are intended to be measured at the same finite number of occasions, but that subjects might have incomplete data across these fixed measurement occasions. A usual linear regression model is posited for describing how the conditional mean of the dependent variable depends on covariates; however, the error variance–covariance matrix is allowed to take on a number of possible forms.

The CPM for the $n_i \times 1$ response vector $\boldsymbol{y}$ for individual i can be written as

$$\underset{n_i \times 1}{\boldsymbol{y}_i} = \underset{n_i \times p}{\boldsymbol{X}_i} \underset{p \times 1}{\boldsymbol{\beta}} + \underset{n_i \times 1}{\boldsymbol{e}_i} \tag{6.1}$$

with $i = 1 \ldots N$ individuals and $j = 1 \ldots n_i$ observations for individual i. Here, $\boldsymbol{y}_i$ is the $n_i \times 1$ dependent variable vector for individual i, $\boldsymbol{X}_i$ is the $n_i \times p$ covariate matrix for individual i, $\boldsymbol{\beta}$ is the $p \times 1$ vector of fixed regression parameters, and $\boldsymbol{e}_i$ is the $n_i \times 1$ error vector. The vector $\boldsymbol{e}_i$ is assumed to be normally distributed with zero mean and variance–covariance matrix $\boldsymbol{\Sigma}_i$. Under these assumptions, the observations $\boldsymbol{y}_i$ are normally distributed with mean $\boldsymbol{X}_i\boldsymbol{\beta}$ and variance–covariance matrix $\boldsymbol{\Sigma}_i$. That is, the mean of the repeated measures is given by $\boldsymbol{X}_i\boldsymbol{\beta}$ (as in an ordinary multiple regression model), and the conditional variance–covariance of $\boldsymbol{y}_i$ given $\boldsymbol{X}_i$ is $\boldsymbol{\Sigma}_i$ (whereas an ordinary multiple regression model has conditional variance of σ^2). Thus, this is an extension of an ordinary multiple regression model by allowing a more general form for the (co)variances of the dependent variable.

Each $\boldsymbol{\Sigma}_i$ is a submatrix of the overall $n \times n$ matrix $\boldsymbol{\Sigma}$, where n is the total number of fixed timepoints. Most of the structures presented in this chapter assume that the n timepoints are equally spaced, though this restriction can be relaxed (see N´uñez-Antón and Woodworth [1994]). A particular individual's $n_i \times n_i$ matrix $\boldsymbol{\Sigma}_i$ is a potentially reduced version of the $n \times n$ matrix $\boldsymbol{\Sigma}$, depending on how many of the n timepoints that individual was measured at. That is, if $n_i < n$ then $\boldsymbol{\Sigma}_i$ has the appropriate rows and columns of $\boldsymbol{\Sigma}$ removed for that individual. The (co)variance matrix $\boldsymbol{\Sigma}$, and thus each $\boldsymbol{\Sigma}_i$, is assumed to be a function of a vector $\boldsymbol{\theta}$ of q (co)variance parameters. The number of parameters depends on the form, or structure, of the variance–covariance matrix. Jennrich and Schluchter [1986] consider several possible forms for $\boldsymbol{\Sigma}$.

6.2.1 Compound Symmetry Structure

A simple form for the variance–covariance matrix is that of compound symmetry (CS), which specifies equal variances and equal covariances. In matrix form, this form can be written as

$$\boldsymbol{\Sigma} = \begin{bmatrix} \sigma^2 + \sigma_1^2 & \sigma_1^2 & \sigma_1^2 & \ldots & \sigma_1^2 \\ \sigma_1^2 & \sigma^2 + \sigma_1^2 & \sigma_1^2 & \ldots & \sigma_1^2 \\ \sigma_1^2 & \sigma_1^2 & \sigma^2 + \sigma_1^2 & \ldots & \sigma_1^2 \\ \cdot & \cdot & \cdot & \ldots & \cdot \\ \sigma_1^2 & \sigma_1^2 & \sigma_1^2 & \ldots & \sigma^2 + \sigma_1^2 \end{bmatrix} . \tag{6.2}$$

Notice that the variance of the dependent variable equals $\sigma^2 + \sigma_1^2$ at every timepoint, and the covariance equals σ_1^2 for the pairwise association of the dependent variable for any two timepoints. The number of variance–covariance parameters $q = 2$ for this structure, and this is the same form that results from a random intercept model, considered earlier in Chapter 4. Thus, the CS structure for the repeated measures results from either specifying a random intercept MRM or by a CPM with the (error) variance–covariance matrix following a CS structure.

6.2.2 First-Order Autoregressive Structure

Another form that only depends on two parameters (*i.e.*, $q = 2$), but that is often better suited to longitudinal data, is the first-order autoregressive (AR1) structure. This is also called a first-order Markov process and is extensively used in time-series analysis [Gottman, 1981]. Here, the (co)variance for timepoints j and j' equals

$$\sigma_{j\,j'} = \sigma^2 \rho^{|j-j'|}, \tag{6.3}$$

where ρ is the AR(1) parameter and σ^2 is the error variance. In terms of the matrix formulation, this structure is given as

$$\Sigma = \sigma^2 \begin{bmatrix} 1 & \rho & \rho^2 & \dots & \rho^{n-1} \\ \rho & 1 & \rho & \dots & \rho^{n-2} \\ \rho^2 & \rho & 1 & \dots & \rho^{n-3} \\ \cdot & \cdot & \cdot & \dots & \cdot \\ \rho^{n-1} & \rho^{n-2} & \rho^{n-3} & \dots & 1 \end{bmatrix}. \tag{6.4}$$

Notice that the correlation decreases exponentially across the lags of the timepoints. For example, if $\rho = .5$, then the correlation of lag-1 or adjacent timepoints is .5, the correlation of lag-2 timepoints is $.5^2 = .25$, the correlation of lag-3 timepoints is $.5^3 = .125$, etc. For longitudinal data, it is common for correlations to diminish as the lag between the timepoints increases, though not always in the exponential manner as AR(1) implies. As described in the next chapter, the above variance–covariance matrix is often parameterized in a slightly different form in the time-series and econometrics literatures, though this difference is not critical.

6.2.3 Toeplitz or Banded Structure

Another useful structure for diminishing correlations across lags that is not as rigid as AR(1) is provided by the Toeplitz or banded structure . Here each lag has its own correlation parameter, namely $\sigma_{j\,j'} = \theta_k$, where $k = | j - j' | + 1$. In matrix form, it specifies

$$\Sigma = \begin{bmatrix} \theta_1 & \theta_2 & \theta_3 & \dots & \theta_n \\ \theta_2 & \theta_1 & \theta_2 & \dots & \theta_{n-1} \\ \theta_3 & \theta_2 & \theta_1 & \dots & \theta_{n-2} \\ \cdot & \cdot & \cdot & \dots & \cdot \\ \theta_n & \theta_{n-1} & \theta_{n-2} & \dots & \theta_1 \end{bmatrix}. \tag{6.5}$$

Here, in this representation of the Toeplitz form, θ_1 equals the variance, θ_2 is the lag-1 covariance, θ_3 is the lag-2 covariance, etc. Whereas the lagged associations are functionally related under AR(1), this is relaxed for the Toeplitz structure. For example, it could be that $\theta_2 = .5$, $\theta_3 = .4$, and $\theta_4 = .3$; instead of the .5, .25, and .125 of the AR(1) form. In general, $q = n$ for the Toeplitz form, though in some cases higher-order lags (*i.e.*, lag n, lag $n - 1$, etc.) can be set to zero and then $q \leq n$. This is usually done in cases where n is large.

6.2.4 Unstructured Form

All of the above structures assume that the variance is constant across time and that the lagged correlations are either all the same (compound symmetry), decrease exponentially (AR-1), or are equal within a lag (Toeplitz). Also, the AR(1) and Toeplitz structures are only reasonable if the time intervals are the same or nearly the same. Clearly, there are cases when one or more of these assumptions are not met. For this, one can assume a general unstructured form that allows all of the parameters of the variance–covariance matrix to be different, namely,

$$\Sigma = \begin{bmatrix} \theta_{11} & \theta_{12} & \theta_{13} & \dots & \theta_{1n} \\ \theta_{21} & \theta_{22} & \theta_{23} & \dots & \theta_{2n} \\ \theta_{31} & \theta_{32} & \theta_{33} & \dots & \theta_{3n} \\ \cdot & \cdot & \cdot & \dots & \cdot \\ \theta_{n1} & \theta_{n2} & \theta_{n3} & \dots & \theta_{nn} \end{bmatrix}. \tag{6.6}$$

Here, because this is a symmetric matrix (and so $\theta_{j\,j'} = \theta_{j'\,j}$), there are $q = n(n+1)/2$ unique parameters. Note that this is the same form that is assumed by the MANOVA model; however, incomplete data across time are allowable under the more general CPM rubric.

6.2.5 Random-Effects Structure

All of the above CPMs are what Jennrich and Schluchter [1986] refer to as "fully specified structures." These are the structures that are typically thought of when one refers to CPMs. For a random-effects structure, an additional specification must be made, namely

$$\boldsymbol{e}_i = \boldsymbol{Z}_i \boldsymbol{v}_i + \boldsymbol{\varepsilon}_i, \tag{6.7}$$

where $\boldsymbol{Z}_i$ is the $n_i \times r$ design matrix for the random effects and $\boldsymbol{v}_i$ is the $r \times 1$ vector of random individual effects. The distribution of the random effects is $\mathcal{N}(0, \boldsymbol{\Sigma}_v)$ and the distribution of the random vector $\boldsymbol{\varepsilon}_i$ is $\mathcal{N}(0, \sigma^2 \boldsymbol{I}_{n_i})$. The variance–covariance matrix of the repeated measures is of the form

$$\boldsymbol{\Sigma}_i = \boldsymbol{Z}_i \boldsymbol{\Sigma}_v \boldsymbol{Z}_i' + \sigma^2 \boldsymbol{I}_{n_i}. \tag{6.8}$$

Notice that, unlike the CPMs, the random-effects structure separates the between-subjects variance $\boldsymbol{\Sigma}_v$ (that part attributable to the random subject effects $\boldsymbol{v}_i$) from the within-subjects variance σ^2 (that part attributable to $\boldsymbol{\varepsilon}_i$). The total number of (co)variance parameters is $q = r(r+1)/2 + 1$, which depends on the number of random effects.

6.3 MODEL SELECTION

An important consideration is determining which of these (co)variance structures to use for a given dataset. In their paper, Jennrich and Schluchter [1986] utilize likelihood ratio tests to compare the various structures to the unstructured form, the latter being a full or saturated model for the variances and covariances. The idea is that if a given structure, which represents some kind of restriction of the general form, does not fit the data statistically worse than the full model (*i.e.*, unstructured), then this structure is a reasonable one. Note that the degrees of freedom for this test equal $(n(n-1)/2)$ - q^*, where $(n(n-1)/2)$ and q^* are the numbers of (co)variance parameters estimated by the full and reduced models, respectively.

There are a few considerations in carrying out model selection. First, the covariates need to be equivalent in the models being compared. Either ML or REML can be used for model estimation and likelihood calculation, but, of course, the method should be the same for a given likelihood ratio test. Also, all covariates of potential interest should be included, since the significance tests of the covariates depends on the (co)variance structure. In other words, model selection of the (co)variance structure is the first stop in a more general two-step procedure of model selection: (1) Including all covariates of potential interest, select an appropriate (co)variance structure; (2) once a (co)variance structure is selected as appropriate, model trimming of the covariates is performed in the usual manner. Because in step (1) these comparisons involve null hypotheses of (co)variance parameters, p-values from the likelihood ratio test need to be adjusted as mentioned in Chapter 4. Namely, as described in detail by Snijders and Bosker [1999] and Berkhof and Snijders [2001], an approximate adjustment that has been shown to work reasonably well is to divide these p-values by two. This adjustment does not apply to use of the likelihood ratio test for comparisons of models with different covariates, only for comparison of models with different variance–covariance parameters.

6.4 EXAMPLE

The published psychiatric dataset from Bock [1983b], which was an early example of a MRM analysis, will be used to illustrate use of CPMs. The dataset consists of 75 depressed patients who received either (a) three weeks of tricyclic antidepressant (TCA) treatment followed by three weeks of no drug treatment (n = 46) or (b) three weeks of no drug treatment followed by three weeks of TCA treatment (n = 29). As this was an observational study, patients were not randomized to these two conditions. Also, all subjects provided data at all six weekly assessments. The dependent variable is the patient's clinical status, as measured by the Weekly Psychiatric Status Scale for Episodic Affective Disorders (WPSS). At each week, patients received a rating on this scale, with scores of: 1, usual self; 2, residual symptomatology; 3, partial remission; 4, marked symptomatology; 5, definite criteria for major depressive disorder; or 6, definite criteria for major depressive disorder with extreme impairment. This type of quasi-continuous scale could be analyzed as an ordinal outcome, using methods described in Chapter 10, here we will treat this as a continuous outcome.

Table 6.1 presents observed WPSS means, standard deviations, and correlations across time. The means suggest that both groups improve to some extent over time. The standard deviations increase across time, and the correlations are very high though they generally diminish as the time interval increases.

Table 6.1. Observed WPSS Means, Standard Deviations, and Correlations Across Time

Treatment group	*N*	Week 1	2	3	4	5	6
		Means					
TCA-None	46	3.76	3.46	3.11	2.89	2.80	2.74
None-TCA	29	4.72	4.62	4.55	4.45	4.21	3.90
		Standard deviations					
		1.30	1.40	1.53	1.61	1.66	1.65
		Correlations					
		1.00					
		0.91	1.00				
		0.75	0.87	1.00			
		0.68	0.82	0.91	1.00		
		0.59	0.70	0.78	0.88	1.00	
		0.60	0.68	0.72	0.84	0.96	1.00

In terms of the fixed effects for the analysis, following Bock [1983b], we will consider both a linear trend across the six timepoints and a change in linear trend between the first and last three-week periods. The latter term is of interest because patients received either drug or no treatment for three-week periods. In Bock's article, these were coded as follows:

	Week 1	Week 2	Week 3	Week 4	Week 5	Week 6
Linear trend	−5/2	−3/2	−1/2	1/2	3/2	5/2
Change of slope	−1/2	0	1/2	1/2	0	−1/2

Notice that these contrasts are expressed in centered form: the linear contrast around its midpoint of week 3.5 and the change of slope contrast is centered within each three-week period. The signs are reversed for the first and last three timepoints of the latter contrast to represent the change in linear slope between the two three-week periods. Additionally, a group dummy-code will be included to represent differences between the two groups (0 = TCA-None and 1 = None-TCA), as will interactions of this dummy-code with the two time-related contrasts (*i.e.*, linear trend and change of slope).

The group by change of slope interaction is of most interest because it tests whether the difference in slopes for each 3-week period depends on drug treatment or not. For example, suppose that the slope equaled -1 during the period when subjects received the drug and equaled 0 when not on drug. This would suggest a beneficial effect of the drug, since lower scores on the WPSS represent less symptomatology. Figure 6.1 presents a plot of WPSS means across time for the two groups under this hypothetical situation. Notice that the change of slope (*i.e.*, the difference in slope between the first and last three-week periods) is exactly opposite for the two groups. Such a pattern would result in a nonzero (and potentially statistically significant) interaction between group and change of slope.

Models with these 5 fixed effects (*i.e.*, linear trend, change of slope, group, group by linear trend, and group by change of slope), plus an intercept, were fit to these data. Table 6.2 lists the deviance values, both under ML and REML estimation, for four CPMs.

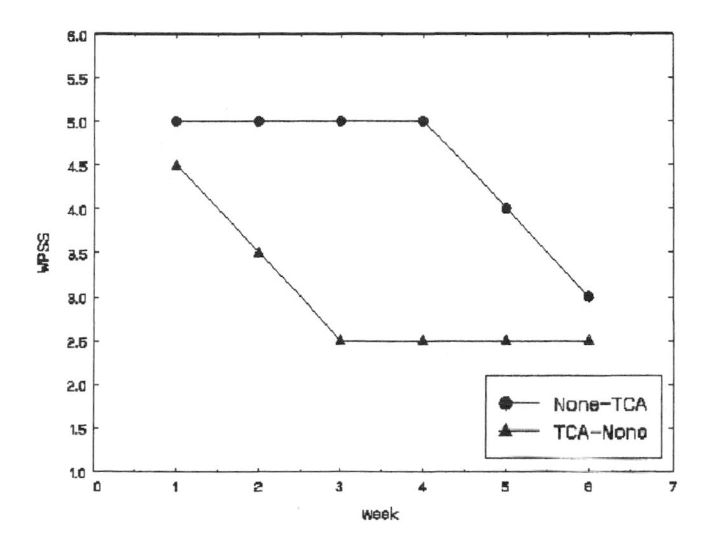

Figure 6.1. Hypothetical WPSS means across time based on group by change of slope interaction.

Table 6.2. Model Deviance Values Under ML and REML Estimation

Structure	q	ML $-2\log L$	REML $-2\log L$
UN	21	945.9	963.1
Toeplitz	6	988.9	1005.3
AR(1)	2	996.3	1013.0
CS	2	1185.8	1204.0

Using likelihood ratio tests to contrast the various CPMs to the unstructured form supports the latter. For example, for the Toeplitz, we get $X^2 = 988.9 - 945.9 = 43.0$ (or $X^2 = 1005.3 - 963.1 = 42.2$ under REML) on 15 degrees of freedom, which is highly significant ($p = .000157/2 = .0000787$). Similarly, for AR(1), it is $X^2 = 996.3 - 945.9 = 50.4$ on 19 degrees of freedom, which is also highly significant ($p = .000114/2 = .0000572$). The compound symmetry form does even worse than AR(1) with 2 parameters, and so it would also fail by the likelihood ratio test. Thus, none of the three restricted variance–covariance structures fit the data reasonably as well as the unstructured form, and so we are obliged to use the latter in our tests of the fixed effects. These are presented in Table 6.3, along with their standard errors and p-values. Here, the ML estimates are presented, which differ negligibly from the REML estimates, since the sample size is of moderate size. Because group interacts with the linear trend and change of slope terms, the "main effects" of these two are for the group coded 0, the TCA-None group. For this group the overall linear trend is clearly significant ($Z = -5.48, p < .0001$) as is the change of slope ($Z = -2.49, p < .015$).

Table 6.3. CPM Results for Unstructured Variance–Covariance Matrix

Parameter	ML Estimate	SE	Z	$p <$
Intercept	3.122	0.179	17.44	.0001
Linear trend	−0.198	0.036	−5.48	.0001
Change of slope	−0.255	0.102	−2.49	.015
Group	1.286	0.288	4.46	.0001
Group by linear trend	0.017	0.058	0.28	.78
Group by change of slope	0.475	0.164	2.89	.005

Note. $-2 \log L = 945.9$. SE = standard error.

To better understand the direction of these effects, we can calculate the estimated means at specific timepoints. For example, the estimated mean at weeks 1 and 3 for the TCA-None group is

$$\begin{aligned}\text{Week 1:} \quad \hat{y} &= (3.122) - 5/2(-.198) - 1/2(-.255), \\ \text{Week 3:} \quad \hat{y} &= (3.122) - 1/2(-.198) + 1/2(-.255),\end{aligned}$$

and so the estimated week 3–week 1 change during this period is $2(-.198) + 1(-.255)$, which translates to an estimated per week change of $-.198 + 1/2(-.255) = -.326$, or approximately a third of a point per week. Note that the fractions in the above equations are simply the values of the linear trend and change of slope contrasts at the indicated timepoints, and the parenthetical values are the estimates for the intercept, linear trend, and change of slope parameters, respectively. Similarly for this group at weeks 4 and 6, we get

$$\begin{aligned}\text{Week 4:} \quad \hat{y} &= (3.122) + 1/2(-.198) + 1/2(-.255), \\ \text{Week 6:} \quad \hat{y} &= (3.122) + 5/2(-.198) - 1/2(-.255),\end{aligned}$$

which yields a per week change of $-.198 - 1/2(-.255) = -.071$, or nearly zero. Thus, for the TCA-None group, negative slopes are estimated for both three-week periods, but the first (when they are on drug) is estimated to be more pronounced than the second (when they are not on drug). This reflects the significant change of slope effect observed in the analysis, which is estimated to be $-.326 - (-.071) = -.255$.

Turning to the group-related effects, we see that the group and group by change of slope terms are significant, and the group by linear trend is not significant. The significant group term indicates that, averaging across time, the None-TCA group is approximately 1.28 points higher on the WPSS and that this a statistically significant difference. The nonsignificant group by linear trend suggests that the two groups have similar overall linear trends across the six timepoints. As mentioned above, the significant group by change of slope is of most interest here. In particular, since it is positive it indicates that whereas the TCA-None group had a more negative slope during the first three-week period, this is reversed for the None-TCA group.

To see this, let's again calculate estimated means. For the None-TCA group, we get

$$\begin{aligned}\text{Week 1:} \quad \hat{y} &= (3.122 + 1.286) - 5/2(-.198 + .017) - 1/2(-.255 + .475), \\ \text{Week 3:} \quad \hat{y} &= (3.122 + 1.286) - 1/2(-.198 + .017) + 1/2(-.255 + .475),\end{aligned}$$

where again the estimates are in parentheses and the contrast coefficients are the fractions. The per week change for this first three-week period, when patients are not on drug, is thus $(-.198 + .017) + 1/2(-.255 + .475) = (-.181) + 1/2(.22) = -.071$, or nearly zero. Conversely, for the second period we get

$$\begin{aligned} \text{Week 4:} \quad \hat{y} &= (3.122 + 1.286) + 1/2(-.198 + .017) + 1/2(-.255 + .475), \\ \text{Week 6:} \quad \hat{y} &= (3.122 + 1.286) + 5/2(-.198 + .017) - 1/2(-.255 + .475), \end{aligned}$$

which yields a per week change of $(-.181) - 1/2(.22) = -.291$, or nearly a third of a point for this period when these patients in the None-TCA group are on drug. The analysis gives evidence that both groups benefit on drug, relative to when they are not on drug.

Figure 6.2 presents a plot of observed and estimated WPSS means across time for the two groups based on this model. As can be seen, the model fits the observed means very well. One can also see the relatively steeper negative slope for the drug period relative to the no drug period for both groups.

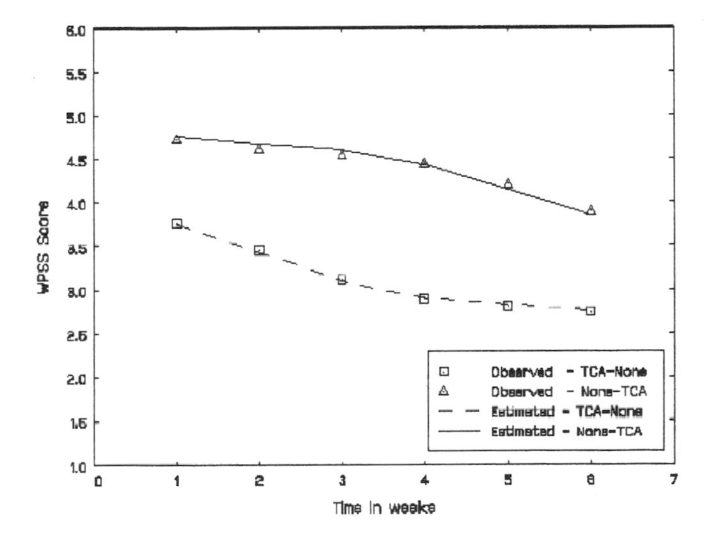

Figure 6.2. Observed and estimated WPSS means across time by group.

In terms of the variance–covariance matrix, this is estimated as

$$\hat{\Sigma} = \begin{bmatrix} 1.443 & 1.361 & 1.129 & 1.058 & 0.939 & 1.003 \\ 1.361 & 1.605 & 1.426 & 1.389 & 1.216 & 1.216 \\ 1.129 & 1.426 & 1.810 & 1.684 & 1.461 & 1.398 \\ 1.058 & 1.389 & 1.684 & 1.995 & 1.792 & 1.788 \\ 0.939 & 1.216 & 1.461 & 1.792 & 2.242 & 2.192 \\ 1.003 & 1.216 & 1.398 & 1.788 & 2.192 & 2.369 \end{bmatrix} . \tag{6.9}$$

Taking the square root of the diagonal entries yields estimated standard deviations across time as 1.201, 1.267, 1.345, 1.413, 1.497, and 1.539. These clearly increase across time indicating the growing spread in WPSS scores across the study timepoints. This gives some sense of why the other CPMs did not fit the data well, since they all assume equal variance across time. Also, converting the (co)variance matrix to a correlation matrix yields

$$\begin{bmatrix} 1.000 & 0.894 & 0.699 & 0.624 & 0.522 & 0.543 \\ 0.894 & 1.000 & 0.836 & 0.776 & 0.641 & 0.624 \\ 0.699 & 0.836 & 1.000 & 0.886 & 0.725 & 0.675 \\ 0.624 & 0.776 & 0.886 & 1.000 & 0.847 & 0.822 \\ 0.522 & 0.641 & 0.725 & 0.847 & 1.000 & 0.951 \\ 0.543 & 0.624 & 0.675 & 0.822 & 0.951 & 1.000 \end{bmatrix} , \tag{6.10}$$

which shows that the correlations decrease as the time interval is increased, and that within a lag the correlations generally increase with time. This latter feature also helps to explain why the AR(1) and Toeplitz structures were not appropriate for these data, since they assume equal correlations within a lag.

One might wonder why these estimated standard deviations and correlations are not closer to the observed values presented earlier, given that the unstructured variance–covariance matrix was estimated. Realize that $\hat{\Sigma}$ is an estimate of the conditional variance–covariance matrix of $\boldsymbol{y}$, given the covariates in $\boldsymbol{X}$, whereas the observed variance–covariance matrix consists of marginal statistics. In the present analysis, the covariates included terms to account for the general linear trend across time, differences in slope between the two periods, and group-related differences across time and change of slope. As Figure 6.2 portrays, these covariates produced a very close fit of the observed group means across time. Thus, $\hat{\Sigma}$ essentially represents the estimated variation and covariation in $\boldsymbol{y}$ accounting for the differential group means across time. This explains why the estimated conditional standard deviations are smaller than the observed marginal standard deviations. Note that this is the same reasoning for why the estimate of the error variance in an ordinary multiple regression (which represents the conditional variance of y given $\boldsymbol{X}$) is generally smaller than the observed variance of y (which is not conditional on $\boldsymbol{X}$).

6.5 SUMMARY

This chapter has presented some of the most common CPMs, but there are more that can be considered. In particular, SAS PROC MIXED can be used to fit a wide variety of forms. Some of the additional structures include generalizations of the AR(1) structure for unequal time intervals and nonstationarity, combined autoregressive moving-average structures, and antedependence forms. Several of the forms are also extended by allowing the error variance to very across time. In particular, SAS PROC MIXED allows heterogeneous error variance versions of the CS, AR(1), and Toeplitz structures. These extended forms can be quite useful in many situations. Further description of many of these additional structures can be found in Wolfinger [1993].

CHAPTER 7

MIXED REGRESSION MODELS WITH AUTOCORRELATED ERRORS

7.1 INTRODUCTION

Chapters 4 and 5 covered mixed-effects regression models (MRMs) where the errors were assumed conditionally independent (conditional on the random effects). In the last chapter, alternatively, covariance pattern models (CPMs) considered direct modeling of the error variance–covariance model in terms of several possible forms. In this chapter these two types of models for longitudinal data will essentially be combined to yield models that allow for subject heterogeneity, via random effects, plus some form of dependency of the errors.

The basic idea is that the model errors are autocorrelated over time, and so are no longer conditionally independent, given the random effects, as was assumed in Chapters 4 and 5. Different types of autocorrelated (AC) errors are possible; the forms that will be considered in this chapter include: the first-order autoregressive process (AR1), the first-order moving average process (MA1), the first-order mixed autoregressive-moving average process (ARMA), and the general Toeplitz autocorrelation stucture. These forms are commonly used in time-series analysis to describe the correlational structure of a univariate series of data, where the number of timepoints is large. Here, we will apply these structures to the errors of MRMs to enhance our modeling of the variance–covariance structure of the repeated observations.

Including autocorrelated errors in regression models has been well-described in the econometrics literature, for example, to model longitudinal earnings data [MaCurdy, 1982]. For MRMs, a key reference is the paper by Chi and Reinsel [1989], which considered a

Longitudinal Data Analysis. By Donald Hedeker and Robert D. Gibbons

MRM with AR(1) errors. Some other related references include Mansour et al. [1985], Hedeker [1989], Jones and Boadi-Boateng [1991], and Rochon [1992].

7.2 MRMS WITH AC ERRORS

Before describing the various models, let us consider some plots. First, Figure 7.1 presents a plot of data from two individuals in which the errors are uncorrelated. In this plot the solid trend line represents the population trend and the two individual trends are depicted by the dashed lines. For a given sample, there would be N such lines, here only two are presented in the figure to give a sense of the individual heterogeneity. Notice that both individuals vary from the population trend in terms of the intercept and linear time trend, so this illustration is of a MRM with random intercepts and trends. For a given individual, the data points in this plot meander around the individual's line in a more or less random manner. In other words, the data are independent conditional on the random subject effects.

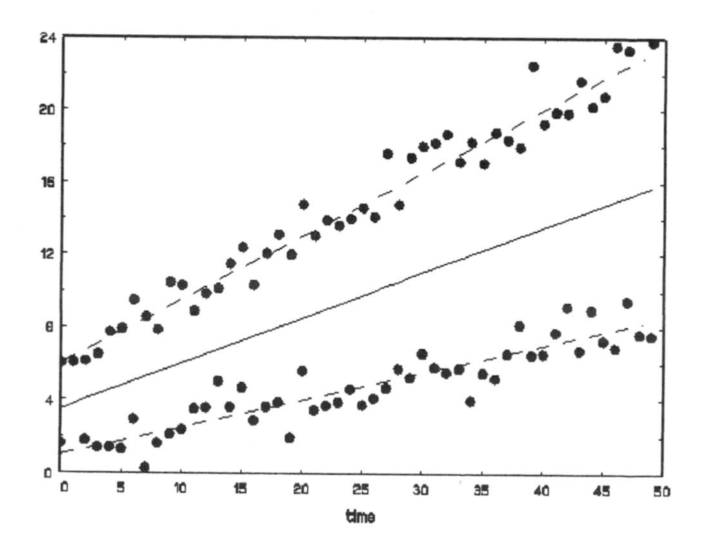

Figure 7.1. MRM for two individuals with uncorrelated errors.

In contrast, Figure 7.2 is a plot of data from the same two individuals (*i.e.*, two individuals with the same intercept and trend deviations from the population trend) in which the errors (the deviations between the data points and the individual lines) are serially correlated according to a first-order autoregressive process. In this plot, the data from each individual display a very systematic pattern of association. Notice that observations that are above (or below) an individual's trend line tend to be followed by another high (or low) observation, not only at the next timepoint, but for the next few timepoints. For illustrative purposes, this plot was generated with the AR(1) parameter equal to .75, which is high, and with 50 timepoints, which is also rather large. Detecting autocorrelations visually with real data is not always so apparent, but these plots give a sense of what the addition of autocorrelated errors into a MRM represent. Verbeke and Molenberghs [2000] present a similar plot in their text, also showing the potential addition of measurement error to the process.

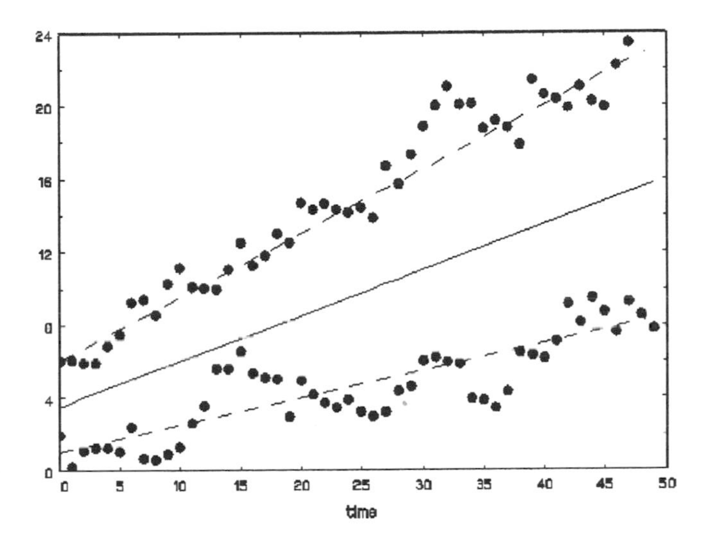

Figure 7.2. MRM for two individuals with AR(1) errors.

As before, the MRM for the $n_i \times 1$ response vector $\boldsymbol{y}$ for individual i is written as

$$\underset{n_i\times 1}{\boldsymbol{y}_i} = \underset{n_i\times p}{\boldsymbol{X}_i}\ \underset{p\times 1}{\boldsymbol{\beta}} + \underset{n_i\times r}{\boldsymbol{Z}_i}\ \underset{r\times 1}{\boldsymbol{v}_i} + \underset{n_i\times 1}{\boldsymbol{\varepsilon}_i} \tag{7.1}$$

with $i = 1 \ldots N$ individuals, $j = 1 \ldots n_i$ observations for individual i, $\boldsymbol{y}_i$ is the $n_i \times 1$ dependent variable vector for individual i, $\boldsymbol{X}_i$ is the $n_i \times p$ covariate matrix for individual i, $\boldsymbol{\beta}$ is the $p \times 1$ vector of fixed regression parameters, $\boldsymbol{Z}_i$ is the $n_i \times r$ design matrix for the random effects, $\boldsymbol{v}_i$ is the $r \times 1$ vector of random individual effects, and $\boldsymbol{\varepsilon}_i$ is the $n_i \times 1$ error vector. Thus far, we have assumed that the errors and the random effects are distributed as

$$\begin{aligned} \boldsymbol{\varepsilon}_i &\sim \mathcal{N}(0, \sigma^2 \boldsymbol{I}_i) \\ \boldsymbol{v}_i &\sim \mathcal{N}(0, \boldsymbol{\Sigma}_v) \end{aligned}$$

And so the variance–covariance matrix of the repeated measures $\boldsymbol{y}$ is of the form

$$V(\boldsymbol{y}_i) = \boldsymbol{Z}_i \boldsymbol{\Sigma}_v \boldsymbol{Z}_i' + \sigma^2 \boldsymbol{I}_i \tag{7.2}$$

This model implies that, conditional on the random effects, the errors are uncorrelated, as is displayed in Figure 7.1. This is seen in the above equation since the error variance is multiplied by the identity matrix (*i.e.*, all correlations of the errors equal zero).

To allow for various forms of autocorrelated errors, we will instead assume that the errors are distributed as $\varepsilon_i \sim \mathcal{N}(0, \sigma^2\boldsymbol{\Omega}_i)$, replacing the identity matrix $\boldsymbol{I}_i$ with the autocorrelation matrix $\boldsymbol{\Omega}_i$. Thus,

$$V(\boldsymbol{y}_i) = \boldsymbol{Z}_i\boldsymbol{\Sigma}_v\boldsymbol{Z}_i' + \sigma^2\boldsymbol{\Omega}_i \tag{7.3}$$

where $\boldsymbol{\Omega}_i$ depends on q autocorrelation parameters, with q varying depending on the type of autocorrelated error structure being considered. The above equation shows that the variance–covariance of the repeated measures is modeled in terms of two components: random individual effects and autocorrelated errors. The first component is concerned with heterogeneity in the population of individuals, whereas the second component posits a correlational structure for the errors that is the same for all individuals. MRMs of Chapters 4 and 5 considered only the first component, whereas the CPMs of Chapter 6 dealt solely with the latter component (with the caveat described in Section 6.2.5). It often happens that models of both types, or with both types combined, can fit the data about the same, so selection of an appropriate variance–covariance structure is an issue. We'll return to this later in the chapter.

The matrix $\boldsymbol{\Omega}$ carries the i subscript to allow for incomplete data across time by individuals. As in the last chapter, here we are assuming that there are a fixed number n of timepoints, however a particular individual might only be measured on n_i of these timepoints (*i.e.*, $n_i \leq n$). The matrix $\boldsymbol{\Omega}_i$ is then of size $n_i \times n_i$ with the appropriate columns and rows of the $n \times n$ matrix $\boldsymbol{\Omega}$ removed. In terms of the spacing of the n timepoints, the autocorrelated error structures presented in this chapter assume that the intervals are the same between timepoints. This restriction can be relaxed; see Jones and Boadi-Boateng [1991] and Jones [1993] for these developments.

As a result of the above assumptions, it can be shown that the observations $\boldsymbol{y}_i$ and random effects $\boldsymbol{v}_i$ have the joint multivariate normal distribution:

$$\begin{bmatrix} \boldsymbol{y}_i \\ \boldsymbol{v}_i \end{bmatrix} \sim \mathcal{N}\left(\begin{bmatrix} \boldsymbol{X}_i\boldsymbol{\beta} \\ \boldsymbol{0} \end{bmatrix}, \begin{bmatrix} \boldsymbol{Z}_i\boldsymbol{\Sigma}_v\boldsymbol{Z}_i' + \sigma^2\boldsymbol{\Omega}_i & \boldsymbol{Z}_i\boldsymbol{\Sigma}_v \\ \boldsymbol{\Sigma}_v\boldsymbol{Z}_i' & \boldsymbol{\Sigma}_v \end{bmatrix} \right). \tag{7.4}$$

Also, the mean of the posterior distribution of $\boldsymbol{v}_i$, given $\boldsymbol{y}_i$, yields the empirical Bayes (EB) estimator of the random effects,

$$\hat{\boldsymbol{v}}_i = \left[\boldsymbol{Z}_i'(\sigma^2\boldsymbol{\Omega}_i)^{-1}\boldsymbol{Z}_i + \boldsymbol{\Sigma}_v^{-1}\right]^{-1} \boldsymbol{Z}_i'(\sigma^2\boldsymbol{\Omega}_i)^{-1}(\boldsymbol{y}_i - \boldsymbol{X}_i\boldsymbol{\beta}). \tag{7.5}$$

Similarly, the corresponding posterior covariance matrix is given by

$$\boldsymbol{\Sigma}_{v|y_i} = \left[\boldsymbol{Z}_i'(\sigma^2\boldsymbol{\Omega}_i)^{-1}\boldsymbol{Z}_i + \boldsymbol{\Sigma}_v^{-1}\right]^{-1}. \tag{7.6}$$

Notice though that these equations are equivalent to the analogous ones in Chapter 4, except that the autocorrelation matrix $\boldsymbol{\Omega}$ has replaced the identity matrix $\boldsymbol{I}$ for the errors.

7.2.1 AR(1) Errors

The first-order autoregressive process (AR1) for the error ε at timepoint j is given as

$$\varepsilon_j = \rho\varepsilon_{j-1} + \xi_j, \tag{7.7}$$

where the disturbances ξ_j are assumed to be distributed $\mathcal{N}(0, \sigma^2)$ and ρ is the autocorrelation coefficient which reflects the degree to which the errors are autocorrelated. It is assumed that $|\rho| < 1$ (*i.e.*, that ρ is a correlation parameter). Notice that the above equation is simply a regression of the errors on themselves, one timepoint removed. This is why it is termed a first-order autoregressive process. If the equation additionally included the errors ε_{j-2} on the right-hand side of the equality, namely,

$$\varepsilon_j = \rho_1 \varepsilon_{j-1} + \rho_2 \varepsilon_{j-2} + \xi_j, \tag{7.8}$$

then it would be a second-order autoregressive process. Here, we will focus on the AR(1) case.

Under AR(1), the variance of the errors at a particular timepoint equals

$$\begin{aligned} V(\varepsilon_j) &= V(\rho \varepsilon_{j-1} + \xi_j) \\ &= \rho^2 V(\varepsilon_{j-1}) + \sigma^2. \end{aligned} \tag{7.9}$$

An assumption that is often made is that of stationarity, which posits that the variance and covariances of the errors are independent of j. This means that the variance of the errors is constant across time and that the correlations are the same within a time lag. With stationarity,

$$\begin{aligned} V(\varepsilon_j) &= \rho^2 V(\varepsilon_j) + \sigma^2 \\ &= \sigma^2/(1-\rho^2) \end{aligned} \tag{7.10}$$

Thus, the diagonal elements of the error variance–covariance matrix are divided by $1 - \rho^2$. Similarly, in terms of the covariance elements, we have

$$\begin{aligned} Cov(\varepsilon_j, \varepsilon_{j-1}) &= E(\varepsilon_j \varepsilon_{j-1}) \\ &= E[(\rho\varepsilon_{j-1} + \xi_j)\, \varepsilon_{j-1}] \\ &= E[(\rho\varepsilon_{j-1})\, \varepsilon_{j-1}] \\ &= \rho V(\varepsilon_{j-1}) \\ &= \rho \sigma^2/(1-\rho^2). \end{aligned} \tag{7.11}$$

More generally,

$$Cov(\varepsilon_j, \varepsilon_{j-s}) = \frac{\rho^s \sigma^2}{1-\rho^2}. \tag{7.12}$$

Taken together, this leads to a variance–covariance matrix of the errors

$$\boldsymbol{\Sigma} = \frac{\sigma^2}{(1-\rho^2)} \begin{bmatrix} 1 & \rho & \rho^2 & \dots & \rho^{n-1} \\ \rho & 1 & \rho & \dots & \rho^{n-2} \\ \rho^2 & \rho & 1 & \dots & \rho^{n-3} \\ . & . & . & \dots & . \\ . & . & . & \dots & . \\ \rho^{n-1} & \rho^{n-2} & \rho^{n-3} & \dots & 1 \end{bmatrix}, \tag{7.13}$$

which is a slightly different parameterization than the AR(1) form given in the previous chapter in equation (6.4). The parameterization above is common in the time-series and econometrics literature, and follows directly from the AR(1) specification in (7.8), whereas the parameterization in (6.4) is perhaps more common in the biostatistics literature. If we denote the the error variance above in (7.13) as σ^{*2} to distinguish it from σ^2 in (6.4) of the previous chapter, then

$$\sigma^{*2} = (1 - \rho^2)\sigma^2 , \tag{7.14}$$

or

$$\sigma^2 = \frac{\sigma^{*2}}{1 - \rho^2} . \tag{7.15}$$

Thus, we see that the only difference is in the scaling of the error variance, and so it doesn't really matter which parameterization is used in data analysis.

7.2.2 MA(1) Errors

Another common form for autocorrelated errors is the first order moving average process, MA(1), which is given as

$$\varepsilon_j = \xi_j - \theta \xi_{j-1} \tag{7.16}$$

with disturbances ξ_j assumed to be $\mathcal{N}(0, \sigma^2)$, and θ is the autocorrelation coefficient for the moving average process. Here, the errors at a particular timepoint equal the disturbances at that timepoint plus a correlated part of the disturbances at the previous timepoint. It can be shown (see Gottman [1981]) that the MA(1) process can be written as an infinite AR process, and likewise that an AR(1) process can be written as an infinite MA process. The practical implication of this is that an MA(1) (or AR(1)) process can often yield a more parsimonious model than a high-order AR (or MA) process.

With stationarity, the error variance covariance matrix, under MA(1), is of the form

$$\sigma^2 \Omega = \sigma^2 \begin{bmatrix} 1+\theta^2 & -\theta & 0 & \ldots & 0 \\ -\theta & 1+\theta^2 & -\theta & \ldots & 0 \\ 0 & -\theta & 1+\theta^2 & \ldots & 0 \\ . & . & . & \ldots & . \\ 0 & 0 & 0 & \ldots & 1+\theta^2 \end{bmatrix} ,$$

that is, a symmetric matrix with $(1 + \theta^2)\sigma^2$ on the main diagonal, $-\theta\sigma^2$ on the first off-diagonal, and 0 everywhere else. This form posits that only the lag-1 errors are correlated. Thus, the errors at a given timepoint are only correlated with those one timepoint apart. This is one reason why this form was not considered in the previous chapter on CPMs. Namely, while the MA(1) form is generally unreasonable for the variance–covariance matrix of e after conditioning on covariates $\boldsymbol{X}$ (as in a CPM), it might well be reasonable for the variance–covariance matrix of ε which is conditional on both covariates $\boldsymbol{X}$ and random effects $\boldsymbol{v}$ (as in a MRM with AC errors).

7.2.3 ARMA(1,1) Errors

A more general form for autocorrelated errors is the first-order mixed autoregressive-moving average (ARMA) process which depends on both the AR parameter ρ and the MA parameter θ and is given as

$$e_k = \rho e_{k-1} + \varepsilon_k - \theta\varepsilon_{k-1} \tag{7.17}$$

with all terms as before. The error variance covariance matrix is now of the form

$$\sigma^2\boldsymbol{\Omega} = \frac{\sigma^2}{(1-\rho^2)}\begin{bmatrix} \gamma_0 & \gamma_1 & \rho\gamma_1 & \cdots & \rho^{n-2}\gamma_1 \\ \gamma_1 & \gamma_0 & \gamma_1 & \cdots & \rho^{n-3}\gamma_1 \\ \rho\gamma_1 & \gamma_1 & \gamma_0 & \cdots & \rho^{n-4}\gamma_1 \\ \rho^2\gamma_1 & \rho\gamma_1 & \gamma_1 & \cdots & \rho^{n-5}\gamma_1 \\ \cdot & \cdot & \cdot & \cdots & \cdot \\ \rho^{n-2}\gamma_1 & \rho^{n-3}\gamma_1 & \rho^{n-4}\gamma_1 & \cdots & \gamma_0 \end{bmatrix},$$

where $\gamma_0 = 1 + \theta^2 - 2\rho\theta$ and $\gamma_1 = (1-\rho\theta)(\rho-\theta)$. Being a combination of the AR(1) and MA(1), this form is similar to AR(1) but with a increased autocorrelation for the lag-1 errors. Thus, it can be useful when the lag-1 errors are relatively large, and the remaining lags diminish more or less exponentially.

7.2.4 Toeplitz Errors

One can also assume that each lag (or each off-diagonal in the error variance covariance matrix) has its own distinct autocorrelation parameter. The error variance covariance matrix is then of the form

$$\sigma^2\boldsymbol{\Omega} = \sigma^2\begin{bmatrix} 1 & \rho_1 & \rho_2 & \cdots & \rho_{n-1} \\ \rho_1 & 1 & \rho_1 & \cdots & \rho_{n-2} \\ \rho_2 & \rho_1 & 1 & \cdots & \rho_{n-3} \\ \cdot & \cdot & \cdot & \cdots & \cdot \\ \rho_{n-1} & \rho_{n-2} & \rho_{n-3} & \cdots & 1 \end{bmatrix}.$$

The matrix $\boldsymbol{\Omega}$ is a symmetric general Toeplitz matrix with $n-1$ unique autocorrelation parameters. It is typical to assume that some of the higher-order lags have zero autocorrelation in a MRM, and so one can define the s-order symmetric Toeplitz matrix to allow only the first s autocorrelations to be nonzero, with the others equal to zero. For instance, a random-intercepts model can only include at most $n-2$ Toeplitz autocorrelations, since this model is equivalent to a CPM with a full Toeplitz structure. To see this, suppose there are three timepoints. Then we get

$$\begin{bmatrix} \theta_1 & \theta_2 & \theta_3 \\ \theta_2 & \theta_1 & \theta_2 \\ \theta_3 & \theta_2 & \theta_1 \end{bmatrix} = \begin{bmatrix} \sigma_v^2 & \sigma_v^2 & \sigma_v^2 \\ \sigma_v^2 & \sigma_v^2 & \sigma_v^2 \\ \sigma_v^2 & \sigma_v^2 & \sigma_v^2 \end{bmatrix} + \begin{bmatrix} \sigma^2 & \rho_1\sigma^2 & 0 \\ \rho_1\sigma^2 & \sigma^2 & \rho_1\sigma^2 \\ 0 & \rho_1\sigma^2 & \sigma^2 \end{bmatrix},$$

where the (conditional) variance–covariance matrix of $\boldsymbol{y}$ is given for (a) the full Toeplitz CPM on the left-hand side of the equality and (b) the random-intercepts model with one

Toeplitz autocorrelation parameter on the right-hand side. Notice that in both cases the number of variance–covariance parameters equals three and that $\theta_1 = \sigma_v^2 + \sigma^2$, $\theta_2 = \sigma_v^2 + \rho_1\sigma^2$, and $\theta_3 = \sigma_v^2$. Thus, these two forms are just reparameterizations of each other.

One trivial, yet confusing, aspect about Toeplitz structures is how the order of the matrix is referred to. In a CPM it is common to count the error variance as one of the elements in the Toeplitz structure and so the above matrix, on the left side of the equality, would be a Toeplitz(3). Alternatively, in some treatments of Toeplitz errors, only the autocorrelation parameters are counted, and so the error variance–covariance matrix above for the model with the random intercepts (*i.e.*, the right side of the equality) would be a Toeplitz(1). Here, we will use the former convention, and so in the above equality the CPM Toeplitz matrix is of order three and the error variance–covariance Toeplitz matrix is of order two. Using this convention, note that the MA(1) form could also be specified as a Toeplitz structure of order two, though the parameterization is different.

7.2.5 Nonstationary AR(1) Errors

The above autocorrelated error forms have all assumed stationarity, and so the error (co)variances are equal within a time-lag in all of these forms. In some cases, it can be advantageous to relax this assumption. In this regard, Mansour et al. [1985] described a two-way mixed analysis of variance model with a first-order nonstationary (NS) AR(1) process; this structure is available within the more general MRM in the MIXREG software program [Hedeker and Gibbons, 1996b]. For this, Mansour et al. [1985] note that, as in (7.9), the variance of the errors at a particular timepoint is given by

$$V(\varepsilon_j) = \rho^2 V(\varepsilon_{j-1}) + \sigma^2. \tag{7.18}$$

Instead of assuming stationarity, assume that the errors have zero variance at time 0 (*i.e.*, one timepoint before the start of the process), namely $V(\varepsilon_0) = 0$. Then one gets the following for the error variance at the first four timepoints:

$$\begin{aligned}
V(\varepsilon_1) &= \sigma^2, \\
V(\varepsilon_2) &= (1+\rho^2)\sigma^2, \\
V(\varepsilon_3) &= (1+\rho^2+\rho^4)\sigma^2, \\
V(\varepsilon_4) &= (1+\rho^2+\rho^4+\rho^6)\sigma^2.
\end{aligned}$$

Notice that the error variance increases across time if $\rho \neq 0$. Similarly, Mansour et al. [1985] show that the correlation of the errors also increase across time under this structure. More generally, the error variance covariance matrix is of the form $\sigma^2\mathbf{\Omega}$, with

$$\mathbf{\Omega} = \begin{bmatrix}
1 & \rho & \rho^2 & \cdots & \rho^{n-1} \\
\rho & (1+\rho^2) & \rho(1+\rho^2) & \cdots & \rho^{n-2}(1+\rho^2) \\
\rho^2 & \rho(1+\rho^2) & (1+\rho^2+\rho^4) & \cdots & \rho^{n-3}(1+\rho^2+\rho^4) \\
\cdot & \cdot & \cdot & \cdots & \cdot \\
\rho^{n-1} & \rho^{n-2}(1+\rho^2) & \rho^{n-3}(1+\rho^2+\rho^4) & \cdots & (1+\sum_{j=1}^{n-1}\rho^{2j})
\end{bmatrix},$$

which depends only on the NS AR(1) parameter ρ and the error variance σ^2. Mansour et al. [1985] point out that the Cholesky factorization $\mathbf{\Omega} = \mathbf{\Upsilon\Upsilon}'$ provides a more convenient form, namely,

$$\mathbf{\Upsilon} = \begin{bmatrix} 1 & 0 & 0 & \dots & 0 \\ \rho & 1 & 0 & \dots & 0 \\ \rho^2 & \rho & 1 & \dots & 0 \\ . & . & . & \dots & . \\ . & . & . & \dots & . \\ \rho^{n-1} & \rho^{n-2} & \rho^{n-3} & \dots & 1 \end{bmatrix}. \tag{7.19}$$

Notice that the lower triangular portion of this matrix is of the ordinary AR(1) form, while the above triangular elements all equal zero.

7.3 MODEL SELECTION

Many possible forms have been described for the (conditional) variance covariance matrix of the repeated measures. Here we discuss tools for helping one decide on a reasonable structure for a given dataset. To review, in Chapter 4, MRMs were considered which posit that this matrix is given by

$$\mathbf{\Sigma}_i = \mathbf{Z}_i \mathbf{\Sigma}_\upsilon \mathbf{Z}_i' + \sigma^2 \mathbf{I}_i. \tag{7.20}$$

The form of $\mathbf{\Sigma}_i$ thus depends on the number of random effects in the model. Alternatively, the CPMs of the last chapter simply posit various forms directly for $\mathbf{\Sigma}_i$. Finally, the models of this chapter have augmented MRMs with autocorrelated errors, namely,

$$\mathbf{\Sigma}_i = \mathbf{Z}_i \mathbf{\Sigma}_\upsilon \mathbf{Z}_i' + \sigma^2 \mathbf{\Omega}_i. \tag{7.21}$$

As already mentioned, because the tests of the fixed effects depend on the variance–covariance structure, it is important to select a reasonable structure for a given dataset. For comparison of nested models, the likelihood ratio test has been described and used in the previous chapters. The likelihood ratio test statistic X^2 is obtained as

$$\begin{aligned} X^2 &= 2(\log L_{full} - \log L_{reduced}) \\ &= -2 \log L_{reduced} - -2 \log L_{full}, \end{aligned} \tag{7.22}$$

where $\log L_{full}$ and $\log L_{reduced}$ are the maximized log-likelihood values for the full and reduced models, respectively. These can be based on ML or REML estimation, but of course for a given test the estimation method needs to be the same. The value $-2 \log L$ is called the deviance, and many computer programs additionally print out this representation of the likelihood value. X^2 is compared to a chi-square distribution with degrees of freedom equal to the number of additional parameters in the full, relative to the reduced, model. As mentioned, use of the likelihood ratio test for variance and covariance parameters suffers from the boundary problem [Verbeke and Molenberghs, 2000] and leads to accepting a more restrictive variance–covariance structure than is correct. As noted by Berkhof and Snijders [2001], this bias can largely be corrected by the simple adjustment of dividing the p-value obtained from the likelihood ratio test (of variance and covariance terms) by two.

For MRMs we have used this test to help determine how many random effects are necessary, building models of increasing complexity, in terms of the random effects, in a sequential manner. In the last chapter, the likelihood-ratio test was used to compare various

CPMs to the totally general unstructured form. These are all examples of comparisons involving nested models.

In some cases, it is necessary to compare models that are not nested. For example, suppose that we want to compare a MRM with multiple random effects to a CPM with a Toeplitz structure. Neither model is a nested version of the other and so the likelihood-ratio test cannot be applied. For this, a simple approach is to use the Akaike Information Criterion (AIC; Akaike [1973]), which is defined for a given model as

$$\begin{aligned} \text{AIC} &= -2(\log L - p) \\ &= -2\log L + 2p \end{aligned} \tag{7.23}$$

where p equals the number of model parameters. Thus, a model's AIC is just its deviance value plus two times the number of estimated parameters in the model. Because lower AIC values are preferred, this adjustment to the deviance is often called the penalty for using (additional) parameters in model fitting. An attractive feature of AIC is that it is the same as Mallows C_p for ordinary linear regression models.

Another criterion that is used for model selection of models that are not nested is the Bayesian Information Criterion (BIC; Schwarz [1978]). This criterion is given by

$$\begin{aligned} \text{BIC} &= -2(\log L - 1/2p\log N) \\ &= -2\log L + p\log N. \end{aligned} \tag{7.24}$$

As with AIC, lower BIC values are preferred. An issue in use of BIC is what N represents, since in longitudinal data one could use the number of level-2 subjects or the number of total level-1 observations as two possibilities. This issue has not been entirely resolved, though Raftery [1995], among others, recommends using the number of subjects. This is also what SAS PROC MIXED, for example, prints out in its output. We will also use this convention here. Notice that the penalty that BIC adds to the deviance is $p\log N$, which will be greater than the AIC penalty of $2p$ if $N > \exp(2) = 7.39$, which it almost always will be. As a result, with moderate sample sizes, BIC extracts a fairly large penalty for the addition of model parameters and leads to simpler models than AIC. For this and other reasons, Fitzmaurice et al. [2004] recommend against use of BIC for model selection of (co)variance structure.

In using likelihood ratio tests, AIC, and/or BIC it is important that the models being compared are fit to the same dataset. This point may seem obvious, but it can be compromised rather easily. For example, suppose that one is interested in comparing models with and without a set of covariates. Suppose further that one or more of these covariates have some missing values in the sample, either at particular timepoints or for particular subjects. Then the models with and without these covariates would be estimated on different datasets and cannot be compared using these model comparison tools. In this case, the data analyst might have to consider imputing the missing values or performing the comparison on the subset of the sample with complete values on these covariates.

7.4 EXAMPLE

Here, we analyze the dataset from Bock [1983b] that was presented in the last chapter. As described, this study consisted of a six-week crossover design where one group of patients received three weeks of anti-depressant medication followed by three weeks of no treatment, while a second group of patients received these two treatment arms for three-week time periods in the reverse order. The variable $Linear$ represents linear change across the six weeks of the study, the $Slope\ Change$ term represents a contrast of the linear trend for the first three weeks versus the last three weeks. As in the Chapter 6, and in Bock's article, these are coded as follows:

	Week 1	Week 2	Week 3	Week 4	Week 5	Week 6
Linear	−5/2	−3/2	−1/2	1/2	−3/2	−5/2
Slope Change	−1/2	0	1/2	1/2	0	−1/2

The variable $Group$ represents the grouping variable (with 0 = drug treatment followed by no drug treatment, and 1 = no drug treatment followed by drug treatment). Two interaction terms will also be included: $Group \times Linear$ and $Group \times Slope\ Change$.

We'll start by considering a model with random intercept, linear trend, and change of slope terms. Specifically, consider the following level-1 model for subject i $(i = 1, \ldots, N)$ at time j $(j = 1, \ldots, n_i)$:

$$WPSS_{ij} = b_{0i} + b_{1i}Linear_{ij} + b_{2i}Slope\ Change_{ij} + \varepsilon_{ij}. \tag{7.25}$$

Also consider the between-subjects (or level-2) model:

$$\begin{aligned} b_{0i} &= \beta_0 + \beta_3 Group_i + \upsilon_{0i}, \\ b_{1i} &= \beta_1 + \beta_4 Group_i + \upsilon_{1i}, \\ b_{2i} &= \beta_2 + \beta_5 Group_i + \upsilon_{2i}. \end{aligned} \tag{7.26}$$

As a minor point, the notation here could actually be slightly simpler. Because all subjects were measured at the same six timepoints, the variables $Linear_{ij}$ and $Slope\ Change_{ij}$ could simply be denoted as $Linear_j$ and $Slope\ Change_j$ (*i.e.*, the values of these time variables do not vary across subjects). Furthermore, because all subjects were measured at all six timepoints (*i.e.*, there is no incomplete data across time), the variable n_i which denotes the number of repeated observations per subject could be n (since it doesn't vary by subjects for these data).

To illustrate application of MRMs with autocorrelated errors, we will contrast models that assume the errors are (conditionally) independent versus following a nonstationary AR(1) process. Table 7.1 presents the results for these two models. Comparing these two models via a likelihood ratio test yields $X_1^2 = 992.5 - 986.7 = 5.8$ which yields a p-value of $.016/2 = .008$. Here, we follow the adjustment of dividing the nominal p-value by 2 for this test of the variance–covariance parameter ρ. Thus, the model with the NS AR(1) parameter is preferred to the model without it. Comparing the estimates, we see that the fixed-effects estimates are nearly identical, as are their standard errors. Clearly, the conclusions regarding these are the same. In fact, the conclusions regarding the fixed effects are the same as those found using the unstructured CPM that was presented in Table 6.3 of the previous chapter.

Turning to the estimates and standard errors of the random-effects variance–covariance parameters, we see that these are affected by the inclusion of the autocorrelated errors. In general, the (co)variance estimates are reduced and their standard errors are increased in the model with NS AR(1) errors. This is not too surprising given the equation for the variance–covariance matrix (for subject i)

$$\boldsymbol{\Sigma}_i = \boldsymbol{Z}_i \boldsymbol{\Sigma}_\upsilon \boldsymbol{Z}_i' + \sigma^2 \boldsymbol{\Omega}_i. \tag{7.27}$$

This equation makes clear that both $\boldsymbol{\Omega}_i$ and $\boldsymbol{\Sigma}_\upsilon$ can be considered to be explanatory determinants of $\boldsymbol{\Sigma}_i$. In other words, variation and covariation in the dependent variable can be explained via either or both $\boldsymbol{\Omega}_i$ and $\boldsymbol{\Sigma}_\upsilon$. Using basic multiple regression logic, it then follows that inclusion of $\boldsymbol{\Omega}_i$ parameters will reduce the effect of $\boldsymbol{\Sigma}_\upsilon$ parameters to the extent that the $\boldsymbol{\Omega}_i$ parameters are related to $\boldsymbol{\Sigma}_i$. Similarly, to the extent that the parameters in $\boldsymbol{\Sigma}_\upsilon$ and $\boldsymbol{\Omega}_i$ are associated with each other, the standard errors of either set will generally increase with the introduction of the other set into the model. Again, this follows ordinary multiple regression logic.

Table 7.1. Results for Two MRMs Applied to Bock (1983) Data

Parameter	Without AC errors			With AC errors		
	Estimate	SE	$p <$	Estimate	SE	$p <$
Fixed Effects						
Constant β_0	3.127	0.180	.0001	3.125	0.180	.0001
Linear (L) β_1	−0.208	0.044	.0001	−0.205	0.040	.0001
Slope Change (SC) β_2	−0.250	0.110	.024	−0.250	0.109	.023
Group (G) β_3	1.281	0.289	.0001	1.282	0.289	.0001
$G \times L$ β_4	0.051	0.070	.46	0.041	0.065	.53
$G \times SC$ β_5	0.440	0.178	.013	0.439	0.176	.013
Variance Terms						
Constant $\sigma^2_{\upsilon 0}$	1.463	0.243		1.214	0.436	
Cons, Lin $\sigma_{\upsilon 0\, \upsilon 1}$	0.099	0.043		0.075	0.070	
Linear $\sigma^2_{\upsilon 1}$	0.079	0.014		0.042	0.024	
Cons, Slope $\sigma_{\upsilon 0\, \upsilon 2}$	0.143	0.107		0.051	0.161	
Lin, Slope $\sigma_{\upsilon 1\, \upsilon 2}$	−0.001	0.026		0.014	0.023	
Slope Change $\sigma^2_{\upsilon 2}$	0.413	0.014		0.190	0.144	
Error σ^2	0.149	0.014		0.315	0.106	
NS AR(1) ρ				0.696	0.324	
$-2 \log L$		992.5			986.7	

Note. SE = standard error.

The upshot of this is that for a given dataset there are often several models of $\boldsymbol{\Sigma}$ that fit the data about equally well. Choice of the final model for $\boldsymbol{\Sigma}$ is then perhaps more a matter of avoiding bad models, and selecting a reasonable model from a number of possible alternative models. This is analogous to the situation in model selection of regressors. For instance, if one performs an all possible regression procedure, one often finds that there are several sets of independent variables that yield similar R^2 values, and so choice of the "best" model is not always clear-cut.

To illustrate these points regarding model selection, consider the results presented in Table 7.2. This table presents results for three types of MRMs: random intercepts; random intercepts and linear trends; and random intercepts, linear trends, and slope changes. For each type, several forms of autocorrelated errors are considered. Additionally, results for the unstructured CPM (model 25), which was deemed the "best" CPM in the last chapter, are also presented. These models were estimated using SAS PROC MIXED, with the exception of the models with NS AR(1) errors which were estimated using MIXREG. In terms of notation and syntax, SAS denotes the random-effects variance–covariance matrix Σ_υ as the $\boldsymbol{G}$ matrix and the error variance–covariance matrix $\sigma^2\Omega$ as the $\boldsymbol{R}$ matrix, and these two are modeled via the `RANDOM` and `REPEATED` statements, respectively.

Table 7.2. Variance–Covariance Structures for Bock (1983) Data

Model	Σ_υ	r	$\sigma^2\Omega_i$	q	$r+q$	$-2\log L$	AIC	BIC
1	Int	1	$\sigma^2 I$	1	2	1185.8	1201.8	1220.4
2	Int	1	AR(1)	2	3	Intercept variance set to zero		
3	Int	1	NS AR(1)	2	3	993.1	1011.1	1032.0
4	Int	1	MA(1)	2	3	1055.1	1073.1	1093.9
5	Int	1	ARMA(1,1)	3	4	Intercept variance set to zero		
6	Int	1	Toeplitz(3)	3	4	1009.1	1029.1	1052.3
7	Int	1	Toeplitz(4)	4	5	988.9	1010.9	1036.4
8	Int	1	Toeplitz(5)	5	6	988.9	1012.9	1040.7
9	Int, Lin	3	$\sigma^2 I$	1	4	1053.0	1073.0	1096.2
10	Int, Lin	3	AR(1)	2	5	Intercept variance set to zero		
11	Int, Lin	3	NS AR(1)	2	5	Intercept variance set to zero		
12	Int, Lin	3	MA(1)	2	5	1006.8	1028.8	1054.3
13	Int, Lin	3	ARMA(1,1)	3	6	Linear variance set to zero		
14	Int, Lin	3	Toeplitz(3)	3	6	990.4	1014.4	1042.2
15	Int, Lin	3	Toeplitz(4)	4	7	Linear variance set to zero		
16	Int, Lin	3	Toeplitz(5)	5	8	980.4	1008.4	1040.9
17	Int, Lin, SC	6	$\sigma^2 I$	1	7	992.5	1018.5	1048.6
18	Int, Lin, SC	6	AR(1)	2	8	Intercept variance set to zero		
19	Int, Lin, SC	6	NS AR(1)	2	8	986.7	1014.7	1047.1
20	Int, Lin, SC	6	MA(1)	2	8	990.2	1018.2	1050.6
21	Int, Lin, SC	6	ARMA(1,1)	3	9	Intercept variance set to zero		
22	Int, Lin, SC	6	Toeplitz(3)	3	9	986.4	1016.4	1051.2
23	Int, Lin, SC	6	Toeplitz(4)	4	10	Unity correlation in Σ_υ		
24	Int, Lin, SC	6	Toeplitz(5)	5	11	Linear variance set to zero		
25		0	UN	21	21	945.9	999.9	1062.5

Int = intercept, Lin = linear, SC = slope change.

Looking over the results in Table 7.2, one notices that many of these models were not fully estimable. For these models, the data did not provide sufficient information for unique estimation of all variance–covariance parameters, and the software program set a particular parameter estimate to zero or yielded a unity correlation for some pair of the random effects. In general, these models would not be reasonable "as is," and would need to be re-estimated with the particular terms omitted, if such a model makes sense, and/or the overall variance–covariance structure simplified.

In selecting the "best" variance–covariance structure for these data, one could adopt the approach suggested by Jennrich and Schluchter [1986] and compare each model to the unstructured form (*i.e.*, model 25, $\sigma^2\Omega_i$ = UN) using a likelihood ratio test. Using this criterion, none of the models in Table 7.2 are statistically competitive with the unstructured CPM. In terms of significance levels, the closest is model 16, which yields a likelihood-ratio $X^2 = 34.5$, df = 13, $p < .001/2 = .0005$ when compared to model 25. Clearly, based on this comparison to the UN form, none of the MRMs or MRMs with AC errors would be selected. Similarly, the UN model yields the lowest AIC value, and so would also be selected using this criterion. If one adopted the BIC criterion, however, then model 3, random intercepts with NS AR(1) errors, would be deemed best. Remember though, as mentioned by Fitzmaurice et al. [2004], that BIC extracts a high penalty for the addition of parameters, leading to variance–covariance structures that are often too simplistic. This certainly is suggested here because model 3 only includes 3 parameters, whereas model 25 uses the full 21 variance–covariance parameters. Therefore, Fitzmaurice et al. [2004] and others warn against its use for model selection of (co)variance structure. Thus, selecting model 25 is the most reasonable course of action.

For the sake of illustration, though, we will examine model 3 a bit closer. First, the estimates of the fixed effects for models 3 and 25 are presented in Table 7.3. One can see that the estimates and standard errors of the fixed effects are quite similar, and certainly no conclusions change between the two models.

Table 7.3. Fixed-Effects Estimates for Models 3 and 25

	Model 3			Model 25		
Parameter	Estimate	SE	$p<$	Estimate	SE	$p<$
Fixed Effects						
Constant β_0	3.125	0.194	.0001	3.122	0.179	.0001
Linear (L) β_1	-0.204	0.038	.0001	-0.198	0.036	.0001
Slope Change (SC) β_2	-0.250	0.101	.013	-0.255	0.102	.015
Group (G) β_3	1.281	0.312	.0001	1.286	0.288	.0001
$G \times L$ β_4	0.039	0.061	.52	0.017	0.058	.78
$G \times SC$ β_5	0.440	0.162	.007	0.475	0.164	.005
$-2 \log L$		993.1			945.9	

Note. SE = standard error.

Next, we will compare the variance–covariance estimates. As presented in the last chapter, the estimated variance–covariance matrix based on the unstructured model 25 is

$$\hat{\Sigma} = \begin{bmatrix} 1.443 & 1.361 & 1.129 & 1.058 & 0.939 & 1.003 \\ 1.361 & 1.605 & 1.426 & 1.389 & 1.216 & 1.216 \\ 1.129 & 1.426 & 1.810 & 1.684 & 1.461 & 1.398 \\ 1.058 & 1.389 & 1.684 & 1.995 & 1.792 & 1.788 \\ 0.939 & 1.216 & 1.461 & 1.792 & 2.242 & 2.192 \\ 1.003 & 1.216 & 1.398 & 1.788 & 2.192 & 2.369 \end{bmatrix} . \quad (7.28)$$

Alternatively, based on model 3, the parameter estimates are $\hat{\sigma}_v^2 = .9473, \hat{\sigma}^2 = .4373$, and $\hat{\rho} = .9145$, so we get

$$\hat{\Sigma} = \begin{bmatrix} .9473 & .9473 & .9473 & .9473 & .9473 & .9473 \\ .9473 & .9473 & .9473 & .9473 & .9473 & .9473 \\ .9473 & .9473 & .9473 & .9473 & .9473 & .9473 \\ .9473 & .9473 & .9473 & .9473 & .9473 & .9473 \\ .9473 & .9473 & .9473 & .9473 & .9473 & .9473 \\ .9473 & .9473 & .9473 & .9473 & .9473 & .9473 \end{bmatrix}$$

$$+ .4373 \begin{bmatrix} 1 & 0 & 0 & 0 & 0 & 0 \\ .9145 & 1 & 0 & 0 & 0 & 0 \\ (.9145)^2 & .9145 & 1 & 0 & 0 & 0 \\ (.9145)^3 & (.9145)^2 & .9145 & 1 & 0 & 0 \\ (.9145)^4 & (.9145)^3 & (.9145)^2 & .9145 & 1 & 0 \\ (.9145)^5 & (.9145)^4 & (.9145)^3 & (.9145)^2 & .9145 & 1 \end{bmatrix}$$

$$\times \begin{bmatrix} 1 & .9145 & (.9145)^2 & (.9145)^3 & (.9145)^4 & (.9145)^5 \\ 0 & 1 & .9145 & (.9145)^2 & (.9145)^3 & (.9145)^4 \\ 0 & 0 & 1 & .9145 & (.9145)^2 & (.9145)^3 \\ 0 & 0 & 0 & 1 & .9145 & (.9145)^2 \\ 0 & 0 & 0 & 0 & 1 & .9145 \\ 0 & 0 & 0 & 0 & 0 & 1 \end{bmatrix}$$

$$= \begin{bmatrix} 1.385 & 1.347 & 1.313 & 1.282 & 1.253 & 1.227 \\ 1.347 & 1.750 & 1.682 & 1.619 & 1.561 & 1.509 \\ 1.313 & 1.682 & 2.056 & 1.961 & 1.875 & 1.795 \\ 1.282 & 1.619 & 1.961 & 2.312 & 2.195 & 2.089 \\ 1.253 & 1.561 & 1.875 & 2.195 & 2.526 & 2.391 \\ 1.227 & 1.509 & 1.795 & 2.089 & 2.391 & 2.705 \end{bmatrix}$$

Given that only three parameters are estimated in model 3, it does a fairly reasonable job of producing the results from the (saturated) unstructured form of model 25. The variances are a bit too spread out in model 3 and the covariances don't quite get small enough towards the lower triangular portion of the matrix.

Expressing these estimated covariance matrices as correlation matrices yields

$$\begin{bmatrix} 1.000 & 0.894 & 0.699 & 0.624 & 0.522 & 0.543 \\ 0.894 & 1.000 & 0.836 & 0.776 & 0.641 & 0.624 \\ 0.699 & 0.836 & 1.000 & 0.886 & 0.725 & 0.675 \\ 0.624 & 0.776 & 0.886 & 1.000 & 0.847 & 0.822 \\ 0.522 & 0.641 & 0.725 & 0.847 & 1.000 & 0.951 \\ 0.543 & 0.624 & 0.675 & 0.822 & 0.951 & 1.000 \end{bmatrix}$$

for model 25, and

$$\begin{bmatrix} 1.000 & 0.865 & 0.778 & 0.716 & 0.670 & 0.634 \\ 0.865 & 1.000 & 0.886 & 0.805 & 0.743 & 0.694 \\ 0.778 & 0.886 & 1.000 & 0.900 & 0.823 & 0.761 \\ 0.716 & 0.805 & 0.900 & 1.000 & 0.908 & 0.835 \\ 0.670 & 0.743 & 0.823 & 0.908 & 1.000 & 0.915 \\ 0.634 & 0.694 & 0.761 & 0.835 & 0.915 & 1.000 \end{bmatrix}$$

for model 3. As can be seen, though the general pattern is similar, model 3's estimated correlations are a bit too close together relative to model 25.

To get a more visual representation, as mentioned in Chapter 4, Grady and Helms [1995] describe graphical techniques to aid in examining model fit of the variance–covariance structure. These authors suggest plots of the (co)variances or correlations as a function of the 'lag' (*i.e.*, the time between measures). For this, Figure 7.3 shows the covariance and correlation plots based on models 25 and 3. Notice that each line in these (co)variance plots in (a) and (b) present the estimates for a given row of the lower triangular portion of the variance–covariance matrix (including the diagonal). This yields a plot of all unique elements of the variance–covariance matrix. For example, for model 25, the first row includes the estimate 1.443 as the only element in the lower triangular portion of this matrix. The first row corresponds to Week 0 and this element corresponds to a lag of 0; this point can be found as the lone point in the lower left-hand corner of plot (a). Continuing with this matrix, the second row, which corresponds to week 1, has estimates of 1.361 (lag 1) and 1.426 (lag 0) for the lower triangular portion of the matrix. Similarly, the correlation plots in (c) and (d) are produced using only the elements in the strictly lower triangular portion (*i.e.*, not including the diagonal) of the respective correlation matrices. Thus, the correlation plots begin with week 1 and not week 0. The careful eye will notice that the week 1 correlation appears missing in plot (c) of Figure 7.3. This is because the week 1 lag 1 correlation of 0.894 is essentially equal to the week 3 lag 1 correlation of 0.886, and so the former is masked by the latter in the plot.

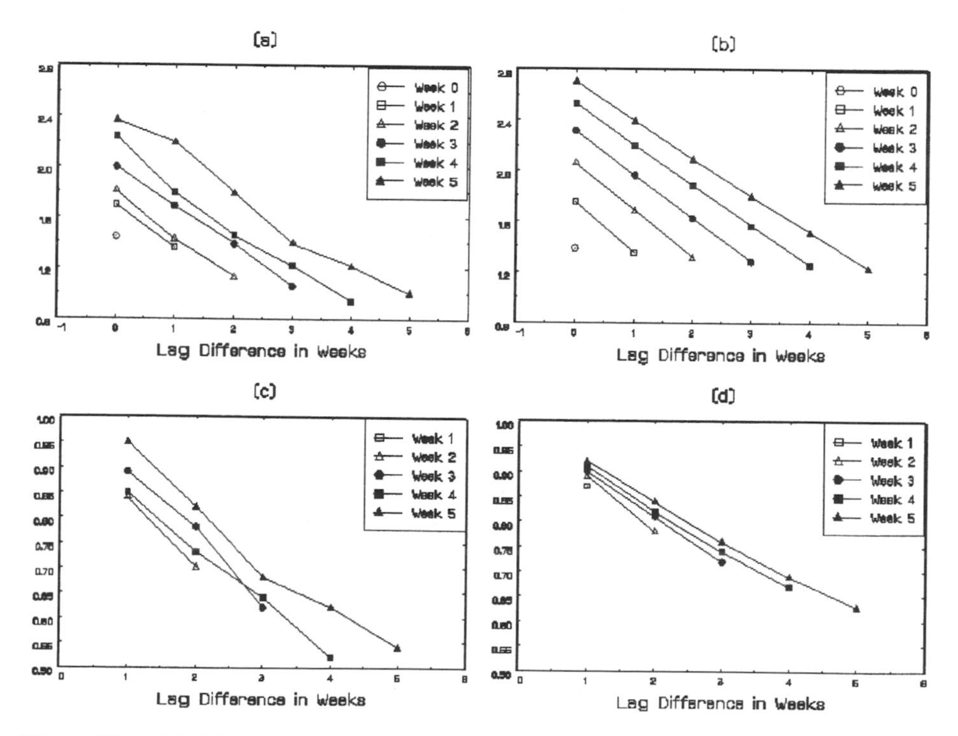

Figure 7.3. Model estimated (co)variance and correlation plots: (a) Model 25 (co)variance; (b) model 3 (co)variance; (c) model 25 correlation; (d) model 3 correlation.

These plots reinforce the notions that model 3 yields variances and covariances that are a bit too spread out and are often too large in value, relative to the unstructured model 25. Similarly, the estimated correlations are too similar based on model 3 relative to model 25. All of this gives some evidence for why model 3 might be considered a relatively good model for the variance–covariance structure, especially given that it only has three parameters, but model 25 is ultimately the best choice for these data, based both on likelihood ratio tests and the AIC and also reinforced by the visual representation.

7.5 SUMMARY

This chapter has presented MRMs with various forms of autocorrelated errors. The forms considered here are largely drawn from the econometrics and time-series literatures, where fixed-effects regression models with autocorrelated errors have a rich history. Of these forms, as noted by Greene [1993] the AR(1) form is easily the most popular for autocorrelated errors in these literatures. For longitudinal data, the AR(1) process makes logical sense in that the correlation of the errors clearly diminishes as the time lag increases. Additionally, because it only depends on one additional parameter it represents a very parsimonious augmentation to the usual conditional independence assumption of MRMs. This is undoubtably why this was the first form of AC errors developed in MRMs [Chi and Reinsel, 1989]. A case can be made, though, that the nonstationary version of AR(1) errors described by Mansour et al. [1985] is worthy of more attention than it has received. This was borne out to some extent in the example presented in this chapter, in which the random intercept model with NS AR(1) errors was seen to be a relatively good model.

As illustrated in this and preceding chapters, there are many possible forms for the variance–covariance Σ of the repeated measures: MRMs, CPMs, and now MRMs with AC errors. Selection of an appropriate model for Σ is important for two potential reasons. First, one might be interested in the form of Σ itself and want to examine hypotheses regarding the variance–covariance parameters. Second, even if one is primarily interested in the fixed effects, it is still important to select a reasonable Σ structure for a given dataset. This is because the variance–covariance structure, while typically not greatly affecting the estimates of the fixed effects, can definitely affect the standard errors of the fixed effects. This is the primary reason why the ANOVA vs. MANOVA choice was once such an important issue in longitudinal data analysis. These days, these two forms (compound symmetry for the ANOVA model and unstructured for the MANOVA model) simply represent the limits of simplest to most general, respectively, with many other possible forms in between.

As already mentioned, when faced with this multitude of choices, and the impact that a given choice might have on the significance levels of the fixed effects, longitudinal data analysis generally proceeds in two stepsmodel selection. First, including all covariates of potential interest in the model, one fits a variety of models for Σ and selects a reasonable one using the tools described in this chapter. Second, using the structure selected in the first step, one tests the significance levels of the model covariates and trims the model in the usual way. In this way, one arrives at reasonable models for both the variance–covariance and the mean structure of the dependent variable.

CHAPTER 8

GENERALIZED ESTIMATING EQUATIONS (GEE) MODELS

8.1 INTRODUCTION

In the 1980s, alongside development of MRMs and CPMs for incomplete longitudinal data, generalized estimating equations (GEE) models were developed [Liang and Zeger, 1986; Zeger and Liang, 1986; Zeger et al., 1988]. Essentially, GEE models extend generalized linear models (GLMs) for the situation of correlated data. Thus, this class of models has become very popular especially for analysis of categorical and count outcomes, though they can be used for continous outcomes as well. One difference between GEE models and MRMs and CPMs is that it uses quasi-likelihood estimation, and so the full likelihood of the data is not specified. GEE models are termed marginal models, and they model the regression of y on X and the within-subject dependence (*i.e.*, the association parameters) separately. As noted in Fitzmaurice et al. [2004], "the term marginal in this context indicates that the model for the mean response depends only on the covariates of interest, and not on any random effects or previous responses." In terms of missing data, it assumes that the missing data are MCAR (as opposed to MAR which is assumed by the models employing full-likelihood estimation). More will be said about this in Chapter 14.

Some useful articles describing GEE models in various literatures include Ballinger [2004], Davis [1993], Dunlop [1994], Fitzmaurice et al. [1993], Norton et al. [1996], Sheu [2000], and Zorn [2001]. Additionally, the texts by Diggle et al. [2002] and Hardin and Hilbe [2003] contain a wealth of information about GEE models. In this chapter we will focus on the class of GEE models originally developed by Liang and Zeger in the aforementioned articles. These are now sometimes termed GEE1 models because of their separation of the estimating equations for the regression parameters and association parameters. In statistical terms, these two parameter vectors are assumed to be orthogonal to each other. GEE1 is the class of models that is most commonly found in statistical software implementations, some of which are reviewed in Horton and Lipsitz [1999]. Subsequent to the development of the GEE1 class of models, GEE2 models were developed that do not make this separation of the regression and association parameters. In other words, GEE2 does not assume orthogonality of these parameter vectors. We will not cover GEE2 models in this chapter; the interested reader is referred to the text by Hardin and Hilbe [2003], which describes GEE1 and GEE2 models, as well as other versions of GEE models, in great detail. In this chapter, GEE models will refer only to the GEE1 class of models.

GEE models have important differences from MRMs, and these are well-described by several authors [Agresti, 2002; Burton et al., 1998; Diggle et al., 2002; Hardin and Hilbe, 2003; Hu et al., 1998; Liang and Zeger, 1993; Neuhaus et al., 1991; Ten Have et al., 1996; Zeger et al., 1988]. A basic premise of the GEE approach is that one is primarily interested in the regression parameters $\boldsymbol{\beta}$ and is not interested in the variance–covariance matrix of the repeated measures. As such, the variance–covariance structure is treated as a nuisance that one must account for in some way in order to get reasonable statistical tests for the regression parameters. Thus, the GEE models are not meant for situations in which scientific interest centers around the variance and/or covariance parameters.

For application of GEE models, as for the CPMs in Chapter 6, one assumes that there are a fixed number of timepoints n that subjects are measured at. A given subject does not have to be measured at all n timepoints, however, it is the full $n \times n$ correlation matrix of the longitudinal data that is considered in a GEE model (as nuisance). The various forms for this matrix are quite similar to the forms considered for CPMs, and so in many ways, GEE models and CPMs share common specifications. A fundamental difference is that CPMs specify the joint distribution and likelihood of the dependent variable vector $\boldsymbol{y}_i$, whereas GEE models only specify the marginal distribution and likelihood of y_{ij} for varying j. Also, the CPMs described in Chapter 6 only apply to continuous normal outcomes, whereas GEE models apply to many different types of outcome variables.

8.2 GENERALIZED LINEAR MODELS (GLMS)

Before describing GEE models, it is useful to review generalized linear models (GLMs), since GEE models can be viewed as an extension of GLMs to the case of correlated data. Here, we present a brief and nontechnical introduction to GLMs; much greater detail and statistical theory can be found in the classic text of McCullagh and Nelder [1989]. GLMs represent a class of models that are used to fit fixed effects regression models to normal and nonnormal data. McCullagh and Nelder [1989] describe this class of models in great detail and point out that the term "generalized linear model" is due to Nelder and Wedderburn [1972], who indicated how linearity could be exploited to unify several diverse statistical techniques. The essential idea is to treat many types of regression models, which differ primarily in terms of the type of dependent variable they model, as special cases of a single family of models. The dependent variable is assumed to come from the class of distributions known as the exponential family, and common GLM family members include linear regression for normally distributed dependent variables, logistic regression for dichotomous dependent varirables, and Poisson regression for counts. There are three specifications in a GLM. First, the linear predictor, denoted as η, of a GLM is of the form

$$\eta_i = \boldsymbol{x}_i'\boldsymbol{\beta}, \tag{8.1}$$

where $\boldsymbol{x}_i$ is the vector of explanatory variables, or covariates, for subject i with fixed effects $\boldsymbol{\beta}$. This first step indicates a linear predictor η_i which is based on covariates $\boldsymbol{x}_i$ and regression coefficients $\boldsymbol{\beta}$. Note that the covariates in $\boldsymbol{x}_i$ can include continuous regressors, dummy variables, interactions, polynomials, etc.

Then, a link function $g(\cdot)$ is specified which converts the expected value μ of the outcome variable y (*i.e.*, $\mu_i = E[y_i]$) to the linear predictor η.

$$g(\mu_i) = \eta_i. \tag{8.2}$$

For example, in ordinary multiple regression, the link function is called the identity link since $g(\mu_i) = \mu_i$ and so $\mu_i = \eta_i$, or

$$E[y_i] = \boldsymbol{x}_i'\boldsymbol{\beta}. \tag{8.3}$$

Under the identity link, the expected value of the dependent variable is simply a linear function of the explanatory variables multiplied by their regression coefficients.

For dichotomous outcomes, logistic regression [Hosmer and Lemeshow, 2000] is a popular choice for analysis. This model is written as

$$\log\left[\frac{P(y_i = 1)}{1 - P(y_i = 1)}\right] = \boldsymbol{x}_i'\boldsymbol{\beta}, \tag{8.4}$$

where y takes on values of 0 or 1. Since $P(y_i = 1) = E[y_i] = \mu_i$ in this case, we see that it is the logit link $g(\mu_i) = \log(\mu_i/(1 - \mu_i))$ which relates the expected value of the outcome variable to the linear predictor.

Similarly, the Poisson regression model [Cameron and Trivedi, 1998], which is used to model count data, is written as

$$\mu_i = \exp(\boldsymbol{x}_i'\boldsymbol{\beta}) \tag{8.5}$$

or

$$\log(\mu_i) = \boldsymbol{x}_i'\boldsymbol{\beta}, \tag{8.6}$$

which shows that it is the log link $g(\mu_i) = \log \mu_i$ that is used for Poisson regression.

So far, we've specified what the covariates are and how they relate to the expectation of the dependent variable. In a GLM, we additionally need to specify the form of the conditional variance of y, given the covariates. This is done as

$$V(y_i) = \phi\, v(\mu_i), \tag{8.7}$$

where $v(\mu_i)$ is a known variance function and ϕ is a scale parameter that may be known or estimated. For example, for ordinary multiple regression $v(\mu_i) = 1$ and ϕ would represent the error variance (*i.e.*, ϕ represents the variance of the conditional normal distribution of y given x) which is estimated. For a dichotomous outcome, the Bernoulli distribution specifies

$$v(\mu_i) = \mu_i(1 - \mu_i) \tag{8.8}$$

and ϕ is typically not estimated but set to 1 in the ordinary GLM. An exception is for models that allow over- or under-dispersion, in which case ϕ is estimated. For a count outcome, the Poisson distribution specifies that the mean equals the variance, and so

$$v(\mu_i) = \mu_i \tag{8.9}$$

and, again, ϕ is set to 1 (*i.e.*, it is not estimated) in the usual GLM. As can be appreciated, the link function and variance specification usually depend on the distribution of the outcome variable y. Here, we have presented three versions of GLMs, but there are many others that are within the GLM family.

With these GLM specifications, one can estimate the regression coefficients $\boldsymbol{\beta}$ by solving the estimating equation

$$U(\boldsymbol{\beta}) = \sum_{i=1}^{N} \left(\frac{\partial \mu_i}{\partial \boldsymbol{\beta}}\right)' (V(y_i))^{-1} [y_i - \mu_i] = 0. \tag{8.10}$$

For example, in the ordinary multiple regression model, we get the usual

$$U(\boldsymbol{\beta}) = \sum_{i=1}^{N} \boldsymbol{x}_i(y_i - \boldsymbol{x}_i'\boldsymbol{\beta}) = 0 \tag{8.11}$$

for solution of the regression coefficients $\boldsymbol{\beta}$.

As noted by Wedderburn [1974], the above estimating equation (8.10) depends only on the mean and variance of y, and therefore the precise distributional form for y is not necessary for estimation of the regression coefficients $\boldsymbol{\beta}$. In this case, solution of this estimating equation provides what are called "quasi-likelihood" estimates.

GLMs are fixed effects models which assume that all observations are independent of each other. Thus, they are not generally appropriate for analysis of longitudinal data. However, they can be extended to account for the correlation inherent in longitudinal data, and this is what Liang and Zeger did in developing GEE models. It should be noted that the models described in later chapters on mixed-effects logistic and Poisson regression, for example, also represent generalizations of GLMs by including random effects, and thus represent generalized linear mixed models (GLMMs). GEE is a different kind of generalization of GLM than that provided by GLMMs.

8.3 GENERALIZED ESTIMATING EQUATIONS (GEE) MODELS

A basic feature of GEE models is that the joint distribution of a subject's response vector $\boldsymbol{y}_i$ does not need to be specified. Instead, it is only the marginal distribution of y_{ij} at each timepoint that needs to be specified. To clarify this further, suppose that there are two timepoints and suppose that we are dealing with a continuous normal outcome. GEE would only require us to assume that the distribution of y_{i1} and y_{i2} are two univariate normals, rather than assuming that y_{i1} and y_{i2} form a (joint) bivariate normal distribution. Thus, GEE avoids the need for multivariate distributions by only assuming a functional form for the marginal distribution at each timepoint.

A related feature of GEE models is that the (co)variance structure is treated as a nuisance. The focus is clearly on the regression of y on X. In this regard, GEE models yield consistent and asymptotically normal solutions for the regression coefficients $\boldsymbol{\beta}$, even with misspecification of the (co)variance structure of the longitudinal data.

Since GEE models can be thought of as an extension of GLMs for correlated data, the GEE specifications involve those of GLM with one addition. So, first, the linear predictor is specified as

$$\eta_{ij} = \boldsymbol{x}'_{ij}\boldsymbol{\beta}, \tag{8.12}$$

where $\boldsymbol{x}_{ij}$ is the covariate vector for subject i at time j. Then a link function

$$g(\mu_{ij}) = \eta_{ij} \tag{8.13}$$

is chosen. As in GLMs, common choices here are the identity, logit, and log link for continuous, binary, and count data, respectively. The variance is then described as a function of the mean, namely,

$$V(y_{ij}) = \phi\, v(\mu_{ij}), \tag{8.14}$$

where, again, $v(\mu_{ij})$ is a known variance function and ϕ is a scale parameter that may be known or estimated.

The additional specification in a GEE model is for the "working" correlation structure of the repeated measures. This working correlation matrix is of size $n \times n$ because one assumes that there are a fixed number of timepoints n that subjects are measured at. A given subject does not have to be measured at all n timepoints; each individual's correlation matrix $\boldsymbol{R}_i$ is of size $n_i \times n_i$ with the appropriate rows and columns removed if $n_i < n$. It is assumed that the correlation matrix $\boldsymbol{R}$, and thus $\boldsymbol{R}_i$, depends on a vector of association parameters denoted $\boldsymbol{a}$. Examples of the various working correlation structures below will make this notion more concrete. These parameters $\boldsymbol{a}$ are assumed to be the same for all subjects. They represent the average dependence among the repeated observations across subjects.

It is generally recommended that choice of $\boldsymbol{R}$ should be consistent with the observed correlations. However, the GEE method yields consistent estimates of the regression coefficients and their standard errors, even with misspecification of the correlational structure. This is a very attractive property of GEE models. Efficiency (*i.e.*, statistical power) is reduced if the choice of $\boldsymbol{R}$ is incorrect; however, the loss of efficiency is lessened as the number of subjects gets large. The usual working correlations considered are independence, exhangeable, AR(1), m-dependent, and unspecified.

8.3.1 Working Correlation Forms

The simplest form is that of independence, namely $\boldsymbol{R}_i(\boldsymbol{a}) = \boldsymbol{I}$, a $n \times n$ identity matrix. This form is equivalent to assuming that the longitudinal data are not correlated. In general, this is not a structure that makes logical sense for longitudinal data, since such data are usually highly correlated. Also, Fitzmaurice [1995] shows that for longitudinal binary outcomes this choice can lead to large efficiency loss for time-varying covariates. Alternatively, Pepe and Anderson [1994] indicate that use of the independence structure does have certain advantages for models that include time-varying covariates.

The next simplest structure is to assume that all of the correlations in $\boldsymbol{R}$ are the same, or "exchangeable." This exchangeable structure specifies that $\boldsymbol{R}_i(\boldsymbol{a}) = \rho$, namely that all of the correlations are equal. This is equivalent to the assumption regarding the correlations in a random intercept MRM or a compound symmetry CPM.

As in CPMs, another useful one parameter model for longitudinal data is the AR(1) structure, namely, $\boldsymbol{R}_i(\boldsymbol{a}) = \rho^{|j-j'|}$. Here, the within-subject correlation over time is an exponential function of the lag. As mentioned, this form is often a very parsimonious one for longitudinal data since it only depends on one term, yet allows the correlations to decline with the order of the time lag.

The m-dependent, or banded, structure is essentially the same as the Toeplitz structure considered in previous chapters. Here, $\boldsymbol{R}_i(\boldsymbol{a}) = \rho_{|j-j'|}$ if $j - j' \leq m$, and $\boldsymbol{R}_i(\boldsymbol{a}) = 0$ if $j - j' > m$. Note that the fullest structure here is $m = n - 1$, in which all of the lagged correlations are estimated. Within a time lag, all of the correlations are assumed to be the same, but there is no functional relationship between lags of different orders, as in the AR(1) structure.

Finally, the unspecified or unstructured form would estimate all $n(n-1)/2$ correlations of $\boldsymbol{R}$. This form is most efficient, but most useful when there are relatively few timepoints. When there are many timepoints, estimation of the $n(n-1)/2$ correlations is not parsimonious. Also, missing data complicates estimation of $\boldsymbol{R}$.

8.4 GEE ESTIMATION

Define $\boldsymbol{A}_i$ to be the $n \times n$ diagonal matrix with $V(\mu_{ij})$ as the jth diagonal element. Also, as indicated above, define $\boldsymbol{R}_i(\boldsymbol{a})$ to be the $n \times n$ "working" correlation matrix (of the n repeated measures) for the i subject. Then, the working variance–covariance matrix for $\boldsymbol{y}_i$ equals

$$V(\boldsymbol{a}) = \phi \boldsymbol{A}_i^{1/2} \boldsymbol{R}_i(\boldsymbol{a}) \boldsymbol{A}_i^{1/2}. \tag{8.15}$$

Notice, that for the case of normally distributed outcomes with homogeneous variance across time, we get

$$V(\boldsymbol{a}) = \phi \boldsymbol{R}_i(\boldsymbol{a}). \tag{8.16}$$

For normal outcomes, Park [1993] extends this to heterogeneous variance across time by allowing the scale parameter ϕ_j to vary across time ($j = 1, \ldots, n$).

The GEE estimator of $\boldsymbol{\beta}$ is the solution of

$$\sum_{i=1}^{N} \boldsymbol{D}_i' \left[V(\hat{\boldsymbol{a}})\right]^{-1} (\boldsymbol{y}_i - \boldsymbol{\mu}_i) \;=\; 0, \tag{8.17}$$

where $\hat{\boldsymbol{a}}$ is a consistent estimate of $\boldsymbol{a}$ and $\boldsymbol{D}_i = \partial \boldsymbol{\mu}_i / \partial \boldsymbol{\beta}$. Notice that this formula is an extension of the estimating equation for $\boldsymbol{\beta}$ in any GLM, which is given in (8.10). Thus, the

GEE solution can be seen as a natural generalization of the GLM solution for correlated data.

As an example, in the normal case

$$\begin{aligned} \boldsymbol{\mu}_i &= \boldsymbol{X}_i\boldsymbol{\beta}, \\ \boldsymbol{D}_i &= \boldsymbol{X}_i, \\ V(\hat{\boldsymbol{a}}) &= \boldsymbol{R}_i(\hat{\boldsymbol{a}}), \end{aligned}$$

and so

$$\sum_{i=1}^{N} \boldsymbol{X}_i' \left[\boldsymbol{R}_i(\hat{\boldsymbol{a}})\right]^{-1} (\boldsymbol{y}_i - \boldsymbol{X}_i\boldsymbol{\beta}) = 0, \tag{8.18}$$

which yields, solving for $\boldsymbol{\beta}$,

$$\hat{\boldsymbol{\beta}} = \left[\sum_{i=1}^{N} \boldsymbol{X}_i' \left[\boldsymbol{R}_i(\hat{\boldsymbol{a}})\right]^{-1} \boldsymbol{X}_i\right]^{-1} \left[\sum_{i-1}^{N} \boldsymbol{X}_i' \left[\boldsymbol{R}_i(\hat{\boldsymbol{a}})\right]^{-1} \boldsymbol{y}_i\right]. \tag{8.19}$$

Notice this is essentially the same as the weighted least-squares (WLS) estimator with the weight matrix being $[\boldsymbol{R}_i(\hat{\boldsymbol{a}})]^{-1}$. A caveat is that in WLS one typically knows the weight matrix, however here it depends on parameters to be estimated (*i.e.*, $\boldsymbol{a}$). In this case, the solution can proceed using iteratively reweighted least squares (IRLS) where iterative estimates of $\boldsymbol{a}$ are used to yield new estimates of $\boldsymbol{\beta}$, with the procedure continuing until convergence. Because equation (8.17) depends only on the mean and variance of y, these are quasi-likelihood estimates.

Solving the GEE involves iterating between the quasi-likelihood solution for estimating $\boldsymbol{\beta}$ and a robust method for estimating $\boldsymbol{a}$ as a function of $\boldsymbol{\beta}$. Essentially, it involves the following steps, which are repeated until convergence.

1. Given estimates of $\boldsymbol{R}_i(\boldsymbol{a})$ and ϕ, calculate estimates of $\boldsymbol{\beta}$ using IRLS.

2. Given estimates of $\boldsymbol{\beta}$, obtain estimates of $\boldsymbol{a}$ and ϕ. For this, calculate Pearson (or standardized) residuals

$$r_{ij} = (y_{ij} - \hat{\mu}_{ij})/\sqrt{[V(\hat{\boldsymbol{a}})]_{jj}} \tag{8.20}$$

   and use these residuals to consistently estimate $\boldsymbol{a}$ and ϕ. Liang and Zeger [1986] present the estimators for several different working correlation structures.

Upon convergence, in order to perform hypothesis tests and construct confidence intervals, it is of interest to obtain standard errors associated with the estimated regression coefficients. These standard errors are obtained as the square root of the diagonal elements of the matrix $V(\hat{\boldsymbol{\beta}})$. The GEE provides two versions of these.

1. Naive or "model-based"

$$V(\hat{\boldsymbol{\beta}}) = \left[\sum_{i}^{N} \boldsymbol{D}_i' \hat{\boldsymbol{V}}_i^{-1} \boldsymbol{D}_i\right]^{-1}. \tag{8.21}$$

2. Robust or "empirical"

$$V(\hat{\beta}) = \boldsymbol{M}_0^{-1} \boldsymbol{M}_1 \boldsymbol{M}_0^{-1}, \tag{8.22}$$

where

$$\begin{aligned} \boldsymbol{M}_0 &= \sum_i^N \boldsymbol{D}_i' \hat{\boldsymbol{V}}_i^{-1} \boldsymbol{D}_i, \\ \boldsymbol{M}_1 &= \sum_i^N \boldsymbol{D}_i' \hat{\boldsymbol{V}}_i^{-1} (\boldsymbol{y}_i - \hat{\boldsymbol{\mu}}_i)(\boldsymbol{y}_i - \hat{\boldsymbol{\mu}}_i)' \hat{\boldsymbol{V}}_i^{-1} \boldsymbol{D}_i. \end{aligned}$$

Here, $\hat{\boldsymbol{V}}_i$ denotes $V_i(\hat{\boldsymbol{a}})$. Notice that if $\hat{\boldsymbol{V}}_i = (\boldsymbol{y}_i - \hat{\boldsymbol{\mu}}_i)(\boldsymbol{y}_i - \hat{\boldsymbol{\mu}}_i)'$ then the two are equal. This occurs only if the true correlation structure is correctly modeled. In the more general case, the robust or "sandwich" estimator, which is due to Royall [1986], provides a consistent estimator of $V(\hat{\boldsymbol{\beta}})$ even if the working correlation structure $\boldsymbol{R}_i(\boldsymbol{a})$ is not the true correlation of $\boldsymbol{y}_i$.

8.5 EXAMPLE

Gruder et al. [1993] describe a smoking-cessation study in which 489 subjects were randomized to either a control, discussion, or social support condition. Control subjects received a self-help manual and were encouraged to watch twenty segments of a daily TV program on smoking cessation, while subjects in the two experimental conditions additionally participated in group meetings and received training in support and relapse prevention. The outcome variable was the subject's smoking status, with 0 coded as smoking and 1 as abstinent. Data were collected at four telephone interviews: post-intervention, and 6, 12, and 24 months later.

Some subjects that were randomized to receive one of two treatment conditions never showed up to any meetings following the phone call informing them of where the group meetings would take place. Thus, these subjects were randomized to receive a treatment (*i.e.*, either discussion or social support group meetings), but in reality only received the same self-help manual that the placebo subjects received. In the Gruder et al. [1993] article, this issue was dealt with by forming four groups in the analysis:

1. Control: randomized to the control condition.
2. No-show: randomized to receive a group treatment, but never showed up to the group meetings.
3. tx1: randomized to and received group meetings (*i.e.*, discussion).
4. tx2: randomized to and received enhanced group meetings (*i.e.*, social support).

In the analysis, these four groups were compared using Helmert contrasts [Bock, 1975]. Specifically, the contrast coefficients for the three group-related Helmert contrasts are:

Group	$H1$	$H2$	$H3$
Control	-1	0	0
No-show	1/3	-1	0
tx1	1/3	1/2	-1
tx2	1/3	1/2	1

These contrasts yield the following interpretation in terms of the four repeated smoking status classifications.

$H1$: test of whether randomization to group versus control influenced subsequent cessation.

$H2$: test of whether showing up to the group meetings influenced subsequent cessation.

$H3$: test of whether the type of meeting influenced cessation.

While $H1$ is an experimental comparison, in that it involves a comparison of subjects randomized to control versus some form of treatment, the $H2$ and $H3$ comparisons are quasi-experimental in nature. As such, the tests concerning $H2$ and $H3$ are open to the possibility of confounding. For example, a person's self-efficacy might be related both to the decision to attend group meetings (*i.e.*, be a no-show or a member of tx1 or tx2) and to the subsequent smoking outcomes. In this case, a finding that $H2$ is significantly related to subsequent smoking outcomes might be due to (a) an effect of the group meetings, (b) an effect due to a subject's self-efficacy, or (c) a combination of (a) and (b). Similar comments apply to $H3$, though it would seem that the potential for confounding variables to bias this comparison of tx1 versus tx2, subject to group attendance, is clearly less. In the Gruder et al. [1993] article, the authors reported that baseline analysis indicated that groups differed in terms of race (white versus nonwhite) and so this variable was included in all analyses.

Smoking abstinence rates, stratified by group, are presented in Table 8.1. These clearly show the general deleterious effect of time on smoking abstinence, though the effect does seem to vary by treatment group. Additionally, in preparation for the GEE analysis, Table 8.2 lists the ordinary Pearson correlations for these four assesments. It is generally advisable to choose a working correlation structure that is similar to the structure of the observed correlations. This is because, although the GEE is robust to misspecification of the correlation structure, efficiency is increased to the extent that the specified structure is correct. In the present case, the exchangeable structure does not appear like a good choice since the correlations are not approximately equal. Also, neither the AR(1) nor the m-dependent structures appear reasonable because the correlations within a time lag vary. Thus, an unspecified structure appears to be the most reasonable choice for these data.

Table 8.1. Point Prevalence Rates (N) of Abstinence over Time by Group

Group	End-of-Program (T1)	6 months (T2)	12 months (T3)	24 months (T4)
No Contact Control	17.4 (109)	7.2 (97)	18.5 (92)	18.2 (77)
No Shows	26.8 (190)	18.9 (175)	18.6 (161)	18.7 (139)
Discussion	33.7 (86)	14.6 (82)	16.3 (80)	22.9 (70)
Social Support	49.0 (104)	20.0 (100)	24.0 (96)	25.6 (86)

Table 8.2. Correlation of Smoking Abstinence (y/n) Across Time

	T1	T2	T3	T4
T1	1.00	0.33	0.29	0.26
T2	0.33	1.00	0.48	0.34
T3	0.29	0.48	1.00	0.49
T4	0.26	0.34	0.49	1.00

For these data, because the dependent variable is dichotomoous, several logistic regression GEE models were fit. In this, it was the probability of smoking abstinence, rather than smoking, across time that was modeled. The regressors include the aforementioned Helmert contrasts for group, time (T_j coded 0, 1, 2, and 4 for the four timepoints), time squared (T_j^2), and race (0, nonwhite; 1, white). Three sets of linear predictors were fit to these data: a main effect models, a model including condition by linear time interactions, and a model including both condition by linear and quadratic time interactions. Specifically, for model 1,

$$\eta_{ij} = \beta_0 + \beta_1 T_j + \beta_2 T_j^2 + \beta_3 H1_i + \beta_4 H2_i + \beta_5 H3_i + \beta_6 Race_i, \tag{8.23}$$

whereas for model 2,

$$\begin{aligned}\eta_{ij} \quad = \quad & \beta_0 + \beta_1 T_j + \beta_2 T_j^2 \\ & + \beta_3 H1_i + \beta_4 H2_i + \beta_5 H3_i + \beta_6 Race_i \\ & + \beta_7 (H1_i \times T_j) + \beta_8 (H2_i \times T_j) + \beta_9 (H3_i \times T_j),\end{aligned} \tag{8.24}$$

and for model 3,

$$\begin{aligned}\eta_{ij} &= \beta_0 + \beta_1 T_j + \beta_2 T_j^2 \\ &\quad + \beta_3 H1_i + \beta_4 H2_i + \beta_5 H3_i + \beta_6 Race_i \\ &\quad + \beta_7 (H1_i \times T_j) + \beta_8 (H2_i \times T_j) + \beta_9 (H3_i \times T_j) \\ &\quad + \beta_{10}(H1_i \times T_j^2) + \beta_{11}(H2_i \times T_j^2) + \beta_{12}(H3_i \times T_j^2). \end{aligned} \tag{8.25}$$

Because the dependent variable is dichotomous, the logit link and Bernoulli distribution are specified. Finally, as mentioned, the unspecified working correlation structure was selected.

Results for these models are presented in Table 8.3. The standard errors presented in this table are based on the "robust" estimates, which are also called the empirical standard errors by SAS, for example. These are obtained under the sandwich estimation of the variance–covariance matrix $V(\hat{\beta})$. The standard errors are the square root of the diagonal elements of this matrix.

Table 8.3. Smoking Status (0, Smoking; 1, Not Smoking) Across Time ($N = 489$)—GEE Logistic Parameter Estimates (Est.), Standard Errors (SE), and p-Values

Parameter	Model 1			Model 2			Model 3		
	Est.	SE	$p <$	Est.	SE	$p <$	Est.	SE	$p <$
Intercept β_0	−.999	.112	.001	−1.014	.117	.001	−1.010	.117	.001
T β_1	−.633	.126	.001	−.610	.127	.001	−.631	.131	.001
T^2 β_2	.132	.029	.001	.130	.029	.001	.135	.030	.001
$H1$ β_3	.583	.170	.001	.811	.214	.001	.865	.224	.001
$H2$ β_4	.288	.121	.018	.366	.142	.010	.431	.150	.004
$H3$ β_5	.202	.119	.091	.271	.141	.055	.268	.148	.070
$Race$ β_6	.358	.200	.074	.353	.200	.078	.354	.200	.078
$H1 \times T$ β_7				−.219	.097	.024	−.466	.226	.039
$H2 \times T$ β_8				.073	.069	.289	−.375	.181	.039
$H3 \times T$ β_9				−.062	.072	.385	−.034	.195	.863
$H1 \times T^2$ β_{10}							.062	.051	.220
$H2 \times T^2$ β_{11}							.079	.042	.060
$H3 \times T^2$ β_{12}							−.007	.046	.878

All three models agree in indicating that the overall change in smoking abstinence across time involves both the linear and quadratic trend components (*i.e.*, both the linear and quadratic trend parameters are highly significant). Because the linear trend estimate is negative and the quadratic trend estimate is positive, a decelerating negative trend is indicated. Abstinence diminishes across time in a curvilinear manner. Also, the effect of *Race* is seen to be marginally significant, suggesting somewhat higher abstinence for whites relative to nonwhites.

8.5.1 Generalized Wald Tests for Model Comparison

In order to interpret the group-related effects, it would be helpful to compare these models statistically to determine if the group by time interaction terms are jointly significant or not.

Because GEE model parameters are estimated using quasi-likelihood procedures, there is no associated likelihood underlying the model. Thus, the usual likelihood ratio tests cannot be applied to compare the above models.

To compare the above GEE models, however, one can construct a multi-parameter Wald test to test the joint null hypothesis that a set of βs equal 0. For this, define a $q \times p$ indicator matrix C of ones and zeros to select the parameters of interest for the multi-parameter test. Here, p equals the number of regressors in the full model (including the intercept) and q equals the number of parameters in the multi-parameter test (*i.e.*, the difference in regressors between the full and reduced models). Each row of C contains a 1 in only one location, and zeros elsewhere, to select one of the parameters that comprise this test. There are q rows in this matrix because each row is used to uniquely select one of the q parameters. For example, suppose that we want to compare models 2 and 3 above. Then $q = 3, p = 13$ and this indicator matrix is given as

$$\boldsymbol{C} = \begin{bmatrix} 0 & 0 & 0 & 0 & 0 & 0 & 0 & 0 & 0 & 0 & 1 & 0 & 0 \\ 0 & 0 & 0 & 0 & 0 & 0 & 0 & 0 & 0 & 0 & 0 & 1 & 0 \\ 0 & 0 & 0 & 0 & 0 & 0 & 0 & 0 & 0 & 0 & 0 & 0 & 1 \end{bmatrix},$$

to select the latter three elements (*i.e.*, the group by time squared terms).

The multi-parameter, or generalized, Wald test then equals

$$X^2 = \hat{\boldsymbol{\beta}}' \, \boldsymbol{C}' \, (\boldsymbol{C} \, V(\hat{\boldsymbol{\beta}}) \, \boldsymbol{C}')^{-1} \, \boldsymbol{C} \, \hat{\boldsymbol{\beta}}, \tag{8.26}$$

which is distributed as χ^2 with q degrees of freedom under the null hypothesis. As a reminder, the prime symbol $'$ indicates the transpose of the matrix or vector. Notice that if C is a $1 \times p$ vector selecting a single regression coefficient β^*, then the above formula simplifies to

$$X^2 = \frac{(\hat{\beta}^*)^2}{V(\hat{\beta}^*)} = \left[\frac{\hat{\beta}^*}{SE(\hat{\beta}^*)} \right]^2$$

on one degree of freedom to test the null hypothesis that β^* equals 0. This is just a re-expression of the Wald test statistic presented in the table above, and elsewhere. Both versions of the one-parameter Wald test are the same (*i.e.*, yield the same p-value) because of the relationship between the standard normal z and the χ^2 distribution on one degree of freedom, namely $z^2 = \chi_1^2$.

In the present example comparing models 2 and 3, the null hypothesis being tested is

$$H_0 = \beta_{10} = \beta_{11} = \beta_{12} = 0.$$

For this, we get $X_3^2 = 4.345$ which yields $p = .23$ and is not significant. Additionally, none of the individual tests of the group by time squared parameters in Table 8.2. are signficant at .05 either. Thus, model 2 is preferred to model 3.

Turning to a comparison of models 1 and 2, and therefore testing the null hypothesis

$$H_0 = \beta_7 = \beta_8 = \beta_9 = 0,$$

we get $X_3^2 = 6.974$ which yields $p = .073$ and is marginally significant. Additionally, turning to the individual interactions, the $H1 \times T$ interaction is significant ($\hat{\beta}_7 = -.219, p < .024$). Thus, there is some evidence for model 2, although it is not quite at the .05 level in terms of the multi-parameter Wald test. Also, inspection of the observed abstinence rates in Table 8.1, supports the notion that there are large group differences at post-intervention, but that these are not maintained across the study.

Model 2 indicates that randomization to group increases abstinence at post-intervention ($\hat{\beta}_3 = .811, p < .001$), but that this benefit goes away across time ($\hat{\beta}_7 = -.219, p < .024$). In terms of an estimated odds ratio at post-intervention, this would equal

$$OR = \exp[4/3(.811)] = 2.95,$$

indicating that the odds of being abstinent at post-intervention is three times that for subjects randomized to one of the groups relative to the control arm. Note that the reason for multiplying the estimated regression coefficient by 4/3 is because this represents the difference between the control and treatment groups in the coding of the H1 contrast (*i.e.*, -1 for the control group and 1/3 for the other treatment conditions). We can also calculate an asymptotic 95% confidence interval for this odds ratio as

$$\exp[4/3(.811) \pm 1.96 \times 4/3(.214)] \;=\; (1.69, 5.16),$$

to give a sense of the uncertainty associated with this odds ratio.

In terms of the second Helmert contrast, Model 2 indicates a significant benefit to attending the group sessions ($\hat{\beta}_4 = .366, p < .010$) at post-intervention, and further suggests that this effect does not vary significantly across time. The estimated odds ratio for this contrast of subjects attending the groups, relative to those not attending the groups, at post-intervention is calculated as

$$OR = \exp[3/2(.366)] = 1.73,$$

with the 95% confidence interval as

$$\exp[3/2(.366) \pm 1.96 \times 3/2(.142)] \;=\; (1.14, 2.63).$$

Thus, there is estimated to be nearly a two-fold increase in the odds of being abstinent at post-intervention for those subjects actually attending the group sessions relative to those not attending. The confidence interval for this contrast is fairly wide indicating the degree of uncertainty in this assessment.

Regarding $H3$, Model 2 posits a marginally significant benefit to the Social Support, relative to the Discussion, at post-intervention which does not vary significantly across time. The estimated odds ratio for this contrast at post-intervention is obtained as

$$OR = \exp[2(.271)] = 1.72,$$

and the 95% confidence interval is gotten as

$$\exp[2(.271) \pm 1.96 \times 2(.141)] = (0.99, 2.99).$$

Similar to the H2 contrast, there is nearly a two-fold increase in the estimated odds of being abstinent for the group receiving the Social Support treatment, relative to the Discussion; however, the confidence interval is quite wide and does include the odds ratio of one.

We can also determine the significance of the Helmert contrasts at any study timepoint using the generalized Wald statistic presented above. Notice that in model 2, the estimated effect of $H1$ is given by

$$(\hat{\beta}_3 + \hat{\beta}_7 \times T).$$

This makes clear that $\hat{\beta}_3$ is the estimated $H1$ contrast at post-intervention (*i.e.*, when T equals 0), and that the degree of the change in the $H1$ contrast across time is estimated by $\hat{\beta}_7$. So, to calculate the estimate of this contrast at the final timepoint (*i.e.*, when T equals 4), we would get

$$.811 + 4 \times -.219 = -.066.$$

To determine the significance, we can use the generalized Wald statistic defined in (8.26). Specifically, (8.26) can be used to test for the significance of any set of linear functions of the model parameters. Above, we used this generalized Wald statistic to perform multi-parameter tests, but it can also be used to assess the significance of one or more linear functions of the parameters. For example, notice that we can define

$$\boldsymbol{C} = \begin{bmatrix} 0 & 0 & 0 & 1 & 0 & 0 & 0 & 4 & 0 & 0 \end{bmatrix}$$

and that if we multiple $\boldsymbol{C}$ by $\hat{\boldsymbol{\beta}}$ from model 2 we would get $-.066$ as the estimate of the H1 contrast at the final timepoint. More generally, applying the generalized Wald statistic in (8.26), we get $X_1^2 = .047$ for this $H1$ contrast at the final timepoint. This is clearly nonsignificant. Similar calculations for $H2$ and $H3$ yield Wald chi-square values of .102 and .009, respectively. Thus, Model 2 suggests that there are no significant group differences by the end of the study. This is supported by the observed proportions in Table 8.1, which suggest minimal group differences at the study's end. It should be noted that performing many of these contrast calculations, and their associated statistical tests, leads to a multiple comparisons issue, and therefore the potential need to correct for the type-I error rate for the family of comparisons. In the present case, it is not so important because these additional contrasts are not significant, even without an adjustment. However, in the more general situation, they should be used judiciously.

8.5.2 Model Fit of Observed Proportions

It is often of interest to examine how well a particular model fits the observed sample proportions. Table 8.4 shows how SAS PROC IML can be used to accomplish this. In doing this, the concatenation operator " | | " is used to combine the necessary columns of the covariate matrix for each of the four groups. Also, in this example we use the estimates from GEE Model 2 to calculate the estimated probabilities across the four timepoints for the four groups. Since *Race* is one of the model covariates, we need to provide a value for this covariate for each group. Here, we used the group mean of this variable. This SAS code yields the following estimated abstinence probabilities:

Group	End-of-Program (T1)	6 months (T2)	12 months (T3)	24 months (T4)
No Contact Control	.152	.121	.121	.231
No Shows	.271	.187	.156	.206
Discussion	.308	.208	.167	.204
Social Support	.441	.292	.218	.217

Figure 8.1 depicts these estimated probabilities and the observed proportions. As can be seen from Figure 8.1, the model fit of the observed proportions is fairly reasonable. The figure and model estimates certainly show how the groups are initially quite distinct, but then converge to an abstinence rate of abut 20% or so.

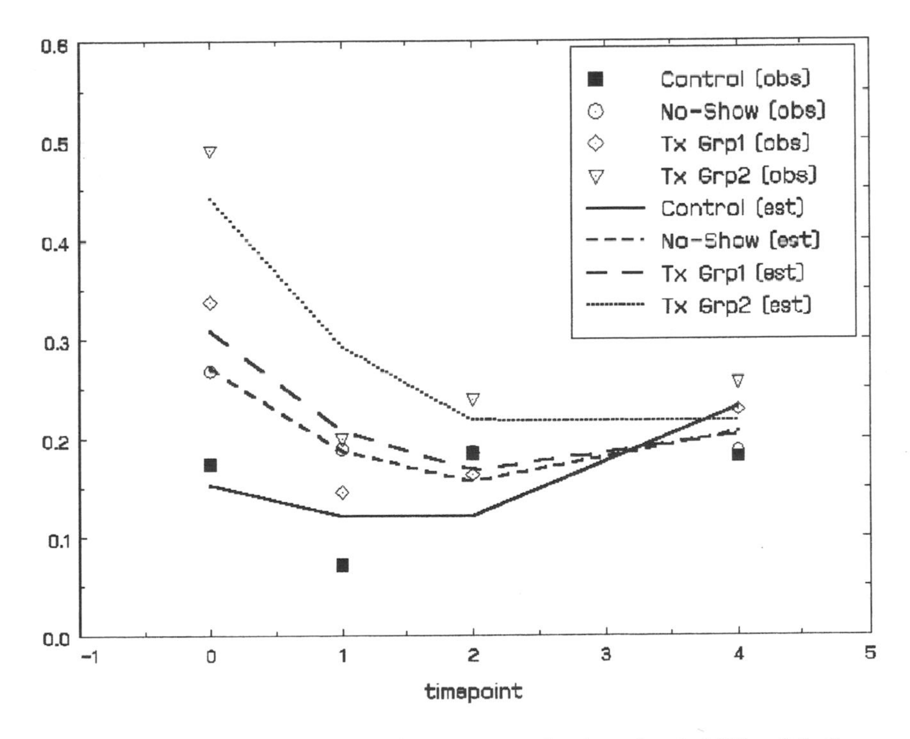

Figure 8.1. Observed point prevalence abstinence rates and estimated probabilities of abstinence across time.

8.6 SUMMARY

This chapter has described the GEE approach for longitudinal data analysis. This approach has several features which makes it particularly useful and popular. Because it is a generalization of GLM, many types of dependent variables can be accommodated within the GEE family of models. Software for performing GEE analysis is readily available in most of the major statistical software packages. Also, the selection of the variance–covariance matrix for the repeated measures is not as critical as with other models, because GEE provides standard errors that are robust to misspecification of the variance–covariance structure. This is an attractive feature, especially for situations where the scientific interest is in estimation and inference of the regression parameters and not of the variance–covariance structure. The converse of this is that if there is scientific interest in the variance–covariance structure of the longitudinal data, then GEE is not appropriate (at least in its GEE1 implementation). GEE does require that the timepoints of measurement are fixed, though subjects are not required to have complete data across time. So, for example, GEE would not be well-suited to problems where subjects vary considerably in terms of the timing of their longitudinal measurements. Also, as will be discussed in greater detail in Chapter 14, GEE makes a rather restrictive assumption (*i.e.*, missing completely at random) about the missing data.

The example in this chapter has illustrated use of the GEE model for longitudinal dichotomous data. As noted, GEE can also be applied to other types of outcomes, for example continuous, counts, ordinal, or nominal dependent variables. For all of these, GEE provides regression estimates that are "population-averaged" rather than the "subject-specific" estimates of the mixed-effects regression models. These two types of estimates agree in terms of scale only for continuous normal outcomes under the identity link. For the other models, there is an important difference in the scaling and interpretation of these two types of estimates. This will be discussed in greater detail in Chapter 9 on mixed-effects models for binary outcomes.

Table 8.4. SAS IML Code: Computing Estimated Probabilities Based on GEE Model 2

```
TITLE1 'Smoking Data - Observed Proportions and Estimated Probabilities';
PROC IML;
/* observed proportions - time by group */;
oprob   =  {   .174 .268 .337 .490,
               .072 .189 .146 .200,
               .185 .186 .163 .240,
               .182 .187 .229 .256 };
ologit = log (oprob / (1-oprob));

/* common part of the covariate matrix - intercept, time, and time squared */;
xmat   =  {   1 0 0,
              1 1 1,
              1 2 4,
              1 4 16};
g1   =  {    -1     0     0  };    /* Helmert contrast coefficients for Control group */
g2   =  {   .333   -1     0  };    /* Helmert contrast coefficients for No-Show group */
g3   =  {   .333   .5    -1  };    /* Helmert contrast coefficients for Tx1 group */
g4   =  {   .333   .5     1  };    /* Helmert contrast coefficients for Tx2 group */
/* use separate group means for race */
raceavg = {.2936, .3474, .0581, .1538};

/* put together the covariate matrices for each of the 4 groups */
xg1 = xmat || xmat[,1]*g1 || xmat[,1]*raceavg[1] || xmat[,2]*g1;
xg2 = xmat || xmat[,1]*g2 || xmat[,1]*raceavg[2] || xmat[,2]*g2;
xg3 = xmat || xmat[,1]*g3 || xmat[,1]*raceavg[3] || xmat[,2]*g3;
xg4 = xmat || xmat[,1]*g4 || xmat[,1]*raceavg[4] || xmat[,2]*g4;

print 'Covariate matrix - Control group', xg1 [FORMAT=8.3];
print 'Covariate matrix - No-Show group', xg2 [FORMAT=8.3];
print 'Covariate matrix - Tx1 group', xg3 [FORMAT=8.3];
print 'Covariate matrix - Tx2 group', xg4 [FORMAT=8.3];

/* estimates from GEE model 2 */
beta = { -1.014, -0.610, 0.130, 0.811, 0.366, 0.271, 0.353, -0.219, -0.073, -0.062};

/* estimated logits for each group */
elogitg1 = xg1*beta;
elogitg2 = xg2*beta;
elogitg3 = xg3*beta;
elogitg4 = xg4*beta;

/* estimated probabilities for each group */
eprobg1 = 1 / (1 + exp( - elogitg1));
eprobg2 = 1 / (1 + exp( - elogitg2));
eprobg3 = 1 / (1 + exp( - elogitg3));
eprobg4 = 1 / (1 + exp( - elogitg4));
/* assemble the group estimates into matrices */
elogit = elogitg1 || elogitg2 || elogitg3 || elogitg4;
eprob = eprobg1 || eprobg2 || eprobg3 || eprobg4;

print 'Observed logits - time by group', ologit [FORMAT=8.3];
print 'Observed probabilities - time by group', oprob [FORMAT=8.3];
print 'Estimated logits - time by group', elogit [FORMAT=8.3];
print 'Estimated probabilities - time by group', eprob [FORMAT=8.3];
```

CHAPTER 9

MIXED-EFFECTS REGRESSION MODELS FOR BINARY OUTCOMES

9.1 INTRODUCTION

Reflecting the usefulness of mixed-effects modeling and the importance of categorical outcomes in many areas of research, generalization of mixed-effects models for categorical outcomes has been an active area of statistical research. For dichotomous response data, several approaches adopting either a logistic or probit regression model and various methods for incorporating and estimating the influence of the random effects have been developed [Gibbons, 1981; Stiratelli et al., 1984; Anderson and Aitkin, 1985; Wong and Mason, 1985; Gibbons and Bock, 1987; Conaway, 1989; Goldstein, 1991]. Several review articles [Fitzmaurice et al., 1993; Rodríguez and Goldman, 1995; Pendergast et al., 1996; Goldstein and Rasbash, 1996] have discussed and compared some of these models and their estimation procedures. Also, Snijders and Bosker [1999, Chapter 14] provide a practical summary of the mixed-effects logistic regression model and the various procedures for estimating its parameters. As these sources indicate, the mixed-effects logistic regression model is a very popular choice for analysis of dichotomous data.

This chapter describes mixed-effects models for binary data that accommodate multiple random effects and allow for a general form for model covariates. Although only 2-level models will be considered here, 3-level generalizations can follow approaches described for 3-level dichotomous response data [Gibbons and Hedeker, 1997; Ten Have et al., 1999] and are discussed further in Chapter 13. We begin by reviewing the mixed-effects logistic regression model for longitudinal (or clustered) binary response data. A full maximum (marginal) likelihood solution is outlined for parameter estimation. In this solution, multi-dimensional quadrature is used to numerically integrate over the distribution of random effects, and an iterative Fisher scoring algorithm is used to solve the likelihood equations. To illustrate application of the mixed-effects logistic regression model, analysis of a longitudinal psychiatric dataset is described.

9.2 LOGISTIC REGRESSION MODEL

Binary outcomes are common in biomedical research, where "success" may indicate that the patient is alive after treatment, develops no particular disease after exposure, or develops no complication after a surgical operation. In health services research a common binary outcome is the use or non-use of services. Logistic regression is a widely accepted method for describing the relationship between a binary or dichotomous outcome and a set of explanatory variables. It is used in many areas such as health care research and biomedical studies [Kramer et al., 1983; Tsutakawa, 1988; Cleary and Angel, 1984; Khuri et al., 1997]. To provide a statistical foundation for mixed-effects generalization of the logistic regression model, we now present an overview of the simpler fixed-effects logistic regression model.

To begin, let p_i represent the probability of a positive outcome (i.e., $Y_i = 1$) for the ith individual. The probability of a negative outcome (i.e., $Y_i = 0$) is then $1-p_i$. Denote the set of covariates as $\boldsymbol{x}_i = (1, x_{i1}, \ldots, x_{ip})'$, where $\boldsymbol{\beta} = (\beta_0, \beta_1, \ldots, \beta_p)'$ is a $(p+1) \times 1$ vector of corresponding regression coefficients. Then the logistic regression model is written as

$$p_i = \Pr(Y_i = 1) = \frac{\exp(\boldsymbol{x}_i'\boldsymbol{\beta})}{1 + \exp(\boldsymbol{x}_i'\boldsymbol{\beta})}. \tag{9.1}$$

With a little algebra, this equation can also be written as

$$p_i = \frac{1}{1 + \exp(-\boldsymbol{x}_i'\boldsymbol{\beta})} = \Psi(\boldsymbol{x}_i'\boldsymbol{\beta}), \tag{9.2}$$

where $\Psi(\cdot)$ is the logistic cumulative distribution function (cdf), namely $\Psi(z) = 1/[1 + \exp(-z)]$. This model can also be represented in terms of the log odds or logit of the probabilities, namely

$$\log\left[\frac{p_i}{1 - p_i}\right] = \boldsymbol{x}_i'\boldsymbol{\beta}. \tag{9.3}$$

The numerator in the logit is the probability of a 1 response, and the denominator equals the probability of a 0 response. The ratio of these probabilities is the odds of a 1 response, and the log of this ratio is the log odds, or logit, or a 1 response. Notice that the log odds is equal to 0 when the probability of a 1 response equals .5 (*i.e.*, equal odds of a response in category 0 and 1), is negative when the probability is less than .5 (*i.e.*, odds favoring a response in category 0), and is positive when the probability is greater than .5 (*i.e.*, odds favoring a response in category 1).

In logistic regression, the logit is called the link function because it maps the (0,1) range of probabilities unto the $(-\infty, \infty)$ range of linear predictors. As (9.3) reveals, the logit is linear in its parameter vector and so has many of the desirable properties of a linear regression model, albeit in terms of the logits. This point is sometimes forgotten, so it is worth emphasizing that the logistic regression model is linear in terms of the logits and not the probabilities. In terms of the probabilities, the model in (9.2) posits an s-shaped logistic relationship between the values of x and the probabilities. For instance, Figure 9.1 depicts this relationship for a simple model with one regressor taking on values from -4 to 4 and whose slope equals one. In contrast, Figure 9.2 depicts the linear relationship between this same x variable and the logit of the response probability.

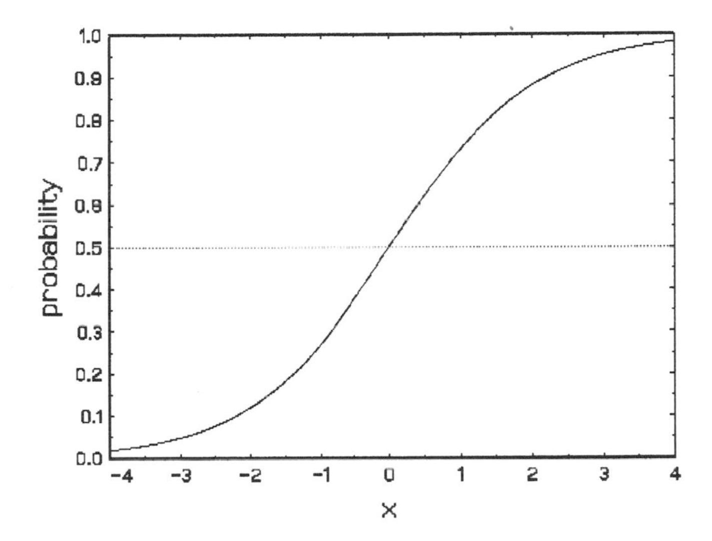

Figure 9.1. Logistic relationship between x and the response probability.

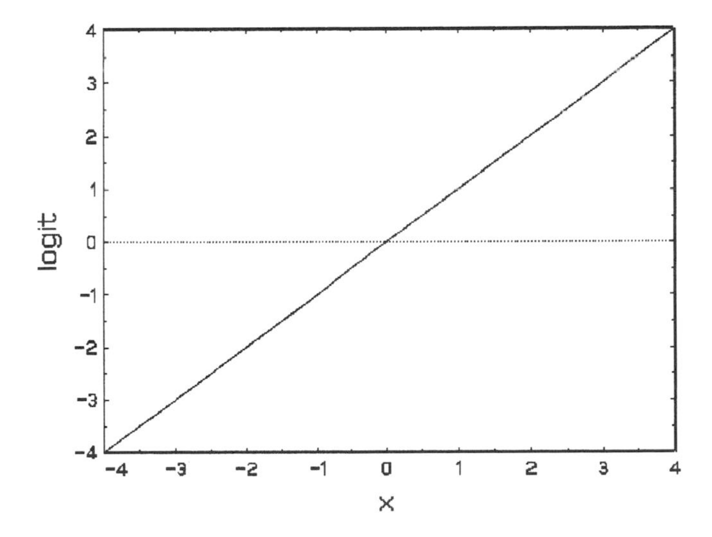

Figure 9.2. Linear relationship between x and the logit.

Since the model is linear in terms of the logits, interpretation of the parameters of the logistic regression model is in terms of the logits. Thus, the intercept β_0 in (9.3) is the log odds of a positive outcome for an individual with a set of covariates $\boldsymbol{x}_i = 0$ and β_p measures the change in the log odds for a unit change in x_p holding all other covariates constant. Often the regression coefficients in logistic regression models are expressed in exponential form, namely $\exp(\beta_p)$. This transformation yields an odds ratio interpretation for the regressors, namely the ratio of the odds of a positive response for a unit change in x.

For estimation, with Y_i as a binary outcome variable from a Bernoulli distribution, we have

$$\Pr(Y_i) = \Psi_i^{Y_i}[1 - \Psi_i]^{1-Y_i} \text{ for } Y_i = 0 \text{ or } 1, \tag{9.4}$$

where $\Psi_i = \Psi_i(\boldsymbol{x}_i'\boldsymbol{\beta})$ as in (9.2). The likelihood function for a sample of N independent observations can be written as the product of equation (9.4) over the N individuals, *i.e.*,

$$L = \prod_{i=1}^{N} \Psi_i^{Y_i}[1 - \Psi_i]^{1-Y_i}. \tag{9.5}$$

Thus the log-likelihood function becomes

$$\log L = \sum_{i=1}^{N} [Y_i \log \Psi_i + (1 - Y_i)\log(1 - \Psi_i)]\,. \tag{9.6}$$

Differentiating the log likelihood function (9.6) with respect to $\boldsymbol{\beta}$ yields the first derivatives for the maximum likelihood (ML) solution:

$$\frac{\partial \log L}{\partial \boldsymbol{\beta}} = \sum_i (Y_i - \Psi_i)\boldsymbol{x}_i. \tag{9.7}$$

This result is due to the fact that for the logistic distribution $\delta\Psi(\cdot) = \Psi(\cdot)(1 - \Psi(\cdot))$. Similarly, the second partial derivatives are obtained as

$$\frac{\partial^2 \log L}{\partial \boldsymbol{\beta} \partial \boldsymbol{\beta}'} = -\sum_i \Psi_i(1 - \Psi_i)\boldsymbol{x}_i\boldsymbol{x}_i', \tag{9.8}$$

Equations (9.7) and (9.8) are nonlinear and require an iterative solution. One such solution is the method of Newton or of Newton–Raphson. In this solution, provisional estimates for the vector of parameters $\boldsymbol{\beta}$, on iteration ι are improved by

$$\boldsymbol{\beta}_{\iota+1} = \boldsymbol{\beta}_\iota - \left[\frac{\partial^2 \log L}{\partial \boldsymbol{\beta}_\iota\, \partial \boldsymbol{\beta}_\iota'}\right]^{-1} \frac{\partial \log L}{\partial \boldsymbol{\beta}_\iota}. \tag{9.9}$$

The iterations continue until the changes in the parameter estimates and/or likelihood value are less than some very small absolute or relative value. At this point the solution is said to have converged, and the large-sample variance-covariance matrix of the maximum likelihood estimator is obtained as the negative inverse of the matrix of second derivatives. Standard errors of the parameter estimates are the square root values of the diagonal entries of this (negative inverse) matrix. The ML estimates and their accompanying standard errors can be used to compute asymptotic Z-statistics (*i.e.*, Wald statistics) or construct confidence intervals. More discussion regarding this ML estimation procedure and the iterative algorithm can be found in McCullagh and Nelder [1989] and Dobson [1990].

9.3 PROBIT REGRESSION MODELS

Probit regression [Finney, 1971] is an alternative to logistic regression that is popular in some fields, for example, genetic studies. In probit regression, the probability of a positive response, p_i, is expressed in terms of the standard normal cumulative distribution function (cdf), namely,

$$p_i = \Pr(Y_i = 1) = \Phi(\boldsymbol{x}_i'\boldsymbol{\beta}), \tag{9.10}$$

where $\Phi(\cdot)$ is the standard normal cdf. Figure 9.3 depicts the standard normal and logistic cdfs. As can be seen, both are symmetric s-shaped curves; however, the logistic places more probability in the tails of the distribution than does the normal. Whereas the standard normal has variance equal to 1, the standard logistic has variance equal to $\pi^2/3$ [Agresti, 2002; Long, 1997]. Thus, the scale of the logistic is greater than the normal.

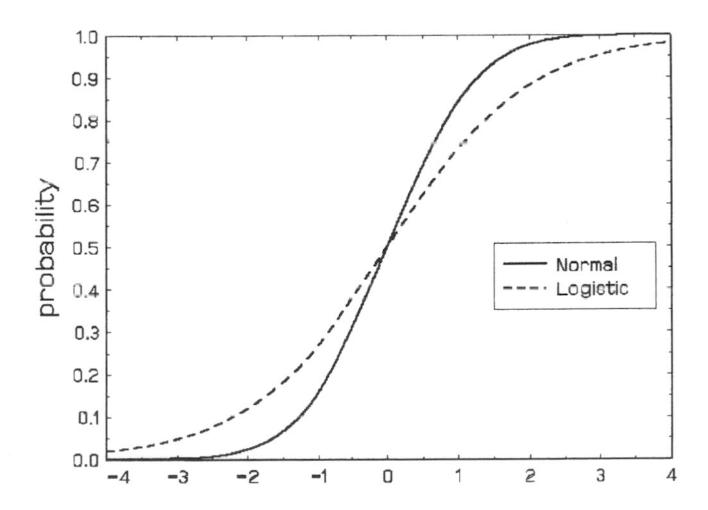

Figure 9.3. Normal and logistic cdfs.

Figure 9.4 depicts the same two cdfs; however, in this figure the logistic is standardized to have variance equal to 1. As is clear from Figure 9.4, the two curves are virtually indistinguishable. Thus, as noted by Doksum and Gasko [1990], large amounts of high-quality data are often necessary for substantive differences to emerge between probit and logistic regression. Since these two typically provide similar fits and conclusions, McCullagh [1980] suggests that choice between the two should be based primarily on ease of interpretation. In this chapter, we will focus primarily on logistic regression, though some discussion of probit regression will be made.

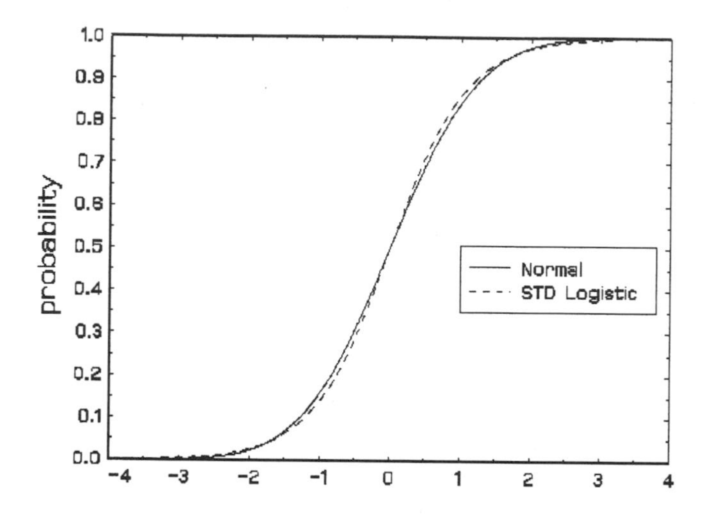

Figure 9.4. Normal and standardized logistic cdfs.

9.4 THRESHOLD CONCEPT

Dichotomous regression models are often motivated and described using the "threshold concept" [Bock, 1975]. This is also termed a latent variable model for dichotomous variables [Long, 1997]. For this, it is assumed that a continuous latent variable y underlies the observed dichotomous response Y. A threshold, denoted γ, then determines if the dichotomous response Y equals 0 ($y \leq \gamma$) or 1 ($y > \gamma$). It is common to fix the location of the underlying latent variable by setting the threshold equal to zero (*i.e.*, $\gamma = 0$). Figure 9.5 illustrates this concept assuming that the continuous latent variable y follows either a normal or logistic probability density function (pdf).

As noted by McCullagh and Nelder [1989], the assumption of a continuous latent distribution, while providing a useful motivating concept, is not a strict model requirement. It does help indicate why different models for dichotomous data (*e.g.*, logistic and probit regression) yield regression coefficients that are on different scales. For this, note that in terms of the continuous latent variable y, the model is written as

$$y_i = \boldsymbol{x}_i' \boldsymbol{\beta} + \epsilon_i. \tag{9.11}$$

Note the inclusion of the errors ϵ_i in this representation of the model. In the logistic regression formulation, the errors ϵ_i are assumed to follow a standard logistic distribution with mean 0 and variance $\pi^2/3$, while for a probit regression formulation the errors are assumed to follow a standard normal distribution with mean 0 and variance 1. The scale of the errors is fixed because y is not observed. Thus, although the above model appears to be the same as an ordinary multiple regression model for continuous outcomes, it is one in which the error variance is fixed and not estimated.

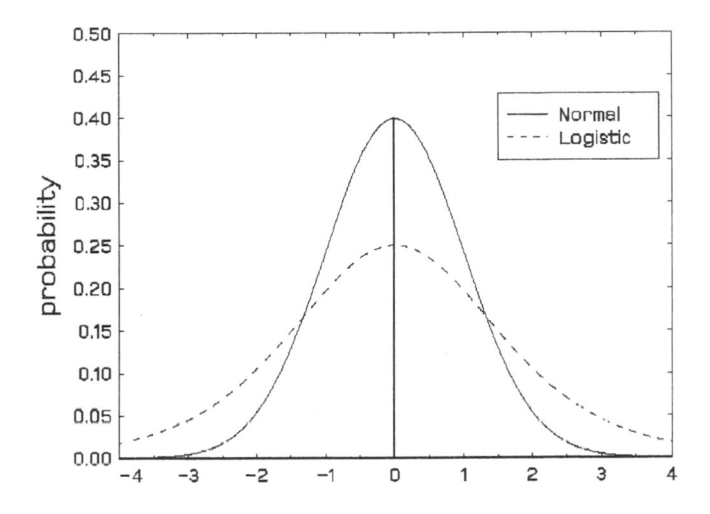

Figure 9.5. Threshold concept for a dichotomous response.

The variance for the underlying variable y is determined by the assumption for the variance of ϵ. Because of this, the scale of the regression coefficients, which indicate the degree of change in y with a one unit change in x, is different between logistic and probit regression. Namely, denoting the coefficients from logistic and probit regression as β_L and β_P, respectively, we get

$$\beta_L \approx \sqrt{\pi^2/3}\beta_P \approx 1.81\beta_P \tag{9.12}$$

by equating the variances of the standard logistic and normal distributions. Alternatively, Amemiya [1981] suggests $\beta_L \approx 1.6\beta_P$, which is based on making the cdfs of the logistic and normal distributions as close as possible, not just equating their variances. Long [1997] posits a factor of 1.7 based on similar calculations. Thus, though it's a bit imprecise, for a given problem the coefficients obtained from a logistic regression will be about 1.6 to 1.8 times as large as those from a probit regression. Similarly for the standard errors. This helps to explain why the Z-statistics (ratio of estimate to standard error) for the regression coefficients are very similar from probit and logistic regression models, whereas the values of these coefficients differ by the approximate scale differnce of 1.6 to 1.8.

9.5 MIXED-EFFECTS LOGISTIC REGRESSION MODEL

When the binary responses are clustered, for example repeatedly measured within individuals, or clustered within clinics, schools, or other social or ecological strata, the fixed-effects logistic regression model fails in its' assumptions to accurately characterize the dependence in the data. Basically, the fixed-effects model assumes that the observations are independent, which they clearly are not when they are clustered within individuals. As described in Chapter 4 for continuous outcomes, one solution to this problem is to generalize the model to the case of a combination of fixed (*e.g.*, treatment) and random (*e.g.*, time-trend coefficients) effects. The random effects allow the correlation between the clustered (*e.g.*, repeated) measurements to be incorporated into the estimates of parameters, standard errors, interval estimates, and tests of hypotheses. One can conceptualize the random effects as representing subject-specific differences in the propensity to respond over time. Those subjects with higher response propensity will exhibit an increased probability of a positive response, conditional on their values of the fixed-effects (*e.g.*, treatment group) included in the model.

Above and beyond the importance of the mixed-effects logistic regression model for statistical practice, the model also provides a foundation for ordinal and nominal responses models, which are considered in the next two chapters. Both ordinal and nominal models can be viewed as different ways of generalizing the dichotomous response model, and so an understanding of the dichotomous model is essential.

To set the notation, let i denote the level-2 units (individuals) and let j denote the level-1 units (nested observations). Assume that there are $i = 1, \ldots N$ level-2 units and $j = 1, \ldots, n_i$ level-1 units nested within each level-2 unit. The total number of level-1 observations across level-2 units is given by $n = \sum_{i=1}^{N} n_i$. Let Y_{ij} be the value of the dichotomous outcome variable, coded 0 or 1, associated with level-1 unit j nested within level-2 unit i. The logistic regression model is written in terms of the log odds (*i.e.*, the logit) of the probability of a response, denoted p_{ij}. Considering first a random intercept model, augmenting the logistic regression model with a single random effect yields

$$\log\left[\frac{p_{ij}}{1 - p_{ij}}\right] = \boldsymbol{x}'_{ij}\boldsymbol{\beta} + v_i, \tag{9.13}$$

where $\boldsymbol{x}_{ij}$ is the $(p+1) \times 1$ covariate vector (includes a 1 for the intercept), $\boldsymbol{\beta}$ is the $(p+1) \times 1$ vector of unknown regression parameters, and v_i is the random subject effect (one for each level-2 subject). These are assumed to be distributed in the population as $\mathcal{N}(0, \sigma_v^2)$. For convenience and computational simplicity, in models for categorical outcomes the random effects are typically expressed in standardized form. For this, $v_i = \sigma_v \theta_i$ and the model is given as

$$\log\left[\frac{p_{ij}}{1 - p_{ij}}\right] = \boldsymbol{x}'_{ij}\boldsymbol{\beta} + \sigma_v \theta_i. \tag{9.14}$$

Notice that the random-effects variance term (*i.e.*, the population standard deviation σ_v) is now explicitly included in the regression model. Thus, it and the regression coefficients are on the same scale, namely, in terms of the log-odds of a response.

In terms of the underlying latent y, the model in (9.14) is written as

$$y_{ij} = \boldsymbol{x}'_{ij}\boldsymbol{\beta} + \sigma_v \theta_i + \epsilon_{ij}. \tag{9.15}$$

This representation helps to explain why the regression coefficients from a mixed-effects logistic regression model do not typically agree with those obtained from a fixed-effects logistic regression model, or for that matter from a GEE logistic regression model which has regression coefficients that agree in scale with the fixed-effects model. In the mixed model (9.15) the conditional variance of the latent y given $\boldsymbol{x}$ equals $\sigma_v^2 + \sigma_\epsilon^2$, whereas in the fixed-effects model (9.11) this conditional variance equals only the latter term σ_ϵ^2 (which equals either $\pi^2/3$ or 1 depending on whether it is a logistic or probit regression model, respectively). As a result, equating the variances of the latent y under these two scenarios yields

$$\boldsymbol{\beta}_M \approx \sqrt{\frac{\sigma_v^2 + \sigma_\epsilon^2}{\sigma_\epsilon^2}}\boldsymbol{\beta}_F \tag{9.16}$$

where $\boldsymbol{\beta}_F$ and $\boldsymbol{\beta}_M$ represent the coefficients from the fixed-effects and (random-intercept) mixed-effects models, respectively. In practice, Zeger et al. [1988] suggest $(15/16)^2\pi^2/3$ works better than $\pi^2/3$ for σ_ϵ^2 in equating results of logistic regression models.

Several authors have commented on the difference in scale and interpretation of the regression coefficients in mixed-models and marginal models, like the fixed-effects and GEE models [Neuhaus et al., 1991; Hu et al., 1998]. Regression estimates from the mixed-model have been termed "subject-specific" to reinforce the notion that they are conditional estimates, conditional on the random (subject) effect. Thus, they represent the effect of a regressor on the outcome controlling for or holding constant the value of the random subject effect. Alternatively, the estimates from the fixed-effects and GEE models are "marginal" or "population-averaged" estimates which indicate the effect of a regressor averaging over the population of subjects.

This difference of scale and interpretation only occurs for nonlinear regression models like the logistic regression model. For the continuous-normal model considered in Chapter 4, this difference does not exist. To see this, we must get a bit more precise with our notation and representation of the mixed-effects logistic model. Specifically, the probability of a response for a subject i at a timepoint j is conditional on the random (subject) effect, and so $p_{ij} = \Pr(Y_{ij} = 1 \mid \theta_i)$. The mixed-effects logistic regression model is

$$\log\left[\frac{\Pr(Y_{ij} = 1 \mid \theta_i)}{1 - \Pr(Y_{ij} = 1 \mid \theta_i)}\right] = \boldsymbol{x}_{ij}'\boldsymbol{\beta} + \sigma_v\theta_i, \tag{9.17}$$

or

$$g[\Pr(Y_{ij} = 1 \mid \theta_i)] = \boldsymbol{x}_{ij}'\boldsymbol{\beta} + \sigma_v\theta_i, \tag{9.18}$$

which yields

$$\Pr(Y_{ij} = 1 \mid \theta_i) = g^{-1}[\boldsymbol{x}_{ij}'\boldsymbol{\beta} + \sigma_v\theta_i], \tag{9.19}$$

where g is the logit link function and g^{-1} is its inverse function, namely the logistic cdf. Taking the expectation, we obtain

$$E(Y_{ij} \mid \theta_i) = g^{-1}[\boldsymbol{x}_{ij}'\boldsymbol{\beta} + \sigma_v\theta_i] \tag{9.20}$$

so that

$$\mu_{ij} = E(Y_{ij}) = E[E(Y_{ij} \mid \theta_i)] = \int_\theta g^{-1}[\boldsymbol{x}_{ij}'\boldsymbol{\beta} + \sigma_v\theta_i] f(\theta)\, d\theta, \tag{9.21}$$

where $\theta \sim N(0, 1)$. When g is a nonlinear function, like the logit link, and if we assume that

$$g(\mu_{ij}) = \boldsymbol{x}'_{ij}\boldsymbol{\beta} + \sigma_v \theta_i \tag{9.22}$$

it is usually not true that

$$g(\mu_{ij}) = \boldsymbol{x}'_{ij}\boldsymbol{\beta} \tag{9.23}$$

unless $\theta_i = 0$ for all i subjects, or g is the identity link (*i.e.*, the normal regression model for y). The idea is really not that difficult to grasp. Essentially it is the same reason behind why the log of the mean of a series of values does not in general equal the mean of the log of those values, namely, the log is a nonlinear function.

9.5.1 Intraclass Correlation

For a random intercept model, it is often of interest to express the level-2 variance (*i.e.*, between-subjects variance) in terms of an intraclass correlation. The intraclass correlation indicates the proportion of unexplained variance that is at the subject level. Thus, it is an indication of the magnitude of the between-subjects variance. For this, one can make reference to the threshold concept and the underlying latent response tendency that determines the observed response. For the logistic model assuming normally distributed random-effects, the estimated intraclass correlation equals $\hat{\sigma}^2_\delta/(\hat{\sigma}^2_\delta + \pi^2/3)$, where the latter term in the denominator represents the variance of the underlying latent response tendency. As mentioned earlier, for the logistic model, this variable is assumed to be distributed as a standard logistic distribution with variance equal to $\pi^2/3$. For a probit model this term is replaced by 1, the variance of the standard normal distribution.

9.5.2 More General Mixed-Effects Models

The model can be easily extended to include multiple random effects. For this, denote $\boldsymbol{z}_{ij}$ as the $r \times 1$ vector of random-effect variables (a column of ones is usually included for the random intercept). The vector of random effects $\boldsymbol{v}_i$ is assumed to follow a multivariate normal distribution with mean vector $\mathbf{0}$ and variance–covariance matrix $\boldsymbol{\Sigma}_v$. To standardize the multiple random effects $\boldsymbol{v}_i = \boldsymbol{T}\boldsymbol{\theta}_i$, where $\boldsymbol{T}\boldsymbol{T}' = \boldsymbol{\Sigma}_v$ is the Cholesky factorization of $\boldsymbol{\Sigma}_v$. The model is now written as

$$\log\left[\frac{p_{ij}}{1 - p_{ij}}\right] = \boldsymbol{x}'_{ij}\boldsymbol{\beta} + \boldsymbol{z}'_{ij}\boldsymbol{T}\boldsymbol{\theta}_i. \tag{9.24}$$

As a result of the transformation, the Cholesky factor $\boldsymbol{T}$ is usually estimated instead of the variance–covariance matrix $\boldsymbol{\Sigma}_v$. As the Cholesky factor is essentially the matrix square-root of the variance–covariance matrix, this allows more stable estimation of near-zero variance terms.

9.5.3 Heterogeneous Variance Terms

Allowing for separate random-effect variance terms for groups of either i or j units is sometimes important. For example, in a twin study it is often necessary to allow the intra-twin correlation to differ between monozygotic and dizygotic twins. In this situation, subjects ($j = 1, 2$) are nested within twin pairs ($i = 1, \ldots, N$). To allow the level-2 variance to vary for these two twin-pair types, the random-effects design vector $\boldsymbol{z}_{ij}$ is specified as a 2×1 vector of dummy codes indicating monozygotic and dizygotic twin-pair status, respectively. $\boldsymbol{T}$ is then a 2×1 vector of independent random-effect standard deviations for monozygotics and dizygotics, and the cluster effect θ_i is a scalar that is pre-multiplied by the vector $\boldsymbol{T}$. For example, for a random intercept logistic model, we would have

$$\log\left[\frac{p_{ij}}{1 - p_{ij}}\right] = \boldsymbol{x}'_{ij}\boldsymbol{\beta} + [MZ_i \;\; DZ_i]\begin{bmatrix} \sigma_{\delta\,(MZ)} \\ \sigma_{\delta\,(DZ)} \end{bmatrix}\theta_i, \tag{9.25}$$

where, MZ_i and DZ_i are dummy codes indicating twin pair status (*i.e.*, if $MZ_i = 1$ then $DZ_i = 0$, and vice versa).

If the probit formulation is used and the model has no covariates (*i.e.*, only an intercept), the resulting intraclass correlations

$$ICC_{MZ} = \frac{\sigma^2_{\delta\,(MZ)}}{\sigma^2_{\delta\,(MZ)} + 1} \quad \text{and} \quad ICC_{DZ} = \frac{\sigma^2_{\delta\,(DZ)}}{\sigma^2_{\delta\,(DZ)} + 1}$$

are tetrachoric correlations for the within twin-pair data. Adding covariates then yields adjusted tetrachoric correlations. Because estimation of tetrachoric correlations is often important in twin and genetic studies, these models are typically formulated in terms of the probit link. Comparing models that allow homogeneous versus heterogeneous subgroup random-effects variance, thus allows examination of whether the tetrachoric correlations are equal across the subgroups.

The use of heterogeneous variance terms can also be found in some item response theory (IRT) models in the educational testing literature [Bock and Aitkin, 1981; Samejima, 1969; Bock, 1972]. Here, item responses ($j = 1, 2, \ldots, m$) are nested within subjects ($i = 1, 2, \ldots, N$) and a separate random-effect standard deviation (*i.e.*, an element of the $m \times 1$ vector $\boldsymbol{T}$) is estimated for each test item (*i.e.*, each j unit). This can be accomplished by specifying $\boldsymbol{z}_{ij}$ as a $m \times 1$ vector of dummy codes indicating the repeated items. To see this, consider the popular two-parameter logistic model for dichotomous responses [Lord, 1980] that specifies the probability of a correct response to item j ($Y_{ij} = 1$) conditional on the ability of subject i (θ_i) as

$$\Pr(Y_{ij} = 1 \mid \theta_i) = \frac{1}{1 + \exp[-a_j(\theta_i - b_j)]}, \tag{9.26}$$

where a_j is the slope parameter for item j (*i.e.*, item discrimination), and b_j is the threshold or difficulty parameter for item j (*i.e.*, item difficulty). The distribution of ability in the population of subjects is assumed to be normal with mean 0 and variance 1 (*i.e.*, the usual assumption for the random effects θ_i). As noted by [Bock and Aitkin, 1981], it is convenient to let $c_j = -a_j b_j$ and write

$$\Pr(Y_{ij} = 1 \mid \theta_i) = \frac{1}{1 + \exp[-(c_j + a_j\theta_i)]}, \tag{9.27}$$

which can be recast in terms of the logit of the response as

$$\log\left[\frac{p_{ij}}{1-p_{ij}}\right] = c_j + a_j\theta_i. \tag{9.28}$$

As an example, suppose that there are four items. Denoting the logit as l, this model can be represented in matrix form as

$$\begin{bmatrix} l_{1i} \\ l_{2i} \\ l_{3i} \\ l_{4i} \end{bmatrix} = \underbrace{\begin{bmatrix} 1 & 0 & 0 & 0 \\ 0 & 1 & 0 & 0 \\ 0 & 0 & 1 & 0 \\ 0 & 0 & 0 & 1 \end{bmatrix}}_{\boldsymbol{X}_j} \underbrace{\begin{bmatrix} c_1 \\ c_2 \\ c_3 \\ c_4 \end{bmatrix}}_{\boldsymbol{c}} + \underbrace{\begin{bmatrix} 1 & 0 & 0 & 0 \\ 0 & 1 & 0 & 0 \\ 0 & 0 & 1 & 0 \\ 0 & 0 & 0 & 1 \end{bmatrix}}_{\boldsymbol{Z}_i} \underbrace{\begin{bmatrix} a_1 \\ a_2 \\ a_3 \\ a_4 \end{bmatrix}}_{\boldsymbol{a}} [\theta_i],$$

showing that this IRT model is simply a mixed-effects model that allows the random effect variance terms to vary across the items at level-1. The usual IRT notation is a bit different than the mixed model notation, but $\boldsymbol{c}$ simply represents the fixed-effects (*i.e.*, $\boldsymbol{\beta}$) and $\boldsymbol{a}$ is the random-effects standard deviation vector $\boldsymbol{T}' = [\sigma_{\delta 1}\ \sigma_{\delta 2}\ \sigma_{\delta 3}\ \sigma_{\delta 4}]$.

The elements of the $\boldsymbol{T}$ vector can also be viewed as the (unscaled) factor loadings of the items on the (unidimensional) underlying ability variable (θ). A simpler IRT model that constrains these factor loadings to be equal is the one-parameter logistic model, the so-called Rasch model [Wright, 1977]. This constraint is achieved by setting $\boldsymbol{Z}_i = \boldsymbol{1}_i$ and $\boldsymbol{a} = a$ in the above model. Thus, the Rasch model is simply a random intercept logistic regression model with item indicators for $\boldsymbol{X}$.

Unlike traditional IRT models, the mixed model formulation easily allows multiple covariates at either level (*i.e.*, items or subjects). This and other advantages of casting IRT models as mixed or multilevel models are described by Adams et al. [1997] and Reise [2000]. In particular, this allows a model for examining whether item parameters vary by subject characteristics, and also for estimating ability in the presence of such item by subject interactions. Interactions between item parameters and subject characteristics, often termed item bias [Camilli and Shepard, 1994], is an area of active psychometric research.

9.5.4 Multilevel Representation

For a multilevel representation of a simple model with only one level-1 covariate x_{ij} and one level-2 covariate x_i, the level-1 model is written in terms of the logit as

$$\log\left[\frac{p_{ij}}{1-p_{ij}}\right] = \beta_{0i} + \beta_{1i}\, x_{ij}, \tag{9.29}$$

or in terms of the latent response variable as

$$y_{ij} = \beta_{0i} + \beta_{1i}\, x_{ij} + \epsilon_{ij}. \tag{9.30}$$

The level-2 model is then (assuming x_{ij} is a random-effects variable):

$$\begin{aligned} \beta_{0i} &= \beta_0 + \beta_2\, x_i + v_{0i}, \\ \beta_{1i} &= \beta_1 + \beta_3\, x_i + v_{1i}. \end{aligned} \tag{9.31}$$

Notice that it's easiest, and in agreement with the normal-theory (continuous) multilevel model, to write the level-2 model in terms of the unstandardized random effects, which are distributed in the population as $\boldsymbol{v} \sim \mathcal{N}(\mathbf{0}, \boldsymbol{\Sigma}_v)$. For models with multiple variables at either level-1 or level-2, the above level-1 and level-2 submodels are generalized in an obvious way.

Because of the fixing of the level-1 variance, the model operates somewhat differently than the more standard normal-theory multilevel model for continuous outcomes. For example, in a continuous mixed model the level-1 variance term is typically reduced as level-1 covariates x_{ij} are added to the model. However, this cannot happen in the above model because the level-1 variance is fixed. As noted by Snijders and Bosker [1999], what happens instead, as level-1 covariates are added, is that the random-effect variance terms tend to become larger as do the other regression coefficients, the latter become larger in absolute value.

9.5.5 Response Functions

The mixed-effects logistic model can also be written in the following way:

$$p_{ij} = \Psi(\boldsymbol{x}'_{ij}\boldsymbol{\beta} + \boldsymbol{z}'_{ij}\boldsymbol{T}\boldsymbol{\theta}_i), \tag{9.32}$$

where $\Psi(\mathrm{z})$ is the logistic cdf, namely

$$\Psi(\mathrm{z}) = \frac{\exp(\mathrm{z})}{1+\exp(\mathrm{z})} = \frac{1}{1+\exp(-\mathrm{z})}. \tag{9.33}$$

The cdf is also termed the response function of the model. An attractive mathematical feature of the logistic distribution is that the probability density function (pdf) is related to the cdf in a simple way, namely, $\psi(\mathrm{z}) = \Psi(\mathrm{z})\,[1 - \Psi(\mathrm{z})]$.

As mentioned, the probit model, which is based on the standard normal distribution, is often proposed as an alternative to the logistic model. For the probit model, the normal cdf $\Phi(\mathrm{z})$ and pdf $\phi(\mathrm{z})$ replace their logistic counterparts, and because the standard normal distribution has variance equal to one, $\epsilon_{ij} \sim \mathcal{N}(0, 1)$. As a result, in the probit model the underlying latent variable vector $\boldsymbol{y}_i$ is distributed normally in the population with mean $\boldsymbol{X}_i\boldsymbol{\beta}$ and variance covariance matrix $\boldsymbol{Z}_i\boldsymbol{T}\boldsymbol{T}'\boldsymbol{Z}'_i + \boldsymbol{I}$. The latter, when converted to a correlation matrix, yields tetrachoric correlations for the underlying latent variable vector $\boldsymbol{y}$, and polychoric correlations for ordinal outcomes (discussed in Chapter 10). For this reason, in some areas, for example familial studies, the probit formulation is preferred to its logistic counterpart.

For the mixed-effects logistic model considered so far, because the errors are assumed to follow a logistic distribution and the random effects a normal distribution, this model and models closely related to it are often referred to as logistic/normal or logit/normit models, especially in the latent trait model literature [Bartholomew and Knott, 1999]. If the errors are assumed to follow a normal distribution, then the resulting model is a mixed-effects probit regression or normal/normal model. Though we focus on the logistic regression formulation in this chapter, interested readers are referred to Gibbons and Hedeker [1994] and Gibbons et al. [1994] for applications of the mixed-effects probit model.

Both the logistic and normal distributions are symmetric around zero and, as mentioned, yield very similar results and conclusions (though the logistic regression parameters and associated standard errors are approximately $\pi/\sqrt{3}$ times as large because of the scale difference between the two distributions). An alternative response function that is sometimes used is the complementary log–log response function, $1 - \exp[-\exp(\mathrm{z})]$. In terms of the mixed-effects regression model, this is written as

$$p_{ij} = 1 - \exp\left[-\exp(\boldsymbol{x}'_{ij}\boldsymbol{\beta} + \boldsymbol{z}'_{ij}\boldsymbol{T}\boldsymbol{\theta}_i)\right] \tag{9.34}$$

or

$$\log\left[-\log(1 - p_{ij})\right] = \boldsymbol{x}'_{ij}\boldsymbol{\beta} + \boldsymbol{z}'_{ij}\boldsymbol{T}\boldsymbol{\theta}_i. \tag{9.35}$$

Unlike the logistic and normal, the distribution that underlies the complementary log–log response function is asymmetric and has variance equal to $\pi^2/6$. Its pdf is given by $\exp(\mathrm{z})(1 - p(\mathrm{z}))$. Use of the complementary log–log response function is most common in the area of survival analysis, since it can be shown to provide a proportional hazards model for grouped-time survival data (see Allison [1995] and Hedeker et al. [2000]).

9.6 ESTIMATION

To set the notation, we assume that there are $i = 1, \ldots, N$ subjects, each with $j = 1, \ldots, n_i$ repeated observations. The outcome variable $Y_{ij} = 1$ for a positive response and $Y_{ij} = 0$ for a negative response. Let us first consider estimation of a random intercept mixed model, that is,

$$\log\left[\frac{p_{ij}}{1 - p_{ij}}\right] = \boldsymbol{x}'_{ij}\boldsymbol{\beta} + \sigma_v\theta_i. \tag{9.36}$$

In this model, we can write the conditional probability of a positive response, conditional on the (standardized) random effect θ_i, as

$$P(Y_{ij} = 1 \mid \theta_i) = \Psi(\mathrm{z}_{ij}), \tag{9.37}$$

where the standard logistic cdf is given in (9.33) and $\mathrm{z}_{ij} = \boldsymbol{x}'_{ij}\boldsymbol{\beta} + \sigma_v\theta_i$. Thus, the probabiity of a negative response is simply

$$P(Y_{ij} = 0 \mid \theta_i) = 1 - \Psi(\mathrm{z}_{ij}). \tag{9.38}$$

The observations within a subject are assumed independent given the random subject effect (*i.e.*, the random effects account completely for the correlation of the data within subjects). This assumption is critical and is known as the conditional independence assumption in the psychometric literature. As a result, we can multiply the conditional probabilities across the n_i timepoints within a subject together to yield the conditional probability for the $n_i \times 1$ response vector $\boldsymbol{Y}_i$:

$$\ell(\boldsymbol{Y}_i \mid \theta) = \prod_{j=1}^{n_i} \Psi(\mathrm{z}_{ij})^{Y_{ij}}[1 - \Psi(\mathrm{z}_{ij})]^{1-Y_{ij}}. \tag{9.39}$$

Note the similarity between this conditional likelihood and the likelihood in equation (9.5). They are functionally of the same form. In the cross-sectional case, for which (9.5) applies, we can multiply the probabilities from each subject together to yield the likelihood of the joint pattern of all N outcomes from the subjects. Similarly, in the longitudinal case we can multiply the probabilities of each timepoint together within a subject to yield the conditional likelihood of the joint pattern of the n_i outcomes across time for that subject. Here, it is a conditional likelihood because these n_i observations are independent (and therefore can be multiplied together as in (9.39)) only conditional on the random effect. To get to the likelihood of the n_i response patterns for all of the N subjects, we need to have an expression for the likelihood of $\boldsymbol{Y}_i$ that doesn't depend on the random effects. We can arrive at such an expression by integrating over the distribution of the random effects. This yields the marginal probability for $\boldsymbol{Y}_i$ in the population of subjects as

$$h(\boldsymbol{Y}_i) = \int_\theta \ell(\boldsymbol{Y}_i \mid \theta)\, g(\theta)\, d\theta, \tag{9.40}$$

where $g(\theta)$ represents the population distribution of the random effects, namely, $N(0, \sigma_\upsilon^2)$. The idea behind this isn't too hard to grasp. Essentially we want to consider the conditional likelihood, which depends on the random effect, for all possible values of the random effect, and thereby obtain an aggregated or marginal likelihood. One can think of equation (9.40) as a weighted average probability, the values of θ modify the response function z_{ij} and thereby modify the conditional likelihood $\ell(\boldsymbol{Y}_i \mid \theta)$, which is weighted by the probability at that point in the distribution $g(\theta)$ as one goes over all values of θ.

We can now form the marginal likelihood of the response patterns $\boldsymbol{Y}_i$ from all subjects, and thus the total sample, by multiplying each of the subject's marginal likelihoods together. Namely,

$$L = \prod_{i=1}^{N} h(\boldsymbol{Y}_i) \tag{9.41}$$

or

$$\log L = \sum_{i=1}^{N} \log h(\boldsymbol{Y}_i). \tag{9.42}$$

As is well known, since the maximum of the likelihod and the maximum of the log-likelihood are at the same point, and since the latter involves summation and not multiplication, it is simpler to differentiate with respect to the log-likelihood. Thus, the log-likelihood in equation (9.42) is what we can differentiate to obtain the maximum (marginal) likelihood solution. For this, let the parameter vector $\boldsymbol{\eta}$ represent either the regressors $\boldsymbol{\beta}$ or the variance parameter σ_υ, then taking derivatives

$$\frac{\partial \log L}{\partial \boldsymbol{\eta}} = \sum_{i=1}^{N} h^{-1}(\boldsymbol{Y}_i) \frac{\partial h(\boldsymbol{Y}_i)}{\partial \boldsymbol{\eta}}. \tag{9.43}$$

Now, express the marginal likelihood in the following way:

$$\begin{aligned}
h(\boldsymbol{Y}_i) &= \int_\theta \ell(\boldsymbol{Y}_i \mid \theta)\, g(\theta)\, d\theta \\
&= \int_\theta \left(\prod_{j=1}^{n_i} \Psi(\mathrm{z}_{ij})^{Y_{ij}} [1 - \Psi(\mathrm{z}_{ij})]^{1-Y_{ij}} \right) g(\theta)\, d\theta \\
&= \int_\theta \left[\exp \left(\log \left\{ \prod_{j=1}^{n_i} \Psi(\mathrm{z}_{ij})^{Y_{ij}} [1 - \Psi(\mathrm{z}_{ij})]^{1-Y_{ij}} \right\} \right) \right] g(\theta)\, d\theta \\
&= \int_\theta \left[\exp \left(\sum_{j=1}^{n_i} Y_{ij} \log[\Psi(\mathrm{z}_{ij})] + (1 - Y_{ij}) \log[1 - \Psi(\mathrm{z}_{ij})] \right) \right] g(\theta)\, d\theta.
\end{aligned}$$

And so, denoting $\ell(\boldsymbol{Y}_i \mid \theta)$ by ℓ_i, we get

$$\begin{aligned}
\frac{\partial h(\boldsymbol{Y}_i)}{\partial \eta} &= \int_\theta \sum_{j=1}^{n_i} \left[\frac{Y_{ij}}{\Psi(\mathrm{z}_{ij})} \partial\Psi(\mathrm{z}_{ij}) + \frac{1 - Y_{ij}}{1 - \Psi(\mathrm{z}_{ij})} (-\partial\Psi(\mathrm{z}_{ij})) \right] \frac{\partial \mathrm{z}_{ij}}{\partial \eta}\, \ell_i\, g(\theta)\, d\theta \\
&= \int_\theta \sum_{j=1}^{n_i} \frac{Y_{ij} - \Psi(\mathrm{z}_{ij})}{\Psi(\mathrm{z}_{ij})(1 - \Psi(\mathrm{z}_{ij}))}\, \partial\Psi(\mathrm{z}_{ij})\, \frac{\partial \mathrm{z}_{ij}}{\partial \eta}\, \ell_i\, g(\theta)\, d\theta,
\end{aligned}$$

yielding

$$\frac{\partial \log L}{\partial \eta} = \sum_{i=1}^{N} h^{-1}(\boldsymbol{Y}_i) \int_\theta \sum_{j=1}^{n_i} \frac{Y_{ij} - \Psi(\mathrm{z}_{ij})}{\Psi(\mathrm{z}_{ij})(1 - \Psi(\mathrm{z}_{ij}))}\, \partial\Psi(\mathrm{z}_{ij})\, \frac{\partial \mathrm{z}_{ij}}{\partial \eta}\, \ell_i\, g(\theta)\, d\theta, \quad (9.44)$$

where $\partial\Psi(\mathrm{z}_{ij})$ equals the pdf, which for the logistic distribution is $\Psi(\mathrm{z}_{ij})[1 - \Psi(\mathrm{z}_{ij})]$, and where

$$\frac{\partial \mathrm{z}_{ij}}{\partial \boldsymbol{\beta}} = \boldsymbol{x}'_{ij}, \quad (9.45)$$

$$\frac{\partial \mathrm{z}_{ij}}{\partial \sigma_\upsilon} = \theta_i. \quad (9.46)$$

For a probit model, the solution is the same except that the normal cdf $\Phi(\mathrm{z}_{ij})$ replaces the logistic cdf $\Psi(\mathrm{z}_{ij})$, and the normal pdf $\phi(\mathrm{z}_{ij})$ replaces the logistic pdf $\Psi(\mathrm{z}_{ij})[1 - \Psi(\mathrm{z}_{ij})]$. As can be appreciated, since the logistic cdf, which equals $1/[1 + \exp(-\mathrm{z}_{ij})]$, is relatively easy to compute and since the logistic pdf is a function of its cdf, the logistic solution is mathematically simpler than the probit. The probit solution relies upon programmed functions for the standard normal cdf and pdf, which are a bit more mathematically involved than the logistic counterparts.

At this point, the second derivatives would need to be obtained to implement the Newton–Raphson procedure outlined for the ordinary logistic regression model in Section 9.2. An alternative procedure that is often simpler to implement is provided by Fisher's method of scoring. For this, provisional estimates for the vector of all parameters $\boldsymbol{\Theta}$, on iteration ι are improved by

$$\boldsymbol{\Theta}_{\iota+1} = \boldsymbol{\Theta}_{\iota} + \boldsymbol{I}(\boldsymbol{\Theta}_{\iota})^{-1} \frac{\partial \log L}{\partial \boldsymbol{\Theta}_{\iota}}, \tag{9.47}$$

where, following Bock and Aitkin [1981], the information matrix $\boldsymbol{I}(\boldsymbol{\Theta}_{\iota})$, or expectation of the negative of second derivatives, is given by

$$\boldsymbol{I}(\boldsymbol{\Theta}_{\iota}) = E\left[-\frac{\partial^2 \log L}{\partial \boldsymbol{\Theta}_{\iota}\, \partial \boldsymbol{\Theta}_{\iota}'}\right] = \sum_{i=1}^{N} h^{-2}(\boldsymbol{Y}_i)\, \frac{\partial h(\boldsymbol{Y}_i)}{\partial \boldsymbol{\Theta}_{\iota}} \left(\frac{\partial h(\boldsymbol{Y}_i)}{\partial \boldsymbol{\Theta}_{\iota}}\right)'.$$

At convergence, the large-sample variance covariance matrix of the parameter estimates is then obtained as the inverse of the information matrix. The square root values of the diagonal elements of this matrix can be used to obtain Wald statistics or construct asymptotic confidence intervals for the model parameters.

9.6.1 Estimation of Random Effects and Probabilities

In many cases, it is useful to obtain estimates of the random effects within the sample. A reasonable choice for this is the expected a posteriori (EAP) or empirical Bayes estimate [Bock and Aitkin, 1981]. For the univariate case, this estimate $\hat{\theta}_i$ can be obtained by

$$\hat{\theta}_i = E(\theta_i \mid \boldsymbol{Y}_i) = h^{-1}(\boldsymbol{Y}_i) \int_{\theta} \theta_i\, \ell(\boldsymbol{Y}_i \mid \theta)\, g(\theta)\, d\theta, \tag{9.48}$$

The variance of the empirical Bayes estimator is similarly obtained as

$$V(\hat{\theta}_i \mid \boldsymbol{Y}_i) = h^{-1}(\boldsymbol{Y}_i) \int_{\theta} (\theta_i - \hat{\theta}_i)^2\, \ell(\boldsymbol{Y}_i \mid \theta)\, g(\theta)\, d\theta. \tag{9.49}$$

These terms may then be used, for example, to evaluate the response probabilities for particular subjects (*e.g.*, person-specific trend estimates). Also Ten Have [1996] describes how they may be used in performing residual diagnostics.

To obtain estimated marginal probabilities (*e.g.*, the estimated response probabilities for the control group at each timepoint of the study), an additional step is required. First, the so-called "subject-specific" probabilities [Neuhaus et al., 1991; Zeger et al., 1988] are estimated for specific values of covariates and the random effect, say θ^*. These subject-specific estimates indicate, for example, the response probability for a subject with random effect level θ^* in the control group at a particular timepoint. Denoting these subject-specific probabilities as $\hat{P}_{ss}$, marginal probabilities $\hat{P}_m$ are then obtained by integrating over the random-effects distribution (this is described in Section 9.6.3), namely, $\hat{P}_m = \int_{\theta} \hat{P}_{ss}\, g(\theta)\, d\theta$. Continuing with our example, the marginalized estimate would indicate the estimated response probability for the entire control group at a particular timepoint. Both subject-specific and marginal estimates have their uses, since they are estimating different quantities, and several authors have characterized the differences between the two [Neuhaus et al., 1991; Hu et al., 1998; Lindsey and Lambert, 1998; Heagerty and Zeger,

2000]. As mentioned earlier, for a random intercept model the marginal probabilities can be approximated in a simpler way. For this, assuming a logistic response function, the estimated regression coefficients from a random intercept model are divided by $\sqrt{k^2\hat{\sigma}_c^2 + 1}$ where $k = 16\sqrt{3}/(15\pi)$. These "marginalized" regression coefficients are then directly used to produce estimated marginal probabilities. We will provide an example of this type of marginalization later in this chapter.

9.6.2 Multiple Random Effects

For the case of multiple random effects, the model is

$$\log\left[\frac{p_{ij}}{1-p_{ij}}\right] = \boldsymbol{x}'_{ij}\boldsymbol{\beta} + \boldsymbol{z}'_{ij}\boldsymbol{T}\boldsymbol{\theta}_i, \tag{9.50}$$

and so estimation of the Cholesky factor $\boldsymbol{T}$ replaces estimation of the scalar term σ_υ. The solution outlined above is only modified insofar as the derivative $\partial z_{ij}/\partial \boldsymbol{T}$ replaces $\partial z_{ij}/\partial \sigma_\upsilon$. In particular we need an expression for the $(r \times (r-1))/2$ unique parameters in the Cholesky factor $\boldsymbol{T}$, which is a lower triangular matrix (in what follows, let r^* denote $(r \times (r-1))/2$).

Such an expression can be obtained using matrix differentiation techniques described in Magnus [1988]. Specifically, vectorizing the part of the model involving $\boldsymbol{T}$ yields:

$$\text{vec}(\boldsymbol{z}'_{ij}\boldsymbol{T}\boldsymbol{\theta}_i) = (\boldsymbol{\theta}'_i \otimes \boldsymbol{z}'_{ij})\ \text{vec}\boldsymbol{T}, \tag{9.51}$$

where the vec operator transforms a matrix into a vector by stacking the columns of the matrix underneath each other, and $\otimes$ represents the Kronecker product. Magnus [1988] further defines the vector $\text{v}(\boldsymbol{A})$ which contains the r^* elements below and on the diagonal of a $r \times r$ matrix $\boldsymbol{A}$, and the relationship between $\text{v}(\boldsymbol{A})$ and $\text{vec}(\boldsymbol{A})$ in the case where, as we have here for $\boldsymbol{T}$, one is dealing with is a lower triangular matrix. Specifically, the two are related by

$$\text{J}'_r\ \text{v}(\boldsymbol{A}) = \text{vec}(\boldsymbol{A}), \tag{9.52}$$

where J_r is the $r^* \times r^2$ elimination matrix, containing 1s and 0s, that removes the elements above the main diagonal. Thus, we can now write

$$\text{vec}(\boldsymbol{z}'_{ij}\boldsymbol{T}\boldsymbol{\theta}_i) = (\boldsymbol{\theta}'_i \otimes \boldsymbol{z}'_{ij})\text{J}'_r\text{v}(\boldsymbol{T}), \tag{9.53}$$

and so

$$\frac{\partial z_{ij}}{\partial \text{v}(\boldsymbol{T})} = (\boldsymbol{\theta}'_i \otimes \boldsymbol{z}'_{ij})\text{J}'_r = [\text{J}_r(\boldsymbol{\theta}_i \otimes \boldsymbol{z}_{ij})]'. \tag{9.54}$$

The Fisher scoring solution outlined above can now proceed substituting the derivative vector (9.54) for the scalar (9.46).

9.6.3 Integration over the Random-Effects Distribution

Various approximations for evaluating the integral over the random-effects distribution have been proposed in the literature. Reviews of many of these approaches can be found in Rodríguez and Goldman [1995], Davidian and Giltinan [1995], and McCulloch and Searle [2001]. Perhaps the most frequently used methods are based on first- or second-order Taylor series expansions. Marginal quasi-likelihood (MQL) involves expansion around the fixed part of the model, whereas penalized or predictive quasi-likelihood (PQL) additionally includes the random part in its expansion [Goldstein and Rasbash, 1996]. Both of these are available in the MLwiN software program [Goldstein et al., 1998]. Unfortunately, several authors [Breslow and Lin, 1995; Rodríguez and Goldman, 1995; Raudenbush et al., 2000b] have reported downwardly biased estimates using these procedures in certain situations, especially for the first-order expansions.

More recently, Raudenbush et al. [2000a] proposed an approach that uses a combination of a fully multivariate Taylor series expansion and a Laplace approximation. Based on the results in Raudenbush et al. [2000b], this method yields accurate results and is computationally fast. Also, as opposed to the MQL and PQL approximations, the deviance obtained from this approximation can be used for likelihood ratio tests. This approach is incorporated in the HLM software program [Raudenbush et al., 2000a].

Alternatively, numerical integration can be used to perform the integration over the random-effects distribution. Specifically, if the assumed distribution is normal, Gauss–Hermite quadrature can be used to approximate the above integral to any practical degree of accuracy [Stroud and Sechrest, 1966]. Additionally, like the Laplace approximation, the numerical quadrature approach yields a deviance that can be readily used for likelihood ratio tests. The integration is approximated by a summation on a specified number of quadrature points Q for each dimension of the integration. Thus, for the transformed $\boldsymbol{\theta}$ space, the summation goes over Q^r points, where r is the number of random effects. For the standard normal univariate density, optimal points and weights (which will be denoted B_q and $A(B_q)$, respectively) are given in Stroud and Sechrest [1966]. For the multivariate density, the r-dimensional vector of quadrature points is denoted by $\boldsymbol{B}'_q = (B_{q1}, B_{q2}, \ldots, B_{qr})$, with its associated (scalar) weight given by the product of the corresponding univariate weights,

$$A(\boldsymbol{B}_q) = \prod_{h=1}^{r} A(B_{qh}). \tag{9.55}$$

Using these points and weights, the response model becomes

$$z_{ijq} = \boldsymbol{x}'_{ij}\boldsymbol{\beta} + \boldsymbol{z}'_{ij}\boldsymbol{T}\boldsymbol{B}_q, \tag{9.56}$$

and so the conditional likelihood is

$$\ell(\boldsymbol{Y}_i \mid \boldsymbol{B}_q) = \prod_{j=1}^{n_i} \Psi(z_{ijq})^{Y_{ij}} [1 - \Psi(z_{ijq})]^{1-Y_{ij}}, \tag{9.57}$$

yielding the approximated marginal likelihood as

$$h(\boldsymbol{Y}_i) \approx \sum_{q=1}^{Q^r} \ell(\boldsymbol{Y}_i \mid \boldsymbol{B}_q)\, A(\boldsymbol{B}_q). \tag{9.58}$$

The first derivatives are then

$$\frac{\partial \log L}{\partial \boldsymbol{\eta}} \approx \sum_{i=1}^{N} h^{-1}(\boldsymbol{Y}_i) \sum_{q=1}^{Q^r} \sum_{j=1}^{n_i} \frac{Y_{ij} - \Psi(\mathrm{z}_{ijq})}{\Psi(\mathrm{z}_{ijq})(1 - \Psi(\mathrm{z}_{ijq}))} \partial\Psi(\mathrm{z}_{ijq}) \frac{\partial \mathrm{z}_{ijq}}{\partial \boldsymbol{\eta}} \ell(\boldsymbol{Y}_i \mid \boldsymbol{B}_q) A(\boldsymbol{B}_q), \tag{9.59}$$

where

$$\frac{\partial \mathrm{z}_{ijq}}{\partial \boldsymbol{\beta}} = \boldsymbol{x}'_{ij} \qquad \text{and} \qquad \frac{\partial \mathrm{z}_{ijq}}{\partial \mathrm{v}(\boldsymbol{T})} = [\mathrm{J}_r(\boldsymbol{B}_q \otimes \boldsymbol{z}_{ij})]'.$$

Although the random effects are typically assumed to be normally distributed, other distributional forms can be used. For example, if a rectangular or uniform distribution is assumed, then Q points may be set at equal intervals over an appropriate range (for each dimension) and the quadrature weights are then set equal to $1/Q$. Other distributions are possible: Bock and Aitkin [1981] discuss the possibility of empirically estimating the random-effect distribution.

Several software packages have implemented Gauss–Hermite quadrature, including EGRET [Corcoran et al., 1999], LIMDEP [Greene, 1998], MIXOR [Hedeker and Gibbons, 1996a], MIXNO [Hedeker, 1999], and SAS PROC NLMIXED. Both EGRET and LIMDEP allow only a univariate solution (*i.e.*, integration over the scalar θ), whereas MIXOR, MIXNO, and NLMIXED permit multiple random effects. Additionally, EGRET only includes logistic regression for dichotomous responses. In contrast to HLM and MLwiN, which permit multiple data levels, all of these programs using the quadrature approach only allow two-level data. A generalization that the quadrature solution allows is that distributions other than the normal can be considered. For this, the points and density weights are substituted for those specified by Gauss–Hermite quadrature. Both MIXOR and MIXNO allow selection of either a normal or uniform distribution for the random effects. Comparison of the results obtained using the normal and uniform distributions provides some information about the sensitivity of the results to the choice of the random-effect distribution.

An issue with the quadrature approach is that it can involve summation over a large number of points, especially as the number of random-effects is increased. For example, if there is only one random effect, the quadrature solution requires only one additional summation over Q points relative to the fixed effects solution. For models with $r > 1$ random effects, however, the quadrature is performed over Q^r points and so becomes computationally burdensome for $r > 5$ or so. An issue, then, is the number of necessary quadrature points to use for accurate estimation of the model parameters. Based on models for dichotomous and ordinal outcomes, respectively, Longford [1993] and Jansen [1990] note that estimation is affected very little when the number of points is 5 or greater for the undimensional solution. Also, as suggested by Bock et al. [1988] in the context of a dichotomous factor analysis model, the number of points in each dimension can be reduced as the dimensionality is increased. These authors noted that as few as three points per dimension were necessary for a five-dimensional solution. Alternatively, both the LIMDEP and EGRET programs use 20 quadrature points by default for their unidimensional models. Similarly, Lesaffre and Spiessens [2001] present an example where the method only gives valid results for a high number of quadrature points. These authors advise practitioners to routinely examine results for the dependence on Q.

More computer-intensive methods, such as the use of Gibbs sampling [Geman and Geman, 1984] and related methods [Tanner, 1996], can also be used to approximate the integration over the random effects distribution. Daniels and Gatsonis [1997] use this approach in their mixed-effects polychotomous regression model, as do Marshall and Spiegelhalter [2001] in their application of a three-level Poisson model. The freeware BUGS software program [Spiegelhalter et al., 1995] can be used to facilitate estimation via Gibbs sampling; see the Marshall and Spiegelhalter [2001] article for some BUGS syntax and discussion. In general, while the quadrature solution is relatively fast and computationally tractable for models with few random effects, Gibbs sampling may be more advantageous for models with many random effects. To address this, several authors have described a method of adaptive quadrature that uses a few number of points per dimension (*e.g.*, 3 or so) that are adapted to the location and dispersion of the distribution to be integrated [Liu and Pierce, 1994; Pinheiro and Bates, 1995; Bock and Shilling, 1997; Rabe-Hesketh et al., 2002]. In this regard, Bock and Shilling [1997] examined dichotomous factor analysis models with 5 and 8 factors (*i.e.*, random effects) and found similar results using adaptive quadrature as compared to a Gibbs sampling approach.

To adapt the Gauss–Hermite quadrature in the case of a single random subject effect, estimation of the posterior mean (*i.e.*, the empirical Bayes estimate $\hat{\theta}_i$) and variance (denote $V(\hat{\theta}_i \mid \boldsymbol{Y}_i)$ as s_i^2) is performed for each subject i at each iteration. Then, following Rabe-Hesketh et al. [2002], the adapted quadrature points are given as

$$B_{iq} = \bar{\theta}_i + s_i B_q \tag{9.60}$$

and the adapted quadrature weights are

$$A_{iq} = \sqrt{2\pi} s_i \exp(B_q^2/2)\phi(B_{iq}) A_q \tag{9.61}$$

where $\phi(\cdot)$ represents the normal pdf.

9.7 ILLUSTRATION

To illustrate application of the mixed-effects logistic regression model, we will present several analyses using data from the National Institute of Mental Health Schizophrenia Collaborative Study. Specifically, we will examine Item 79 of the Inpatient Multidimensional Psychiatric Scale (IMPS; Lorr and Klett [1966]). Item 79, "Severity of Illness," was originally scored on a 7-point scale ranging from *normal, not at all ill* (0) to *among the most extremely ill* (7). For the purpose of this chapter, we dichotomize the score between *mildly ill* (3) to *moderately ill* (4). For these data, we have previously illustrated application of the mixed-effects probit regression model in Gibbons and Hedeker [1994]. Here, we will present results from the logistic counterpart. In this study, patients were randomly assigned to receive one of four medications: placebo, chlorpromazine, fluphenazine, or thioridazine. Since our previous analyses revealed similar effects for the three anti-psychotic drug groups, they were combined in the present analysis. The experimental design and corresponding sample sizes are presented in Table 9.1. In this study, the protocol called for subjects to be measured at weeks 0, 1, 3, and 6; however, a few subjects were additionally measured at weeks 2, 4, and 5. There is a fair amount of attrition; this issue will be examined in more detail for these data in Chapter 14.

Table 9.1. Experimental Design and Weekly Sample Sizes Across Time

	Sample Size at Week						
Group	0	1	2	3	4	5	6
Placebo (N=108)	107	105	5	87	2	2	70
Drug (N=329)	327	321	9	287	9	7	265

Drug = Chlorpromazine, fluphenazine, or thioridazine.

The main question of interest is addressing whether there is differential change across time for the drug groups, here considered jointly, relative to the control group. To prepare for the analyses, Table 9.2 presents the observed proportions, odds, and log odds for the two groups at the four primary timepoints. It should be made clear that although these statistics are presented only for the four primary study timepoints, because of the sparseness of data at the other timepoints, the analysis will consider the data from all timepoints in the study.

As these descriptive statistics reveal, almost all patients were initially scored greater than moderately ill. The proportion of moderately ill patients diminishes across time for both groups, though the decrease appears more pronounced among the drug patients. Table 9.2 makes clear that proportions above .5 equate to odds greater than unity and positive logit values. Conversely, proportions less than .5 yield odds less than unity and negative logits. Additionally, Table 9.2 illustrates that the odds ratio, for the group comparison at a given timepoint, is equal to the exponential of the difference in logit values.

Table 9.2. Descriptive Statistics

	Week 0	Week 1	Week 3	Week 6
Observed proportions $\geq$ "moderately ill"				
Placebo	.98	.91	.89	.71
Drug	.99	.82	.66	.42
Observed odds $\geq$ "moderately ill"				
Placebo	52.5	9.50	7.70	2.50
Drug	80.8	4.63	1.93	.73
Ratio	.65	2.05	3.99	3.42
Observed log odds, logits, $\geq$ "moderately ill"				
Placebo	3.96	2.25	2.04	.92
Drug	4.39	1.53	.66	−.31
Difference	−.43	.72	1.38	1.23
exp(*difference*)	.65	2.05	3.99	3.42

Since it is the logits that are modeled in the logistic regression model, Figure 9.6 presents the observed logits for these two groups at the four primary timepoints. Figure 9.6 suggests that the trend across time, measured in terms of weeks, is not linear on the logit scale. There is an appreciable drop after the first timepoint, after which a more or less linear relationship ensues. Part of the reason for this is that the logits at the first timepoint are very large because most subjects are ill at this timepoint. A curvilinear relationship such as this could be modeled using polynomial trend models as described in Chapter 5. Here, we will instead linearize the relationship between the logit and time by transforming the time variable. Figure 9.7 shows the approximate linear relationship that is observed when time is expressed as the square root of week. An advantage of transforming the metric of time, as in Figure Figure 9.7 , is that only one term is needed in the model to express the effect of time on the logit outcome. This yields a parsimonious model, though admittedly it is somewhat more difficult to describe since the time effects are in terms of the square root of week, rather than the week value.

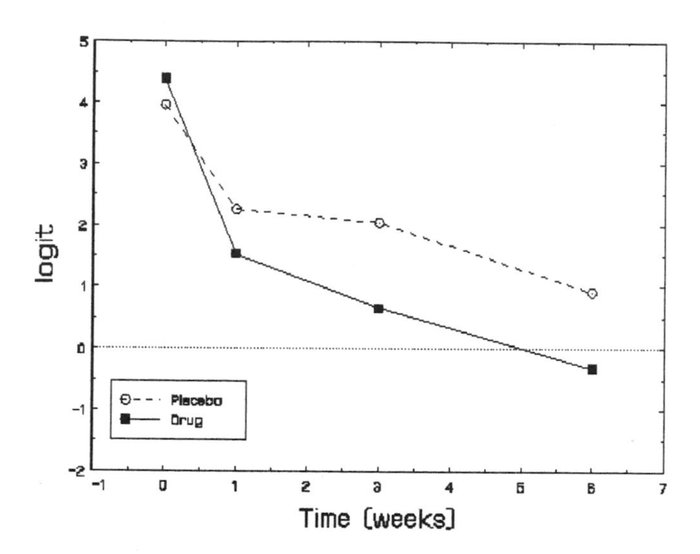

Figure 9.6. Observed logits across time (week) by condition.

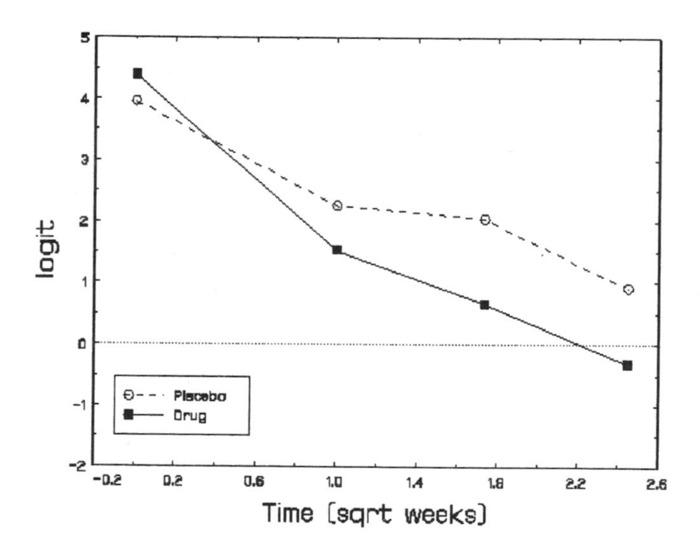

Figure 9.7. Observed logits across time (square root of week) by condition.

9.7.1 Fixed-Effects Logistic Regression Model

As a first analysis, and for later comparison, we present a fixed-effects logistic regression including Drug (0 = placebo, 1 = drug), Time (square root of week, with week coded 0 to 6), and the Drug by Time interaction. Table 9.3 lists the results of this analysis. Note that this model is being presented solely for comparison purposes. It might be an appropriate model if the data were cross-sectional, but it is not appropriate here because the data are longitudinal and this model assumes all observations are independent.

Table 9.3. NIMH Schizophrenia Collaborative Study: Severity of Illness Across Time (N = 437)—Logistic Regression Results

	Estimate	SE	Z	$p <$
Intercept	3.703	0.442	8.38	.001
Drug (0 = placebo; 1 = drug)	−0.405	0.494	−0.82	.41
Time ($\sqrt{\text{week}}$)	−1.113	0.234	−4.76	.001
Drug by time	−0.418	0.262	−1.60	.11
$-2 \log L = 1362.06$				

SE = standard error.

From these estimates we can easily calculate the fit of the model to the observed logits. This is presented in Figure 9.8. These estimated logits are obtained by substituting in the values of Drug (0 for placebo and 1 for drug), and Time (square root of weeks 0, 1, 3, and 6) into the estimated regression equation:

$$\log \left[\frac{\hat{p}_{ij}}{1 - \hat{p}_{ij}} \right] = 3.70 - .41\, Drug_i - 1.11\, Time_j - .42\, (Drug_i \times Time_j). \quad (9.62)$$

As Figure 9.8 attests, the observed logits appear to be fit well by the ordinary fixed-effects logistic regression model. Figure 9.9 presents the model fit in terms of the observed proportions. These are obtained by using the aforementioned logistic cdf formula, namely $p_{ij}(\hat{z}) = 1/[1 + \exp(-\hat{z})]$, where $\hat{z}$ is the estimated logit obtained from (9.62).

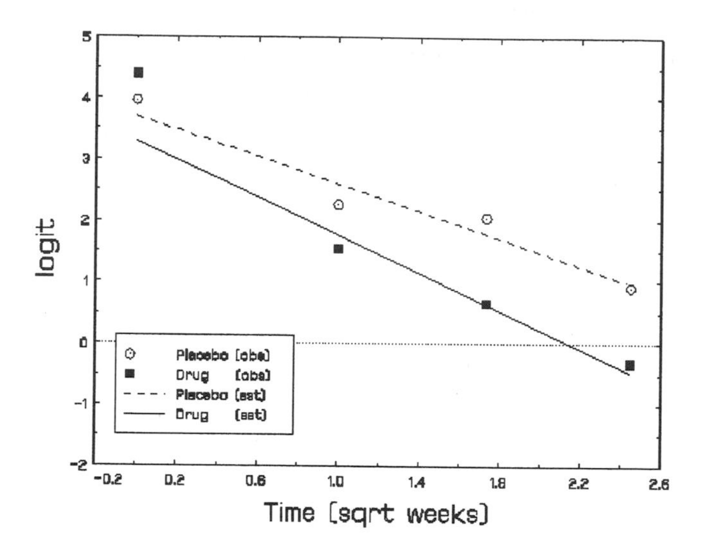

Figure 9.8. Fitted logits across time by condition: Fixed-effects logistic regression model.

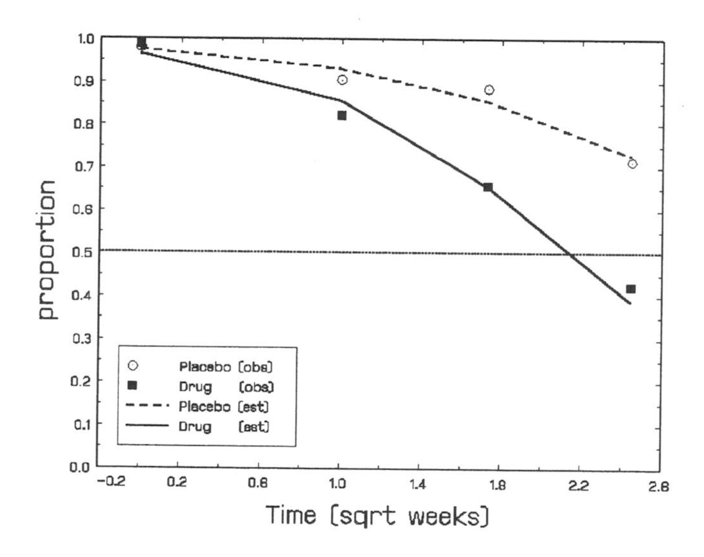

Figure 9.9. Fitted proportions across time by condition: Fixed-effects logistic regression model.

9.7.2 Random Intercept Logistic Regression Model

We now consider mixed-effects logistic regression modeling of these data, starting first with a random intercept model. The within-subjects, or level-1, model is

$$\log\left[\frac{p_{ij}}{1-p_{ij}}\right] = b_{0i} + b_{1i}\sqrt{Week_j} \tag{9.63}$$

and the between-subjects, or level-2, model is

$$\begin{aligned} b_{0i} &= \beta_0 + \beta_2 Drug_i + v_{0i}, \\ b_{1i} &= \beta_1 + \beta_3 Drug_i, \end{aligned} \tag{9.64}$$

where the random subject effects v_{0i} are distributed as $\mathcal{N}(0, \sigma_v^2)$. In this model, β_0 represents the week 0 IMPS79 logit for placebo patients (Drug = 0), β_1 is the logit change in IMPS79 due to time ($\sqrt{\text{week}}$) for placebo patients (Drug = 0), β_2 is the difference in week 0 IMPS79 logit for drug patients (Drug = 1), and β_3 is the difference in the logit due to time ($\sqrt{\text{week}}$) for drug patients (Drug = 1). This latter term indicates whether the drug effect changes across the study timepoints, and therefore is of primary interest here. The random subject effects v_{0i} indicate each individual's deviation from their group trend. It is important to keep in mind, as detailed earlier, that all of these regression coefficients are "subject-specific" terms, meaning that they are conditional on the random subject effect. For example, the drug-related terms indicate the effectiveness of the drug relative to placebo for the sub-population of individuals with the same level on the random subject effect. Table 9.4 lists the results of this analysis.

Table 9.4. NIMH Schizophrenia Collaborative Study: Severity of Illness Across Time (N = 437)—Random Intercept Logistic Regression Results

	Estimate	SE	Z	$p <$
Intercept	5.387	0.535	10.07	.001
Drug (0 = placebo; 1 = drug)	−0.025	0.601	−0.04	.97
Time ($\sqrt{\text{week}}$)	−1.500	0.228	−6.59	.001
Drug by time	−1.015	0.274	−3.70	.001
Intercept sd	2.116	0.215		
$-2\log L = 1249.74$				

SE = standard error.

Comparing this model with the fixed-effects model in Table 9.3, using a a likelihood ratio test, yields $X_1^2 = 112.3$. This is highly significant, indicating that the data are definitely correlated within subjects. Expressing the subject-variation as an intraclass correlation yields

$$r = \frac{2.116^2}{2.116^2 + \pi^2/3} = .58, \tag{9.65}$$

showing that a sizable proportion of the variation is attributable to subjects.

Inspecting the results for the regression coefficients indicates that the drug and placebo groups do not differ significantly at week 0 (drug term), that the placebo group does improve significantly across time (time term), but that the drug group improves at a more pronounced rate across time than the placebo group (drug by time interaction). Notice that the conclusions from this analysis are somewhat different than those in the fixed-effects analysis, primarily for the test of most interest: the drug by time interaction term. Remember, that the fixed-effects analysis completely ignores the correlation in the data and is therefore not an appropriate model for these data—we are presenting it in this chapter for comparative purposes only.

At this point, one might be interested in examining how well the estimated model fits the observed data. Because of the inclusion of the random effect, this is a bit more involved than in the ordinary fixed effects model. Here, one can calculate estimated trend lines for the two treatment groups (drug and placebo) for specific values of the random effect. For example, Figure 9.10 illustrates these trends, in terms of the probability of response, for two values of the random subject effects: $-1\ \hat{\sigma}_\upsilon$ and $1\ \hat{\sigma}_\upsilon$, where $\hat{\sigma}_\upsilon = 2.12$ from the current analysis.

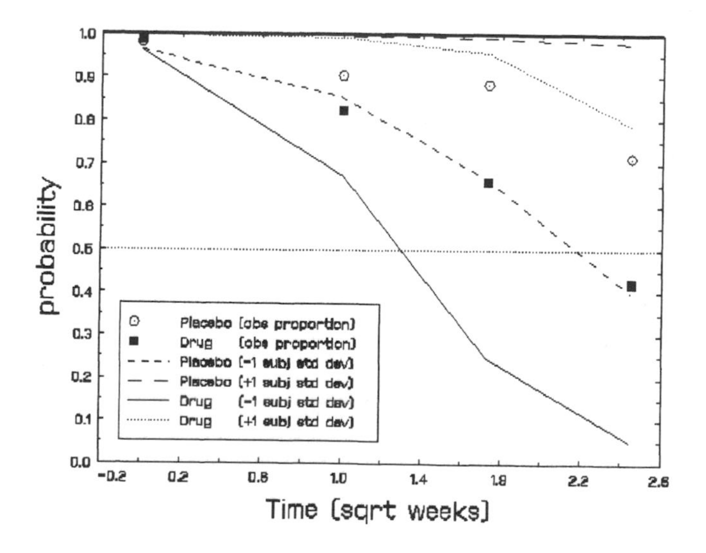

Figure 9.10. Observed proportions and estimated probabilities across time by condition.

These lines are generated using the formula (letting D and W denote $Drug$ and $Week$, respectively):

$$P(Y_{ij} = 1) = \\ 1/\left\{1 + \exp\left[-\left(5.39 - .03\,D_i - 1.50\,\sqrt{W_j} - 1.02\,D_i \times \sqrt{W_j} + \upsilon_{0i}\right)\right]\right\},$$

substituting values for $Drug$ (0 or 1), $Week$ (0 to 6), and where υ_{0i} is set equal to -2.12 and 2.12, respectively. This figure shows the benefit of the drug relative to placebo for both of these levels of the random subject effect. The probability of a positive response (*i.e.*, being sick) is clearly less for the drug relative to placebo group, and this benefit grows over time. What is also enlightening is the depiction of the individual effect on response. As the figure indicates, placebo subjects with a random effect equal to -1 standard deviation units actually have a greater probability of being well, rather than sick, at the final timepoint (the dashed line goes below .5 at the final timepoint). Also, such subjects have a better probability of being well than drug patients with a +1 (standard deviation unit) random effect values. Clearly, subjects have great influence on their change across time.

Figures 9.11 and 9.12 present histograms of the empirical Bayes estimates of the random subject effects υ_{0i} for the control and drug patients, respectively. As these histograms reveal, while not the majority, there are some placebo patients with random effect estimates of minus one standard deviation unit (*i.e.*, -2.12) or less, and some drug patients with random effect estimates of approximately one standard deviation unit above zero. Computing percentiles for these distributions indicates that slightly more than 10% of the placebo patients have estimates of -2.12 or less, while approximately 25% of the drug patients have estimates greater than 1.93 (the maximum estimate for the drug group is 2.04). Thus, the analysis suggests that there is a fair proportion of placebo patients who do get well and drug patients who do not get well.

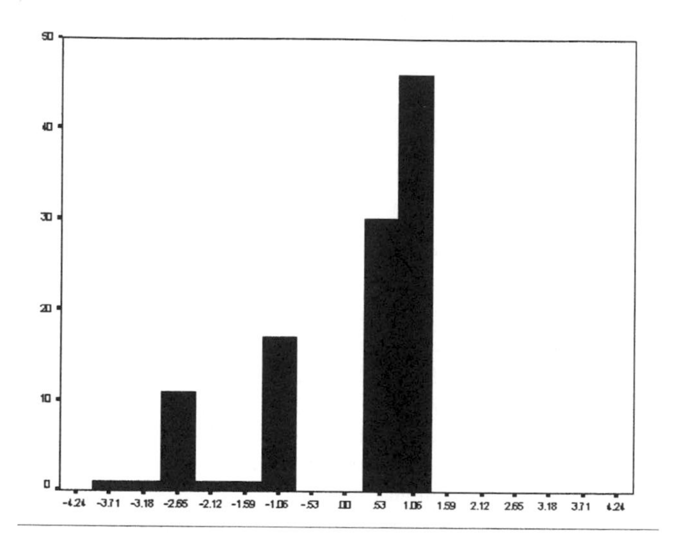

Figure 9.11. Schizophrenia data: Empirical Bayes estimates for control patients.

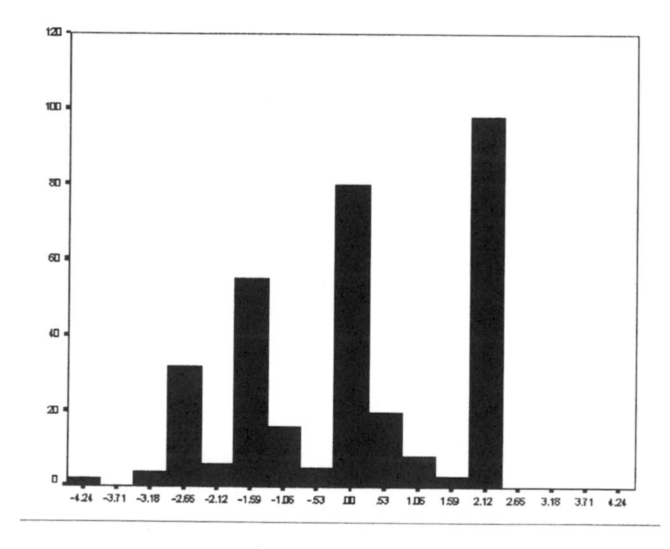

Figure 9.12. Schizophrenia data: Empirical Bayes estimates for drug patients.

The previous Figure 9.10 presents the observed proportions for the two groups across the four primary study timepoint. A natural question to ask is how well does the model fit these marginal proportions. Since the parameter estimates from the logistic mixed model represent "subject-specific" quantities, they must be "marginalized" to address this point. This can be done in the following way.

1. Calculate $\hat{\boldsymbol{y}}_i = \boldsymbol{X}_i \hat{\boldsymbol{\beta}}$.

2. Calculate the "marginalization" vector:

$$\hat{\boldsymbol{s}} = \frac{1}{\sigma} \left[\text{Diag}(\hat{V}(\boldsymbol{y}_i)) \right]^{1/2}$$

   where $\hat{V}(\boldsymbol{y}_i) = \boldsymbol{Z}_i \hat{\boldsymbol{\Sigma}} \boldsymbol{Z}_i' + \sigma^2 \boldsymbol{I}_i$, with $\sigma = 1$ for probit and $\sigma = \pi/\sqrt{3}$ for logistic. $\boldsymbol{Z}_i$ = the design matrix for random effects; for a random intercept model $\boldsymbol{Z}_i = \boldsymbol{1}_i$, and so, $\hat{s} = \sqrt{\hat{\sigma}_v^2/\sigma^2 + 1}$.

3. Perform element-wise division: $\hat{\mathbf{z}}_i = \hat{\boldsymbol{y}}_i \; /. \; \hat{\boldsymbol{s}}$.

4. Obtain the estimated probabilities as $\hat{\boldsymbol{p}}_i = \Phi(\hat{\mathbf{z}}_i)$ for probit and $\hat{\boldsymbol{p}}_i = \Psi(\hat{\mathbf{z}}_i)$ for logistic, where $\Phi(\cdot)$ and $\Psi(\cdot)$ represent the normal and logistic cdf, respectively.

In practice, as mentioned for the logistic, Zeger et al. [1988] suggests that $(15\pi)/(16\sqrt{3})$ works better than $\pi/\sqrt{3}$ as σ. Also, for the logistic the results are approximate and rely on the cumulative normal approximation to the logistic function.

Table 9.5 shows how SAS PROC IML can be used to accomplish this marginalization for the two treatment groups and four primary study timepoints. In this code, `x0` and `x1` represent the covariate matrices for the placebo and drug groups, respectively (not including the intercept). The first column of these matrices is the drug code, the next column is the square root of week, and the third column is the drug by time interaction. The estimated parameters are then given in `int` (intercept), `sd` (standard deviation of the random effects), and `beta` (regression coefficients). Several SAS PROC IML functions are utilized including `J` (matrix initialization), `diag` (construction of a diagonal matrix), `T` (matrix transpose), and `vecdiag` (extraction of the diagonal of a matrix).

Table 9.5. SAS PROC IML Code: Computing Marginal Probabilities

```
PROC IML;
/* Results from random intercept model */;
x0   =  {   0 0.00000 0,
            0 1.00000 0,
            0 1.73205 0,
            0 2.44949 0};
x1   =  {   1 0.00000 0.00000,
            1 1.00000 1.00000,
            1 1.73205 1.73205,
            1 2.44949 2.44949};
int    =   {5.387};
sd     =   {2.116};
beta   =   {-.025, -1.500, -1.015};
/* Approximate Marginalization Method */;
pi     =   3.141592654;
nt     =   4;
ivec   =   J(nt,1,1);
zvec   =   J(nt,1,1);
evec   =   (15/16)**2 * (pi**2)/3 * ivec;
/* nt by nt matrix with evec on the diagonal and zeros elsewhere */;
emat = diag(evec);
/* variance--covariance matrix of underlying latent variable */;
vary = zvec * sd * T(sd) * T(zvec) + emat;
sdy = sqrt(vecdiag(vary) / vecdiag(emat));
z0 = (int + x0*beta) / sdy ;
z1 = (int + x1*beta) / sdy;
grp0 = 1 / ( 1 + EXP(0 - z0));
grp1 = 1 / ( 1 + EXP(0 - z1));
print 'Random intercept model';
print 'Approximate Marginalization Method';
print 'marginal prob for group 0 - response' grp0 [FORMAT=8.4];
print 'marginal prob for group 1 - response' grp1 [FORMAT=8.4];
```

Figure 9.13 displays the observed proportions and marginalized probabilities that are obtained using this SAS PROC IML code (based on the random intercept model). As can be seen, the marginalized estimates fit the observed proportions very well. Thus, the use of the square root of week does a reasonable job of fitting the trends across time in these two groups (utilizing only a single parameter for time for each group).

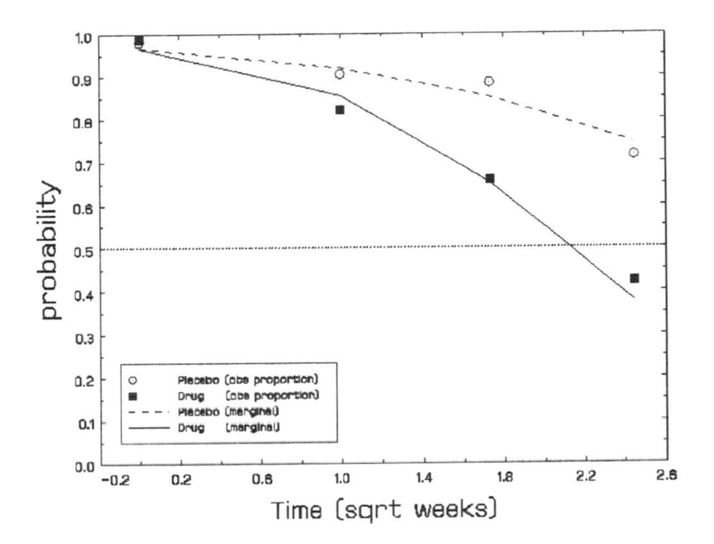

Figure 9.13. Observed proportions and estimated marginalized probabilities across time by condition.

9.7.3 Random Intercept and Trend Logistic Regression Model

As in the continuous case, it is natural to include both random subject intercepts and trends in the model. For this, the within-subjects, or level-1, model is as before:

$$\log\left[\frac{p_{ij}}{1-p_{ij}}\right] = b_{0i} + b_{1i}\sqrt{Week_j}, \tag{9.66}$$

while the between-subjects, or level-2, model is now

$$\begin{aligned} b_{0i} &= \beta_0 + \beta_2 Drug_i + v_{0i}, \\ b_{1i} &= \beta_1 + \beta_3 Drug_i + v_{1i}, \end{aligned} \tag{9.67}$$

where the random subject effects vector $\boldsymbol{v}_i$ is distributed as a bivariate normal $\mathcal{N}(\mathbf{0}, \boldsymbol{\Sigma}_v)$. A probit version of this model was first described by Gibbons and Bock [1987] and applied to these data in Gibbons and Hedeker [1994]. Here, we will utilize the logistic model.

Table 9.6 contains the results of this analysis. Comparing this model to the previous random intercept model using a likelihood ratio test yields $X_2^2 = 22.31, p < .001$, indicating that there is appreciable heterogeneity in the individual trends. This test has two degrees of freedom because it is testing the null hypothesis that the slope variance $\sigma_{v_1}^2$ and slope intercept covariance $\sigma_{v_0 v_1}$ are jointly equal to 0. This null hypothesis is clearly rejected in the present case.

Table 9.6. NIMH Schizophrenia Collaborative Study: Severity of Illness Across Time (N = 437)—Random Intercept and Trend Logistic Regression Results

	Estimate	SE	Z	$p <$
Intercept	6.025	0.918	6.56	.001
Drug (0 = placebo; 1 = drug)	0.281	0.761	0.37	.71
Time ($\sqrt{\text{week}}$)	−1.477	0.451	−3.27	.001
Drug by time	−1.587	0.479	−3.31	.001
Cholesky elements				
Intercept	2.726	0.597		
Intercept–time covariance	−0.829	0.352		
Time	1.561	0.248		
$-2 \log L = 1227.43$				

SE = standard error.

In terms of the fixed effets, the conclusions from this analysis are not different from the previous random intercept analysis. Namely, the groups do not differ significantly at week 0 or baseline (the drug term), the placebo group does have a significant negative trend across time (the time term), though the drug group has an even greater negative trend across time (the drug by time interaction). These statements are conditional on, or adjusted for, the random subject intercepts and trends. Table 9.6 also lists the estimated elements, and standard errors, in the Cholesky factorization of the random-effect variance–covariance matrix. Using these, the estimated variance–covariance matrix is obtained as

$$\hat{\Sigma}_\upsilon = \begin{bmatrix} 2.73 & 0 \\ -.83 & 1.56 \end{bmatrix} \begin{bmatrix} 2.73 & -.83 \\ 0 & 1.56 \end{bmatrix} = \begin{bmatrix} 7.43 & -2.27 \\ -2.27 & 3.12 \end{bmatrix}. \quad (9.68)$$

Additionally, converting the estimated covariance to a correlation yields $r_{\upsilon_0 \upsilon_1} = -.47$. These results make clear that there is considerable individual heterogeneity in terms of both the intercepts and slopes. For example, the mean intercept is estimated as 6.03 and 6.31 for placebo and drug groups, respectively, but the standard deviation in the population distribution of individual intercepts is $\sqrt{7.43} = 2.73$. Similarly, while the average slopes are -1.48 and -3.06 for placebo and drug groups, the standard deviation in the slopes equals $\sqrt{3.12} = 1.77$.

Figure 9.14, which presents the empirical Bayes intercept and slope estimates for each subject, bears this heterogeneity out. As can be seen from Figure 9.14 there are both drug and placebo patients with more negative slopes than their group averages (to the left of the zero on the x axis). There are also some patients who do not improve across time, namely those to the right in the figure. The figure also reveals the nature of the negative covariance between the random intercepts and slopes, namely subjects with more negative intercepts (more likely to be well at baseline) have more positive slopes (less likely to improve over time).

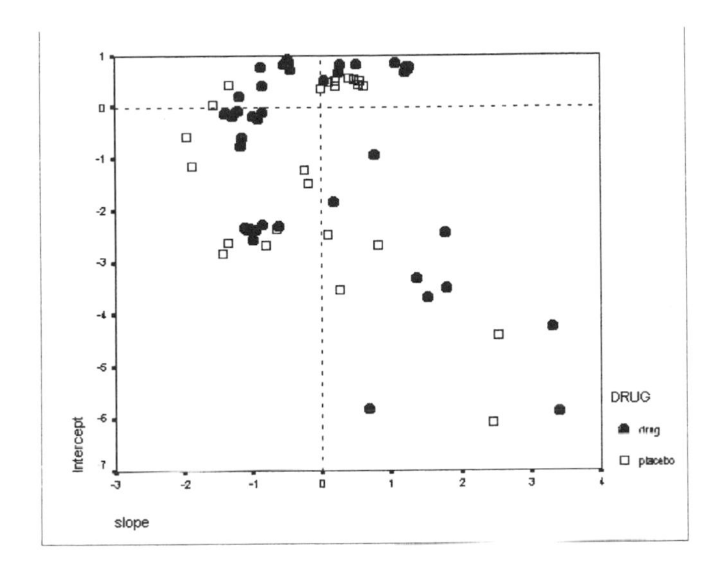

Figure 9.14. Schizophrenia data: Empirical Bayes estimates.

Figures 9.15 and 9.16 use these empirical Bayes estimates to yield the estimated trend lines for the probability of being sick across time for each of the control and drug patients, respectively. These lines are generated by (letting D and W denote $Drug$ and $Week$, respectively):

$$Pr(Y_{ij} = 1) = 1/\{1 + \exp[-(6.03 + .28\, D_i - 1.48\, \sqrt{W_j} - 1.59\, D_i \times \sqrt{W_j} + \upsilon_{0i} + \upsilon_{1i}\, \sqrt{W_j})]\},$$

with values for each subject's group, $Drug$ (0 or 1), the seven values of $Week$ (0 to 6), and each subject's empirical Bayes estimates of υ_{0i} and υ_{1i}. As these figures reveal, there is a fair proportion of placebo patients in the sample with high probability of improving across time. Likewise, there is a fair proportion of drug patients with high probability of being sick consistently across time. Thus, while the results presented in Table 9.6 yield important information for how the groups differ, controlling for the subject effects, the subject effects themselves provide useful information for indicating the broad range of response trends across subjects within groups.

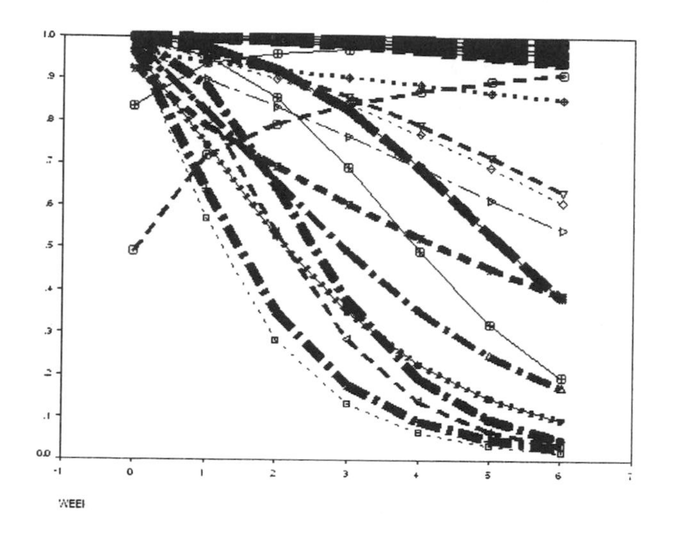

Figure 9.15. Estimated probabilities across time for placebo patients.

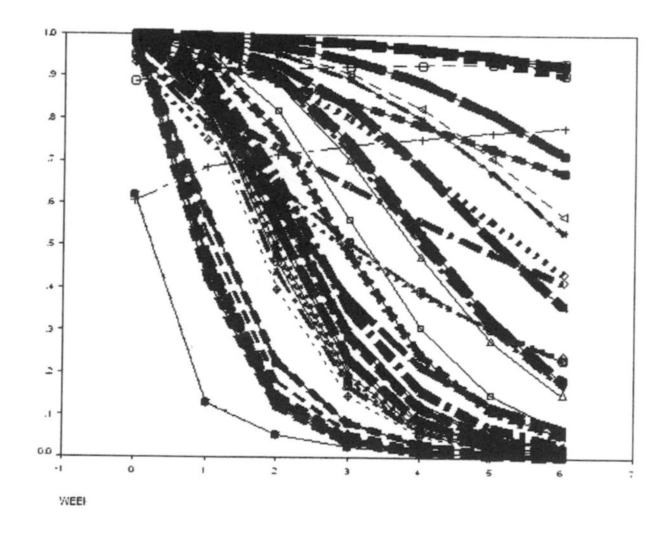

Figure 9.16. Estimated probabilities across time for drug patients.

Finally, for this model it is useful to calculate the estimated marginal group trends across time. Table 9.7 presents SAS PROC IML code for generating these estimates based on the parameter estimates in Table 9.6. This code augments the code presented earlier for the random intercept model in Table 9.5 for the inclusion of the random subject trend. Thus the logic is the same, though the particulars are model-specific. As before, several SAS PROC IML functions are utilized including J (matrix initialization), diag (construction of a diagonal matrix), T (matrix transpose), and vecdiag (extraction of the diagonal of a matrix).

Table 9.7. SAS PROC IML Code: Computing Marginal Probabilities

```
PROC IML;
/* Results from random intercept & trend model */;
x0   =  {   0 0.00000 0,
            0 1.00000 0,
            0 1.73205 0,
            0 2.44949 0};
x1   =  {   1 0.00000 0.00000,
            1 1.00000 1.00000,
            1 1.73205 1.73205,
            1 2.44949 2.44949};
int    =   {6.025};
chol   =   {2.726 0,
            -.829 1.561};
beta   =   { .281, -1.477, -1.587};
/* Approximate Marginalization Method */;
pi     =   3.141592654;
nt     =   4;
ivec   =   J(nt,1,1);
zmat   =   {1 0.00000,
            1 1.00000,
            1 1.73205,
            1 2.44949};
evec   =   (15/16)**2 * (pi**2)/3 * ivec;
/* nt by nt matrix with evec on the diagonal and zeros elsewhere */,
emat = diag(evec);
/* variance--covariance matrix of underlying latent variable */;
vary = zmat * chol * T(chol) * T(zmat) + emat;
sdy = sqrt(vecdiag(vary) / vecdiag(emat));
z0 = (int + x0*beta) / sdy ;
z1 = (int + x1*beta) / sdy;
grp0 = 1 / ( 1 + EXP(0 - za));
grp1 = 1 / ( 1 + EXP(0 - za));
print 'Random intercept and trend model';
print 'Approximate Marginalization Method';
print 'marginal prob for group 0 - response' grp0 [FORMAT=8.4];
print 'marginal prob for group 1 - response' grp1 [FORMAT=8.4];
```

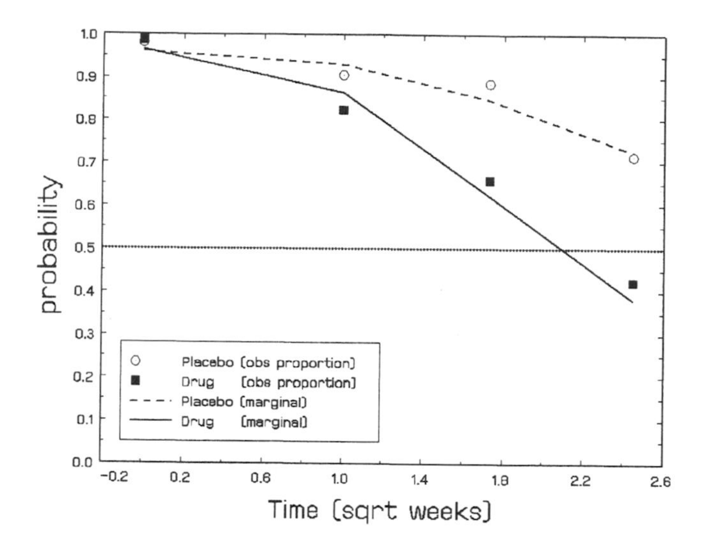

Figure 9.17. Observed proportions and marginalized probabilities by group for the random intercept and trend model.

Figure 9.17 displays these marginalized estimates along with the observed proportions across time for the two groups. As can be seen, the model fits the marginal proportions well. This isn't surprising given that the previous random intercept model also fit these marginal proportions well. What is interesting to see in the current model is the wide range of response trends in the data, for both drug and placebo groups, and the importance of the subject effects on the probability of being sick across time.

9.8 SUMMARY

This chapter has focused on random-effects logistic regression models for longitudinal dichotomous responses. As mentioned, one could also use a probit response function to yield random effects probit regression models. The logistic version of the model is particuarly popular in many fields, though not all. The probit version is often preferred in genetic studies because of its reliance on an underlying normal distribution.

In this chapter we have represented time in its raw metric, but use of orthogonal polynomials can also be applied here. This would follow the ideas and presentation in Chapter 5 for normal outcomes. In many respects, the orthogonal polynomial representation of time is preferable because it usually provides less computational problems, since the time variables are made nearly independent for the usual case of some incomplete data across time. Since estimation of dichotomous mixed models is challenging in and of itself, researchers should consider the orthogonal polynomial representation of the model if computational difficulties arise in estimation. Of course, in some cases even this won't eliminate computational problems, thereby suggesting model simplification. Since computational difficulties are more often due to the random effects than the fixed effects, simplication of the random effects is a likely possibility.

CHAPTER 10

MIXED-EFFECTS REGRESSION MODELS FOR ORDINAL OUTCOMES

10.1 INTRODUCTION

Often, the response measure of interest is measured in a series of ordered categories. Such measures are termed "ordinal" and can represent a variety of graded responses such as Likert scales, psychiatric ratings of severity (*e.g.*, none, mild, moderate, severe), agreement ratings (disagree, undecided, agree), to cite a few examples. In some cases, the response measure of interest may represent a count (*e.g.*, number of heath service visits) that has large probability mass at zero (*i.e.*, no service use), a majority of values in the one to two-visit range, and a few extreme values. Such data are sometimes not amenable to analysis using methods suitable for Poisson random variables (see Chapter 12). In these cases, an ordinal variable can be constructed with ordered categories of 0, 1, 2, and 3 or more visits. The relative frequency with which the categories are endorsed is not a factor for the ordinal regression model, whereas quite strict requirements are imposed under the assumption of a Poisson process.

Extending the methods for dichotomous responses to ordinal response data has been actively pursued [Agresti and Lang, 1993; Dos Santos and Berridge, 2000; Ezzet and Whitehead, 1991; Fielding, 1999; Harville and Mee, 1984; Hedeker and Gibbons, 1994; Jansen, 1990; Ten Have, 1996; Tutz and Hennevogl, 1996]. Again, developments have been mainly in terms of logistic and probit regression models. In particular, because the proportional odds assumption described by McCullagh [1980], which is based on the logistic regression formulation, is a common choice for analysis of ordinal data, many of the mixed models for ordinal data are generalizations of this model. The proportional odds model characterizes the ordinal responses in C categories in terms of $C-1$ cumulative category comparisons, specifically, $C-1$ cumulative logits (*i.e.*, log odds). In the proportional odds model, the covariate effects are assumed to be the same across these cumulative logits, or proportional across the cumulative odds. As noted by Peterson and Harrell [1990], however, examples of nonproportional odds are not difficult to find. To overcome this limitation, Hedeker and Mermelstein [1998] described an extension of the mixed-effects ordinal logistic regression model to allow for nonproportional odds for a set of regressors.

This chapter describes mixed-effects models for ordinal data that accommodate multiple random effects and a general form for model covariates. For ordinal outcomes, proportional odds, partial proportional odds, and related survival analysis models for discrete or grouped-time survival data are described. To illustrate application of the various mixed-effects ordinal regression models analysis of a longitudinal psychiatric dataset is described.

10.2 MIXED-EFFECTS PROPORTIONAL ODDS MODEL

Let the C ordered response categories be coded as $c = 1, 2, \ldots, C$. As ordinal models often utilize cumulative comparisons of the categories, define the cumulative probabilities for the C categories of the outcome Y as $P_{ijc} = \Pr(Y_{ij} \leq c) = \sum_{k=1}^{c} p_{ijk}$, where p_{ijk} represents the probability of response in category k. The mixed-effects logistic regression model for the cumulative probabilities is given in terms of the cumulative logits as

$$\log\left[\frac{P_{ijc}}{1-P_{ijc}}\right] = \gamma_c - \left[\boldsymbol{x}'_{ij}\boldsymbol{\beta} + \boldsymbol{z}'_{ij}\boldsymbol{T}\boldsymbol{\theta}_i\right] \quad (c = 1, \ldots, C-1), \tag{10.1}$$

with $C-1$ strictly increasing model thresholds γ_c (*i.e.*, $\gamma_1 < \gamma_2 \cdots < \gamma_{C-1}$). As before, $\boldsymbol{x}_{ij}$ is the $(p+1) \times 1$ covariate vector (including the intercept), and $\boldsymbol{z}_{ij}$ is the design vector for the r random effects, both vectors being for the jth timepoint nested within subject i. Also, $\boldsymbol{\beta}$ is the $(p+1) \times 1$ vector of unknown fixed regression parameters, and $\boldsymbol{\theta}_i$ is the standardized $r \times 1$ vector of unknown random effects for subject i. This standardization is exactly as in the dichotomous case discussed in the last chapter, that is, we let $\boldsymbol{v} = \boldsymbol{T}\boldsymbol{\theta}$, where $\boldsymbol{T}\boldsymbol{T}' = \boldsymbol{\Sigma}_v$ is the Cholesky factorization of random-effect variance covariance matrix $\boldsymbol{\Sigma}_v$. Thus, the unstandardized random effects $\boldsymbol{v}_i$ are assumed to be multivariate normally distributed with mean vector $\boldsymbol{0}$ and variance–covariance matrix $\boldsymbol{\Sigma}_v$, whereas $\boldsymbol{\theta}_i$ are assumed to be multivariate *standard* normally distributed with mean vector $\boldsymbol{0}$ and variance–covariance matrix $\boldsymbol{I}$.

The relationship between the latent continuous variable y and an ordinal outcome with three categories is depicted in Figure 10.1. In this case, the observed ordinal outcome $Y_{ij} = c$ if $\gamma_{c-1} \leq y_{ij} < \gamma_c$ for the latent variable (with $\gamma_0 = -\infty$ and $\gamma_C = \infty$). As in the dichotomous case, it is common to set a threshold to zero to set the location of the latent variable. Typically, this is done in terms of the first threshold (*i.e.*, $\gamma_1 = 0$). In the figure above, setting $\gamma_1 = 0$ implies that $\gamma_2 = 2$. These threshold parameters, in addition to the model intercept, represent the marginal response probabilities in the C categories. For example, for this case with $C = 3$, $0 - \beta_0$ represents the log odds for a response in the first category, relative to categories 2 and 3; $\gamma_2 - \beta_0$ represents the log odds for a response in the first two categories, relative to the third category. An alternative specification is to set the model intercept $\beta_0 = 0$ and to estimate $C-1$ thresholds. Denoting these $C-1$ thresholds as γ^*, we would then have the following relationship between these two parameterizations: $\gamma_1^* = 0 - \beta_0$ and $\gamma_2^* = \gamma_2 - \beta_0$.

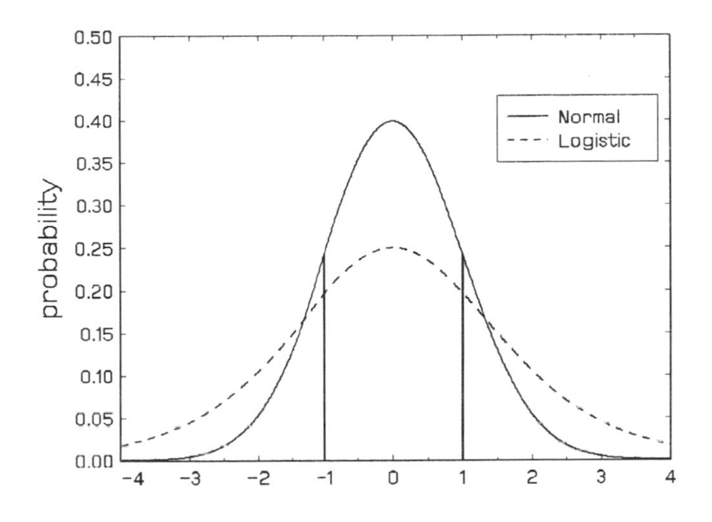

Figure 10.1. Threshold concept for an ordinal reponse with three categories.

At first glance, it may appear that the parameterization of the model in (10.1) is not consistent with the dichotomous model in the previous chapter. To see the connection, notice that for a dichotomous outcome (coded 0 and 1), the model is written as

$$\log\left[\frac{P_{ij0}}{1-P_{ij0}}\right] = 0 - \left[\boldsymbol{x}'_{ij}\boldsymbol{\beta} + \boldsymbol{z}'_{ij}\boldsymbol{T}\boldsymbol{\theta}_i\right], \tag{10.2}$$

and since for a dichotomous outcome $P_{ij0} = p_{ij0}$ and $1 - P_{ij0} = p_{ij1}$,

$$\log\left[\frac{1-P_{ij0}}{P_{ij0}}\right] = \log\left[\frac{p_{ij1}}{1-p_{ij1}}\right] = \boldsymbol{x}'_{ij}\boldsymbol{\beta} + \boldsymbol{z}'_{ij}\boldsymbol{T}\boldsymbol{\theta}_i, \tag{10.3}$$

which is the same as before.

In terms of the underlying latent variable y, the multilevel representation of the ordinal model is identical to the dichotomous version presented in the previous chapter. If the multilevel model is written in terms of the observed response variable Y, then the level-1 model is written instead in terms of the cumulative logits, namely,

$$\log\left[\frac{P_{ijc}}{1-P_{ijc}}\right] = \gamma_c - [b_{0i} + b_{1i}x_{ij}] \quad (c = 1, \ldots C-1), \tag{10.4}$$

for the case of a model with one within-subjects or level-1 covariate x_{ij}. Because the between-subjects or level-2 model does not really depend on the response function or variable, it would be the same as given for the dichotomous model in the previous chapter, namely,

$$\begin{aligned} b_{0i} &= \beta_0 + \beta_2 x_i + v_{0i}, \\ b_{1i} &= \beta_1 + \beta_3 x_i + v_{1i}. \end{aligned} \tag{10.5}$$

for the case of one between-subjects covariate x_i that influences both the intercept ("main effect" of x_i) and the effect of the level-1 covariate ($x_i \times x_{ij}$ interaction). Also, the above model includes random effects for both the intercept and slope of x_{ij}.

Since the regression coefficients $\boldsymbol{\beta}$ do not carry the c subscript, they do not vary across categories. Thus, the relationship between the explanatory variables and the cumulative logits does not depend on c. McCullagh [1980] calls this assumption of identical odds ratios across the $C-1$ cut-offs the proportional odds assumption.

Figure 10.2 illustrates the proportional odds assumption for a single regressor x and an ordinal outcome variable with three categories. Here, we are considering only the fixed-effects part of the model and only considering one regressor x which takes on values of either 0 or 1. The model for the two cumulative logits is therefore

$$\begin{aligned} \log\left[\frac{Pr(Y_{ij} \le 1)}{1 - Pr(Y_{ij} \le 1)}\right] &= \log\left[\frac{Pr(Y_{ij} = 1)}{Pr(Y_{ij} = 2 \text{ or } 3)}\right] = \gamma_1 - (x\,\beta_1)\,, \\ \log\left[\frac{Pr(Y_{ij} \le 2)}{1 - Pr(Y_{ij} \le 2)}\right] &= \log\left[\frac{Pr(Y_{ij} = 1 \text{ or } 2)}{Pr(Y_{ij} = 3)}\right] = \gamma_2 - (x\,\beta_1)\,. \end{aligned}$$

In this illustration, $\gamma_1 = -1$, $\gamma_2 = 1$, and $\beta_1 = -.5$, so that the cumulative logits equal -1 and 1, respectively, for the group with $x = 0$, and $-.5$ and 1.5 for the group with $x = 1$. Since the covariate effect β is equal across both cumulative logits, there is a parallel shift in logit values (of .5) for the group with $x = 1$, relative to the group with $x = 0$, which is depicted above. Tracing these logit values onto the logistic cdf indicates the response probabilities (y axis) for the three categories for each of the two groupsproportional odds assumption.

As written above in (10.1), a positive coefficient for a regressor indicates that as values of the regressor increase so do the odds that the response is greater than or equal to c. Although this is a natural way of writing the model, because it means that for a positive β as x increases so does the value of Y, it is not the only way of writing the model. In particular, the model is sometimes written as

$$\log\left[\frac{P_{ijc}}{1 - P_{ijc}}\right] = \gamma_c + \boldsymbol{x}'_{ij}\boldsymbol{\beta} + \boldsymbol{z}'_{ij}\boldsymbol{T}\boldsymbol{\theta}_i \quad (c = 1, \ldots, C-1), \tag{10.6}$$

in which case the regression parameters $\boldsymbol{\beta}$ are identical but of opposite sign. This alternate specification is commonly used in survival analysis models (see Section 10.2.3).

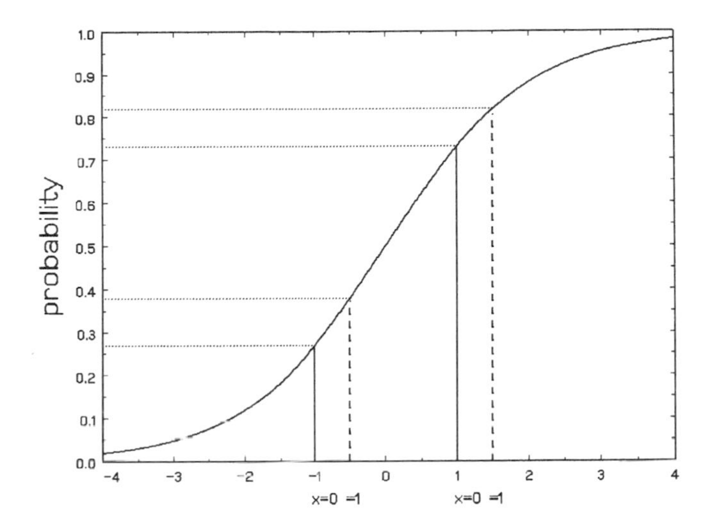

Figure 10.2. Proportional odds assumption.

10.2.1 Partial Proportional Odds

As noted by Peterson and Harrell [1990], violation of the proportional odds assumption is not uncommon. Thus, they described a (fixed-effects) partial proportional odds model in which covariates are allowed to have differential effects on the $C - 1$ cumulative logits. Similarly, Terza [1985] developed a similar extension of the (fixed-effects) ordinal probit model. Hedeker and Mermelstein [1998, 2000] utilize this extension within the context of a mixed-effects ordinal regression model. For this, the model for the $C - 1$ cumulative logits can be written as

$$\log\left[\frac{P_{ijc}}{1 - P_{ijc}}\right] = \gamma_c - (\boldsymbol{x}'_{ij}\boldsymbol{\beta} + \boldsymbol{u}'_{ij}\boldsymbol{\alpha}_c + \boldsymbol{z}'_{ij}\boldsymbol{T}\boldsymbol{\theta}_i) \quad (c = 1, \ldots, C-1), \tag{10.7}$$

where $\boldsymbol{u}_{ij}$ is a $h \times 1$ vector containing the values of observation ij on the set of h covariates for which proportional odds is not assumed. In this model, $\boldsymbol{\alpha}_c$ is a $h \times 1$ vector of regression coefficients associated with these h covariates. Because $\boldsymbol{\alpha}_c$ carries the c subscript, the effects of these h covariates are allowed to vary across the $C - 1$ cumulative logits. This extended model has recently been applied successfully in several articles [Wakefield et al., 2001; Xie et al., 2001; Freels et al., 2002], and a similar Bayesian hierarchical model is described in Ishwaran [2000].

Figure 10.3 illustrates a single regressor u, with values 0 or 1, that has a nonproportional effect on the ordinal outcome variable with three categories. Again, we are considering only the fixed-effects part of the model and only considering one regressor u which takes on values of either 0 or 1. The model for the two cumulative logits is therefore

$$\log\left[\frac{Pr(Y_{ij} \le 1)}{1 - Pr(Y_{ij} \le 1)}\right] = \log\left[\frac{Pr(Y_{ij} = 1)}{Pr(Y_{ij} = 2 \text{ or } 3)}\right] = \gamma_1 - (u\,\alpha_1),$$

$$\log\left[\frac{Pr(Y_{ij} \le 2)}{1 - Pr(Y_{ij} \le 2)}\right] = \log\left[\frac{Pr(Y_{ij} = 1 \text{ or } 2)}{Pr(Y_{ij} = 3)}\right] = \gamma_2 - (u\,\alpha_2).$$

As in the previous Figure 10.2, $\gamma_1 = -1$, and $\gamma_2 = 1$, however now $\alpha_1 = -.05$ and $\alpha_2 = -1$. Thus, the covariate has very little effect on the first cumulative logit and a large effect on the second cumulative logit. Thus, as the logistic cdf indicates, these two groups have very similar response probabilities in the first category, but differ quite a bit in the response probabilities for categories 2 and 3.

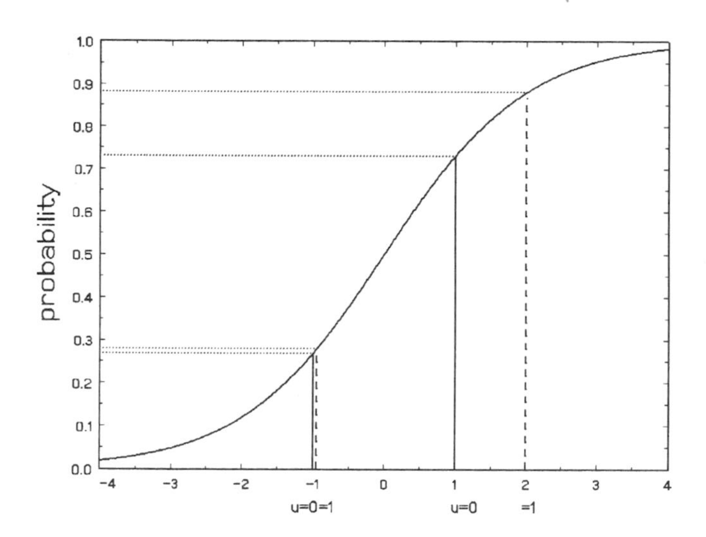

Figure 10.3. Nonproportional odds.

A caveat about this extension of the proportional odds model should be mentioned. The effects on the cumulative log odds, namely $\boldsymbol{u}'_{ij}\boldsymbol{\alpha}_c$, result in $C - 1$ nonparallel regression lines. These regression lines inevitably cross for some values of $\boldsymbol{u}$, leading to negative fitted values for the response probabilities. For $\boldsymbol{u}$ variables contrasting two levels of a regressor (*e.g.*, gender coded as 0 or 1), this crossing of regression lines occurs outside the range of admissible values (*i.e.*, < 0 or > 1). However, if the regressor is continuous, this crossing can occur within the range of the data, and so, allowing for nonproportional odds can lead to illogical results. Figure 10.4 illustrates and contrasts the proportional and nonproportional odds effect for a continuous variable on an ordinal outcome with three categories.

In the top figures, the effect of x is assumed to be the same for both cumulative logits (left- and right-side figures, respectively), namely $\beta = -.5$. Here, $\gamma_1 = -1$ and $\gamma_2 = 1$, which equal the two cumulative logit values for $x = 0$. For $x = 1$, we get $\gamma_1 - \beta = -.5$ and $\gamma_2 - \beta = 1.5$, for the two cumulative logits, and for $x = 5$ we get $\gamma_1 - 5\beta = 1.5$ and $\gamma_2 - 5\beta = 5$. In the top figures, the first cumulative logit is always less than the second cumulative logit for all values of x. Since the effect of x does not vary across the cumulative logit, this is always ensured in the proportional odds model. Alternatively, the bottom figures illustrates what can happen by allowing a continuous covariate u to have a nonproportional odds effect on the cumulative logits. For the first cumulative logit (bottom left figure) the effect of u is $\alpha_1 = -.75$, while for the second cumulative logit (bottom right figure) the effect is $\alpha_2 = -.25$; thus the covariate has less effect on the second cumulative logit, relative to the first. This could happen if the distribution of responses in the third category was relatively similar across levels of u, while the relative distribution of responses in the first two categories varied across levels of u. Here, for $u = 5$, the first cumulative logit ($\gamma_1 - 5\alpha_1 = 2.75$) is actually greater than the second cumulative logit ($\gamma_2 - 5\alpha_2 = 2.25$). This would imply the untenable result that for $u = 5$ the probability of a response in categories 2 and 3 (y axis probability from the upper horizontal dashed line to 1.0 in the lower-left figure) is less than the probability of a response in category 3 alone (y axis probability from the upper horizontal dashed line to 1.0 in the lower-right figure). This is clearly impossible, but there is nothing in the model that prevents this from happening in the case of a continuous regressor.

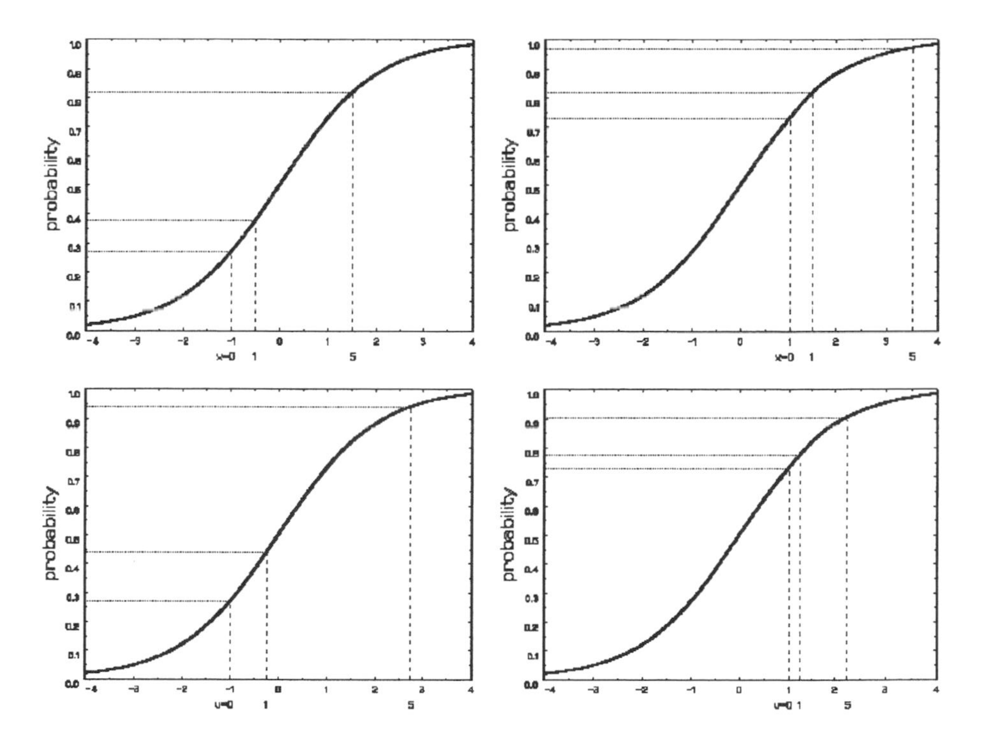

Figure 10.4. Upper left and right: First and second cumulative logit values, respectively, for a proportional odds effect of a continuous regressor x. Lower left and right: First and second cumulative logit values, respectively, for a nonproportional odds effect of a continuous regressor u.

For continuous explanatory regressors, other than requiring proportional odds, a solution to this dilemma is sometimes possible if the variable has, say, m levels with a reasonable number of observations at each of these m levels. In this case $m-1$ dummy-coded variables can be created and substituted into the model in place of the continuous variable. The idea is simply to replace the continuous variable with a categorical version of it. Admittedly this is a somewhat ad-hoc solution; a more sophisticated alternative is to specify a family of nonlinear response functions for the continuous regressors so that they can be nonparallel but do not cross [Kauermann and Tutz, 2003].

10.2.2 Models with Scaling Terms

A somewhat different extention of the proportional odds model is described by Tosteson and Begg [1988]. Here, in the context of ROC analysis, the *scale* of the regressor effects are allowed to vary. McCullagh and Nelder [1989] refer to this extended model for ordinal data as a generalized "rational" model. Ishwaran and Gatsonis [2000] build upon this approach in describing analysis of correlated ROC data using Bayesian methods. For cross-sectional data, Cox [1995] brought together these extensions of the proportional odds model into what he termed location-scale cumulative odds models. For longitudinal data, Hedeker et al. [2006] describe this approach within a mixed-effects model, and show how this extended model can be used to allow both heterogeneous within-subjects and between-subjects variation. The scaling terms allow modeling of the within-subjects variance, just like the random effects model the between-subjects variance. This extended model is

$$\log\left[\frac{P_{ijc}}{1-P_{ijc}}\right] = \frac{\gamma_c - (\boldsymbol{x}'_{ij}\boldsymbol{\beta} + \boldsymbol{u}'_{ij}\boldsymbol{\alpha}_c + \boldsymbol{z}'_{ij}\boldsymbol{T}\boldsymbol{\theta}_i)}{\exp(\boldsymbol{w}'_{ij}\boldsymbol{\tau})} \qquad (c = 1, \ldots, C-1), \qquad (10.8)$$

where $\boldsymbol{w}_{ij} = k \times 1$ vector for the set of k regressors which influence the scale, and $\boldsymbol{\tau}$ are their corresponding effects. Note that the $\boldsymbol{w}$ variables can be the same as the $\boldsymbol{x}$ variables. The result is a nonproportional odds model, using only one additional parameter for each regressor.

Figure 10.5 depicts a model that has a scaling effect for a single explanatory variable w, coded 0 or 1 (but with no explanatory variables in $\boldsymbol{x}$ or $\boldsymbol{u}$). As can be seen, here, the explanatory variable changes the scale of the logistic cdf. Since for this illustration $\tau = -.5$, the scale has been reduced for the group with $w = 1$, relative to the group with $w = 0$. Thus, the former has lower response probabilities in the extreme categories, and a higher probability of response in the middle category.

From this general model, the proportional odds model is obtained only if $\boldsymbol{\alpha}_c = \boldsymbol{\tau} = 0$, otherwise a nonproportional odds model results. To summarize, the two ways nonproportionality occurs are by the inclusion of threshold interactions $\boldsymbol{\alpha}_c \neq 0$ (change of location) or by inclusion of scale terms $\boldsymbol{\tau} \neq 0$ (change of scale).

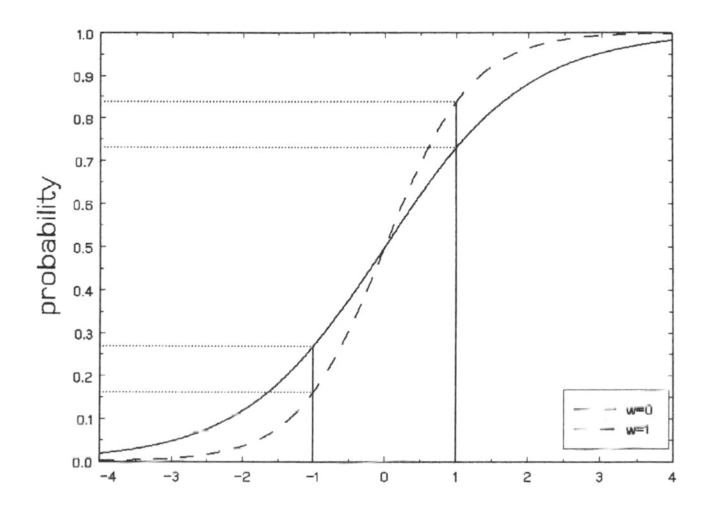

Figure 10.5. Nonproportional odds model with scaling effect.

10.2.2.1 Intraclass Correlation and Partitioning of Between- and Within-Cluster Variance For a random intercept model (*i.e.*, $z_i = \mathbf{1}_i$) it is often of interest to express the subject (level-2) variance in terms of an intraclass correlation. The intraclass correlation indicates the proportion of unexplained variance at the subject level; in other words, it reflects the magnitude of the between-subjects variance. Reference is made to the threshold concept and the underlying latent response tendency that determines the observed ordinal response. For the logistic model assuming normally distributed random effects, the estimated intraclass correlation equals $\hat{\sigma}_v^2/(\hat{\sigma}_v^2 + \pi^2/3)$, where the latter term in the denominator represents the variance of the underlying latent response tendency. As mentioned earlier, for the logistic model, this variable is assumed to be distributed as a standard logistic distribution with variance equal to $\pi^2/3$.

If one allows the random intercept to vary across groups of subjects (level-2 units), then separate intraclass correlations are obtained, one for each subject group. For example, in Hedeker et al. [2006], adolescent smokers were classified in terms of their lifetime cigarette smoking experience, and there was interest in examining whether the variation in responses from more experienced smokers differed from those less experienced. In that article, subjects were grouped into three levels depending on their lifetime smoking experience. Here, for simplicity, suppose there are two groups. For this, define $z_{ij} = [\text{HI}_i \ \ \text{LO}_i]$ (where HI_i and LO_i are subject-level indicator variables for these two groups), $\boldsymbol{T} = [\sigma_{v\,(HI)} \ \ \sigma_{v\,(LO)}]$, and $\boldsymbol{\theta}_i = \theta_i$. The between-subjects variance is then allowed to be different for these two groups, and the resulting intraclass correlations are

$$ICC_{HI} = \frac{\hat{\sigma}^2_{v\,(HI)}}{\hat{\sigma}^2_{v\,(HI)} + \pi^2/3} \quad \text{and} \quad ICC_{LO} = \frac{\hat{\sigma}^2_{v\,(LO)}}{\hat{\sigma}^2_{v\,(LO)} + \pi^2/3}.$$

The inclusion of scaling terms in the ordinal mixed model additionally allows an examination of heterogeneity in terms of the within-subjects (level-1) variance. For instance, continuing with the smoking groups example, suppose that a model with $w_{ij} = LO_i$ was fit. In this case, the within-subjects variance for the HI group would be $\pi^2/3$, whereas it would be $(\exp \tau)^2 \, \pi^2/3$ for the LO group. The resulting intraclass correlations are then

$$ICC_{HI} = \frac{\hat{\sigma}^2_{v(HI)}}{\hat{\sigma}^2_{v(HI)} + \pi^2/3} \quad \text{and} \quad ICC_{LO} = \frac{\hat{\sigma}^2_{v(LO)}}{\hat{\sigma}^2_{v(LO)} + [(\exp \tau)^2 \, \pi^2/3]}.$$

Thus, the model allows both heterogeneous between-subjects (level-2) and within-subjects (level-1) variance. This feature can be useful in many applications.

As an aside, in some research areas it is common to recast the model in terms of a probit formulation, instead of the logistic, because these intraclass correlations are then equivalent to polychoric correlations (for ordinal responses) or tetrachoric correlations (for binary responses). Fortunately, modification of the response function and associated likelihood equations is trivial. As has been described, to recast the model in terms of the probit response function, we simply replace the logistic response function (cdf) with the normal cdf, and likewise the logistic pdf is replaced by the normal pdf. Also, in the probit specification, the level-1 variance corresponding to the standard normal distribution is 1 (instead of the logistic value of $\pi^2/3$).

10.2.3 Survival Analysis Models

Several authors have noted the connection between survival analysis models and binary and ordinal regression models for survival data that are discrete or grouped within time intervals (for practical introductions see Allison [1982, 1995], D'Agostino et al. [1990], and Singer and Willett [1993, 2003]). This connection has been utilized in the context of categorical mixed-effects regression models by many authors as well [Han and Hausman, 1990; Ten Have and Uttal, 1994; Ten Have, 1996; Scheike and Jensen, 1997; Hedeker et al., 2000; Reardon et al., 2002; Muthén and Masyn, 2005]. For this, assume that time (of assessment) can take on only discrete positive values $c = 1, 2, \ldots, C$. To make the connection to ordinal models more direct, time is here denoted as c, however more commonly it is denoted as t in the survival analysis literature. For each level-1 unit, observation continues until time Y_{ij} at which point either an event occurs ($d_{ij} = 1$) or the observation is censored ($d_{ij} = 0$), where censoring indicates being observed at c but not at $c + 1$. Define P_{ijc} to be the probability of failure, up to and including time interval c, that is,

$$P_{ijc} = \Pr\,[Y_{ij} \leq c] \tag{10.9}$$

and so the probability of survival beyond time interval c is simply $1 - P_{ijc}$.

Because $1 - P_{ijc}$ represents the survivor function, McCullagh (1980) proposed the following grouped-time version of the continuous-time proportional hazards model:

$$\log[-\log(1 - P_{ijc})] = \gamma_c + \boldsymbol{x}'_{ij}\boldsymbol{\beta}. \tag{10.10}$$

This is the aforementioned complementary log–log response function, which can be re-expressed in terms of the cumulative failure probability, $P_{ijc} = 1 - \exp(-\exp(\gamma_c + \boldsymbol{x}'_{ij}\boldsymbol{\beta}))$.

In this model, $\boldsymbol{x}_{ij}$ includes covariates that vary either at level 1 or 2, however they do not vary with time (*i.e.*, they do not vary across the ordered response categories). They may, however, represent the average of a variable across time or the value of the covariate at the time of the event.

The covariate effects in this model are identical to those in the grouped-time version of the proportional hazards model described by Prentice and Gloeckler [1978]. As such, the $\boldsymbol{\beta}$ coefficients are also identical to the coefficients in the underlying continuous-time proportional hazards model. Furthermore, as noted by Allison [1982], the regression coefficients of the model are invariant to interval length. Augmenting the coefficients $\boldsymbol{\beta}$, the threshold terms γ_c represent the logarithm of the integrated baseline hazard (*i.e.*, when $\boldsymbol{x} = \mathbf{0}$). While the above model is the same as that described in McCullagh [1980], it is written so that the covariate effects are of the same sign as the Cox proportional hazards model. A positive coefficient for a regressor then reflects increasing hazard (*i.e.*, lower values of Y) with greater values of the regressor. Adding (standardized) random effects, we get

$$\log[-\log(1 - P_{ijt})] = \gamma_c + \boldsymbol{x}'_{ij}\boldsymbol{\beta} + \boldsymbol{z}'_{ij}\boldsymbol{T}\boldsymbol{\theta}_i. \tag{10.11}$$

This model is a mixed-effects ordinal regression model with a complementary log–log response function instead of the logistic. Though the logistic model has also been proposed for analysis of grouped and/or discrete time survival data, its regression coefficients are not invariant to time interval length and it requires the intervals to be of equal length [Allison, 1982].

In the ordinal treatment, survival time is represented by the ordered outcome Y_{ij}, which is designated as being censored or not. Alternatively, each survival time can be represented as a set of dichotomous dummy codes indicating whether or not the observation failed in each time interval that was experienced [Allison, 1982; D'Agostino et al., 1990; Singer and Willett, 1993]. Specifically, each survival time Y_{ij} is represented as a vector with all zeros except for its last element, which is equal to d_{ij} (*i.e.*, =0 if censored and =1 for an event). Alternatively, one can code Y_{ij} as a vector with all ones except for its last element, which is equal to $1 - d_{ij}$; the results are the same in this case but of opposite sign. The length of the vector for observation ij equals the observed value of Y_{ij} (assuming that the survival times are coded as $1, 2, \ldots, C$). These multiple time indicators are then treated as distinct observations in a dichotomous regression model. In a mixed model, a given subject's repeated survival time response vector $\boldsymbol{Y}_i$ is then of size $(\sum_{j=1}^{n_i} Y_{ij}) \times 1$. This method has been called the pooling of repeated observations method by Cupples et al. [1985] and has been well-described in a multilevel context by Reardon et al. [2002]. It is particularly useful for handling time-dependent covariates and fitting nonproportional hazards models because the covariate values can change across time.

Table 10.1 indicates the difference in data structures between the ordinal and dichotomous approaches for a situation with three timepoints. Notice that in the dichotomous representation there would be up to three records depending on when the event or censoring took place. While the table denotes the values of the one dependent variable for the three potential records as $Y1$, $Y2$, and $Y3$, this not a multivariate approach, as there is only a single dependent variable Y. Also, the coding of the dependent variable is such that lower values indicate less survival time.

Table 10.1. Ordinal and Dichotomous Representations of Grouped-Time Survival Data

	Ordinal		*Dichotomous*
Outcome	Dependent Variable	Event Indicator	Dependent Variable
Censor at T1	$Y = 1$	$d = 0$	$Y1 = 1$
Event at T1	$Y = 1$	$d = 1$	$Y1 = 0$
Censor at T2	$Y = 2$	$d = 0$	$Y1 = 1$ $Y2 = 1$
Event at T2	$Y = 2$	$d = 1$	$Y1 = 1$ $Y2 = 0$
Censor at T3	$Y = 3$	$d = 0$	$Y1 = 1$ $Y2 = 1$ $Y3 = 1$
Event at T3	$Y = 3$	$d = 1$	$Y1 = 1$ $Y2 = 1$ $Y3 = 0$

For the dichotomous approach, define p_{ijc} to be the probability of failure in time interval c, conditional on survival prior to c:

$$\mathrm{p}_{ijc} = \Pr\left[Y_{ij} = c \mid Y_{ij} \geq c\right]. \tag{10.12}$$

Similarly, $1 - \mathrm{p}_{ijc}$ is the probability of survival beyond time interval c, conditional on survival prior to c. The mixed proportional hazards model is then written as

$$\log[-\log(1 - \mathrm{p}_{ijc})] = \boldsymbol{x}'_{ijc}\boldsymbol{\beta} + \boldsymbol{z}'_{ij}\boldsymbol{T}\boldsymbol{\theta}_i, \tag{10.13}$$

where now the covariates $\boldsymbol{x}$ can vary across time and so are denoted as $\boldsymbol{x}_{ijc}$. The first elements of $\boldsymbol{x}$ are usually timepoint dummy codes. Because the covariate vector $\boldsymbol{x}$ now varies with c, this approach automatically allows for time-dependent covariates, and relaxing the proportional hazards assumption only involves including interactions of covariates with the timepoint dummy codes.

Under the complementary log–log link function, the two approaches characterized by (10.11) and (10.13) yield identical results for the parameters that do not depend on c [Läärä and Matthews, 1985; Engel, 1993]. Comparing these two approaches, notice that for the ordinal approach each observation consists of only two pieces of data: the (ordinal) time of the event and whether it was censored or not. Alternatively, in the dichotomous approach each survival time is represented as a vector of dichotomous indicators, where the size of the vector depends upon the timing of the event (or censoring). Thus, the ordinal approach can be easier to implement and offers savings in terms of the dataset size, especially as the number of timepoints gets large, while the dichotomous approach is superior in its treatment of time-dependent covariates and relaxing of the proportional hazards assumption.

10.2.3.1 Intraclass Correlation In the case of a random intercept model (*i.e.*, $z_i = 1_i$), it was previously noted that under the logistic formulation assuming normally distributed random effects, the estimated intraclass correlation equals $\hat{\sigma}_v^2/(\hat{\sigma}_v^2 + \pi^2/3)$, where the latter term in the denominator represents the variance of the underlying latent response tendency. For models, like the survival models presented in this section, using the complementary log–log link, the corresponding variance of the underlying latent response tendency equals $\pi^2/6$ [Agresti, 2002]. Thus, the estimated intraclass correlation is calculated as $\hat{\sigma}_v^2/(\hat{\sigma}_v^2 + \pi^2/6)$ under the complementary log–log link.

10.2.4 Estimation

For the ordinal models presented, the probability of a response in category c for a given level-2 unit i, conditional on the random effects $\boldsymbol{\theta}$ is equal to

$$\Pr(Y_{ij} = c \mid \boldsymbol{\theta}) = P_{ijc} - P_{ij,c-1}, \tag{10.14}$$

where $P_{ijc} = 1/\left[1 + \exp\left(-z_{ijc}\right)\right]$ under the logistic response function (formulas for other response functions are given in Section 9.5.3). Because $\gamma_0 = -\infty$ and $\gamma_C = \infty$, $P_{ij0} = 0$ and $P_{ijC} = 1$. Here, z_{ijc} denotes the response model, for example,

$$\log\left[\frac{P_{ijc}}{1 - P_{ijc}}\right] = z_{ijc} = \frac{\gamma_c - (\boldsymbol{x}'_{ij}\boldsymbol{\beta} + \boldsymbol{u}'_{ij}\boldsymbol{\alpha}_c + \boldsymbol{z}'_{ij}\boldsymbol{T}\boldsymbol{\theta}_i)}{\exp(\boldsymbol{w}'_{ij}\boldsymbol{\tau})} \qquad (c = 1, \ldots, C-1), \tag{10.15}$$

or one of the other variants of z_{ijc} given in this chapter. In what follows, the general model allowing for nonproportional odds will be considered, since the more restrictive proportional odds model is just a special case (*i.e.*, when $\boldsymbol{\alpha}_c = 0$ and $\boldsymbol{\tau} = 0$).

Let $\boldsymbol{Y}_i$ denote the vector of ordinal responses from subject i (for the n_i repeated observations nested within). The probability of any pattern $\boldsymbol{Y}_i$ conditional on $\boldsymbol{\theta}$ is equal to the product of the probabilities of the level-1 responses:

$$\ell(\boldsymbol{Y}_i \mid \boldsymbol{\theta}) = \prod_{j=1}^{n_i} \prod_{c=1}^{C} (P_{ijc} - P_{ij,c-1})^{y_{ijc}}, \tag{10.16}$$

where $y_{ijc} = 1$ if $Y_{ij} = c$ and 0 otherwise (*i.e.*, for each ijth observation, $y_{ijc} = 1$ for only one of the C categories). For the ordinal representation of the survival model, where right-censoring is present, the above likelihood is generalized to

$$\ell(\boldsymbol{Y}_i \mid \boldsymbol{\theta}) = \prod_{j=1}^{n_i} \prod_{c=1}^{C} \left[(P_{ijc} - P_{ij,c-1})^{d_{ij}} \left(1 - P_{ijc}\right)^{1-d_{ij}}\right]^{y_{ijc}}, \tag{10.17}$$

where $d_{ij} = 1$ if Y_{ij} represents an event, or $d_{ij} = 0$ if Y_{ij} represents a censored observation. Notice that (10.17) is equivalent to (10.16) when $d_{ij} = 1$ for all observations. With right-censoring, because there is essentially one additional response category (for those censored at the last category C), it is $\gamma_{C+1} = \infty$ and so $P_{ij,C+1} = 1$. In this case, parameters γ_c and $\boldsymbol{\beta}_c$ with $c = 1, \ldots, C$ are estimable; otherwise c only goes to $C - 1$. In what follows, we'll solve for the likelihood in (10.16); for the solution corresponding to (10.17) see Hedeker et al. [2000].

The marginal density of $\boldsymbol{Y}_i$ in the population is expressed as the following integral of the likelihood, $\ell(\cdot)$, weighted by the prior density $g(\cdot)$:

$$h(\boldsymbol{Y}_i) = \int_{\boldsymbol{\theta}} \ell(\boldsymbol{Y}_i \mid \boldsymbol{\theta})\, g(\boldsymbol{\theta})\, d\boldsymbol{\theta}, \tag{10.18}$$

where $g(\boldsymbol{\theta})$ represents the multivariate standard normal density. The marginal log-likelihood from the N level-2 units, $\log L = \sum_i^N \log h(\boldsymbol{Y}_i)$, is then maximized to yield maximum (marginal) likelihood estimates.

For this, denote the parameter vector as $\boldsymbol{\eta}' = \left[\gamma_c \,\vdots\, \boldsymbol{\alpha}_c \,\vdots\, \boldsymbol{\beta} \,\vdots\, \boldsymbol{\tau} \,\vdots\, \boldsymbol{T}\right]$. Also, note that the first derivatives equal the following:

$$\frac{\partial \log L}{\partial \boldsymbol{\eta}} = \sum_{i=1}^{N} [h(\boldsymbol{Y}_i)]^{-1} \int_{\boldsymbol{\theta}} \frac{\partial \ell_i}{\partial \boldsymbol{\eta}}\, g(\boldsymbol{\theta})\, d\boldsymbol{\theta}, \tag{10.19}$$

where

$$\ell_i = \ell(\boldsymbol{Y}_i \mid \boldsymbol{\theta}) = \prod_{j=1}^{n_i} \prod_{c=1}^{C} (p_{ijc})^{y_{ijc}}, \tag{10.20}$$

and

$$p_{ijc} = P_{ijc} - P_{ij,c-1}. \tag{10.21}$$

Furthermore,

$$\frac{\partial L_i}{\partial \boldsymbol{\eta}} = \frac{\partial(\exp \log L_i)}{\partial \boldsymbol{\eta}} = \sum_{j=1}^{n_i} \sum_{c=1}^{C} y_{ijc} \frac{\partial \log(p_{ijc})}{\partial \boldsymbol{\eta}} L_i$$

and

$$\frac{\partial \log(p_{ijc})}{\partial \boldsymbol{\eta}} = \frac{\partial p_{ijc}/\partial \boldsymbol{\eta}}{p_{ijc}} = \frac{\partial\left[\Psi(z_{ijc}) - \Psi(z_{ij,c-1})\right]/\partial \boldsymbol{\eta}}{\Psi(z_{ijc}) - \Psi(z_{ij,c-1})},$$

where, again, $\Psi(\cdot)$ denotes the logistic cdf.

By the chain rule:

$$\frac{\partial \Psi(z_{ijc})}{\partial \boldsymbol{\eta}} = \psi(z_{ijc}) \frac{\partial z_{ijc}}{\partial \boldsymbol{\eta}},$$

where $\psi(\cdot)$ is the logistic probability density function (pdf). As a result, for parameters whose derivatives do not vary with c, namely the regression coefficients $\boldsymbol{\beta}$ and the random-effects variance terms in $\boldsymbol{T}$, we get

$$\frac{\partial \log(p_{ijc})}{\partial \boldsymbol{\eta}} = \frac{\psi(z_{ijc}) - \psi(z_{ij,c-1})}{\Psi(z_{ijc}) - \Psi(z_{ij,c-1})} \frac{\partial z_{ijc}}{\partial \boldsymbol{\eta}},$$

where

$$\frac{\partial z_{ijc}}{\partial \boldsymbol{\beta}} = \frac{-\boldsymbol{x}_{ij}}{\exp(\boldsymbol{w}'_{ij}\boldsymbol{\tau})}, \qquad \frac{\partial z_{ijc}}{\partial(\mathrm{v}(\boldsymbol{T}))} = \frac{-\mathrm{J}_r(\boldsymbol{\theta} \otimes \boldsymbol{z}_{ij})}{\exp(\boldsymbol{w}'_{ij}\boldsymbol{\tau})},$$

$\mathrm{v}(\boldsymbol{T})$ is the vector containing the $r(r+1)/2$ unique elements of the Cholesky factor $\boldsymbol{T}$, and J_r is the transformation matrix of Magnus [1988] that eliminates the elements above the main diagonal, and $\otimes$ is the Kronecker product. If $\boldsymbol{T}$ is a $r \times 1$ vector multiplying a scalar random effect θ (*i.e.*, if $\boldsymbol{z}_{ij}$ is a $r \times 1$ vector of level-1 or level-2 grouping variables) then the numerator of the latter derivative is simply $\boldsymbol{z}_{ij}\theta$, while if a random intercept model is specified then the numerator is simply θ. Notice that to avoid the negative sign in the derivatives, one can write

$$\frac{\partial \log(p_{ijc})}{\partial \boldsymbol{\eta}} = \frac{\psi(z_{ij,c-1}) - \psi(z_{ijc})}{\Psi(z_{ijc}) - \Psi(z_{ij,c-1})} \left(-\frac{\partial z_{ijc}}{\partial \boldsymbol{\eta}}\right).$$

For parameters that vary with c, the $C-1$ thresholds γ_c and the corresponding threshold interaction terms $\boldsymbol{\alpha}_c$, we get

$$\frac{\partial \log(p_{ijc})}{\partial \boldsymbol{\eta}_{c'}} = \frac{\delta_{cc'}\, \psi(z_{ijc}) \frac{\partial z_{ijc}}{\partial \boldsymbol{\eta}_{c'}} - \delta_{c-1,c'}\, \psi(z_{ij,c-1}) \frac{\partial z_{ij,c-1}}{\partial \boldsymbol{\eta}_{c'}}}{\Psi(z_{ijc}) - \Psi(z_{ij,c-1})},$$

where

$$\delta_{cc'} = \begin{cases} 1 & \text{if } c = c' \\ 0 & \text{if } c \neq c' \end{cases}$$

with

$$\frac{\partial z_{ijc}}{\partial \gamma_{c'}} = \frac{1}{\exp(\boldsymbol{w}'_{ij}\boldsymbol{\tau})}, \quad \text{and} \quad \frac{\partial z_{ijc}}{\partial \boldsymbol{\alpha}_{c'}} = \frac{-\boldsymbol{u}_{ij}}{\exp(\boldsymbol{w}'_{ij}\boldsymbol{\tau})}.$$

Finally, for the scaling terms $\boldsymbol{\tau}$ note that $\partial \exp(\boldsymbol{w}'_{ij}\boldsymbol{\tau})/\partial \boldsymbol{\tau} = \exp(\boldsymbol{w}'_{ij}\boldsymbol{\tau})\, \boldsymbol{w}_{ij}$, and

$$\frac{\partial z_{ijc}}{\partial \boldsymbol{\tau}} = \frac{-\left[\exp(\boldsymbol{w}'_{ij}\boldsymbol{\tau})\, \boldsymbol{w}_{ij}\right]\left[\gamma_c - (\boldsymbol{u}'_{ij}\boldsymbol{\alpha}_c + \boldsymbol{x}'_{ij}\boldsymbol{\beta} + \boldsymbol{z}'_{ij}\boldsymbol{T}\boldsymbol{\theta}_i)\right]}{\left[\exp(\boldsymbol{w}'_{ij}\boldsymbol{\tau})\right]^2} = -\boldsymbol{w}'_{ij}\, z_{ijc}.$$

As a result,

$$\frac{\partial \log(p_{ijc})}{\partial \boldsymbol{\tau}} = \frac{\psi(z_{ijc}) \frac{\partial z_{ijc}}{\partial \boldsymbol{\tau}} - \psi(z_{ij,c-1}) \frac{\partial z_{ij,c-1}}{\partial \boldsymbol{\tau}}}{\Psi(z_{ijc}) - \Psi(z_{ij,c-1})}$$

$$= \frac{\psi(z_{ij,c-1}) z_{ij,c-1} - \psi(z_{ijc}) z_{ijc}}{\Psi(z_{ijc}) - \Psi(z_{ij,c-1})} \boldsymbol{w}_{ij}.$$

As in the binary case, Fisher's method of scoring can be used to provide the solution to these likelihood equations. Again, provisional estimates for the vector of parameters $\boldsymbol{\Theta}$, on iteration ι are improved by

$$\boldsymbol{\Theta}_{\iota+1} = \boldsymbol{\Theta}_{\iota} + \boldsymbol{I}(\boldsymbol{\Theta}_{\iota})^{-1} \frac{\partial \log L}{\partial \boldsymbol{\Theta}_{\iota}} \tag{10.22}$$

with

$$\boldsymbol{I}(\boldsymbol{\Theta}_{\iota}) = \sum_{i=1}^{N} h_i^{-2} \frac{\partial h_i}{\partial \boldsymbol{\Theta}_{\iota}} \left(\frac{\partial h_i}{\partial \boldsymbol{\Theta}_{\iota}} \right)'.$$

Also, as in the binary case described in the last chapter, numerical quadrature can be used to perform the integration over the random effects distribution $\boldsymbol{\theta}$.

10.3 PSYCHIATRIC EXAMPLE

For our first application of the ordinal mixed model, we will use the same data from the National Institute of Mental Health Schizophrenia Collaborative Study that was presented in Chapter 9. Here, though, we will analyze the outcome as an ordinal variable, specifically recoding the original seven ordered categories of the IMPS 79 severity score into four: (1) normal or borderline mentally ill, (2) mildly or moderately ill, (3) markedly ill, and (4) severely or among the most extremely ill.

Figures 9.6 and 9.7 present the observed proportions across the four primary study timepoints for the control and drug groups, respectively. As can be seen, the proportion in the normal category increase across time for both groups, though more dramatically for the drug group. Similarly, the proportion of responses in the most severe category diminish across time, especially for the drug group.

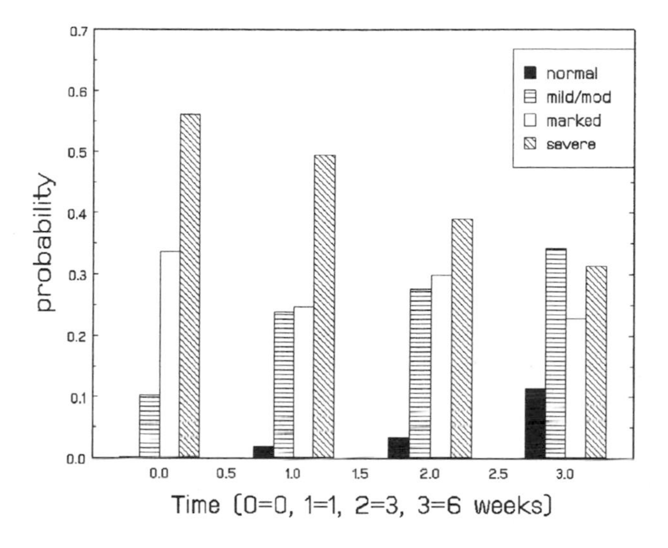

Figure 10.6. Ordinal proportions across time for the placebo group.

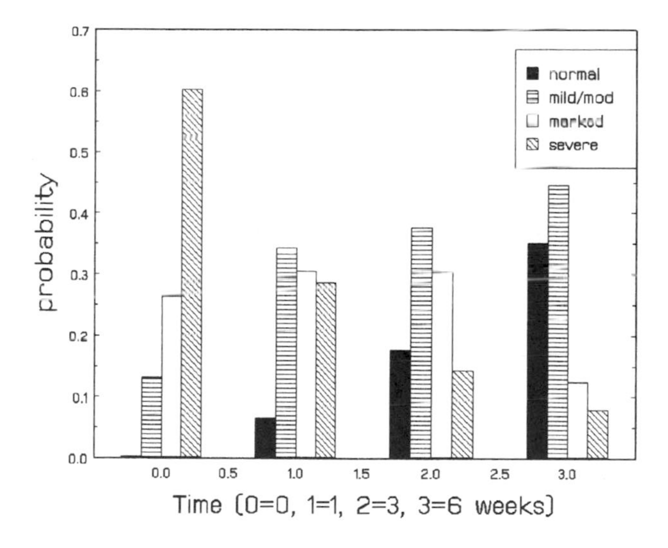

Figure 10.7. Ordinal proportions across time for the drug group.

To prepare for the analysis, these observed proportions are plotted in terms of the three cumulative logits that will be modeled. Figure 10.8 presents these three cumulative logits for the two treatment groups. As in Chapter 9, we will utilize the square root of week to linearize the relationship over time. These observed logits are formed in the following way. The first, in the top plot, expresses the log odds of a response in the severe category (category 4) relative to the non-severe category (categories 1, 2, or 3). The middle plot illustrates the log odds of a response in the marked or severe categories (category 3 or 4) relative to the lower two categories (1 or 2). Finally the lower plot has the log odds of a response in mild, marked, or severe categories (2, 3, or 4) relative to the normal category (1). This logit is "off the map" for the first timepoint because very few subjects were classified as normals at the baseline visit.

Since the proportional odds model posits that the covariate effects are the same across the cumulative logits, in terms of the plots in Figure 10.8, this would imply that the group trend lines are parallel across the three plots. That is to say, the lines for each group are parallel across the three plots. Of course, the drug and placebo lines can be different when contrasted to each other; the difference between these two is of primary interest in the analysis. A visual inspection of the proportional odds assumption is inconclusive (as most visual inspections of this generally are), since for each group the three trend lines look similar but not exactly the same. A statistical test of this assumption, below, will address this issue more conclusively.

We'll first consider a random intercept model for these data. Specifically, for subject i at timepoint j, the within-subjects model is (for the $C - 1$ cumulative logits):

$$\log\left[\frac{P_{ijc}}{1 - P_{ijc}}\right] = \gamma_c - [b_{0i} + b_{1i}\sqrt{Week_j}] \tag{10.23}$$

and the between-subjects model (for the $i = 1, \ldots, N$ subjects) is

$$\begin{aligned} b_{0i} &= \beta_0 + \beta_2 Drug_i + \upsilon_{0i}, \\ b_{1i} &= \beta_1 + \beta_3 Drug_i, \end{aligned} \tag{10.24}$$

where $Drug$ is coded 0 or 1 for placebo or drug patients, respectively. Also, the random subject effects υ_{0i} are assumed to be normally distributed in the population of subjects with 0 mean and variance σ_υ^2.

In this model, $0 - \beta_0$ represents the week 0 IMPS79 1st logit (*i.e.*, category 1 vs. 2–4), $\gamma_1 - \beta_0$ the week 0 IMPS79 2nd logit (1–2 vs. 3–4), and $\gamma_2 - \beta_0$ the week 0 IMPS79 3rd logit (1–3 vs. 4) for the placebo group. These terms reflect the response proportions in the four categories. In terms of the regression parameters, β_1 represents the weekly (in square root units) logit change for placebo patients, β_2 the difference in the week 0 logit for drug patients, and β_3 is the difference in the weekly (square root) logit change between drug and placebo groups. Note, that the latter two parameters represent drug differences relative to the placebo group. Finally, the random subject effect υ_{0i} represents the deviation of individual i from their group trend. Since this is only a random intercept model, this deviation is assumed to be the same across time. Also, the regression coefficients are conditional effects; they represent the effects of the regressors conditional on, or adjusted for, the level of the random effect. Table 10.2 lists the results of this analysis.

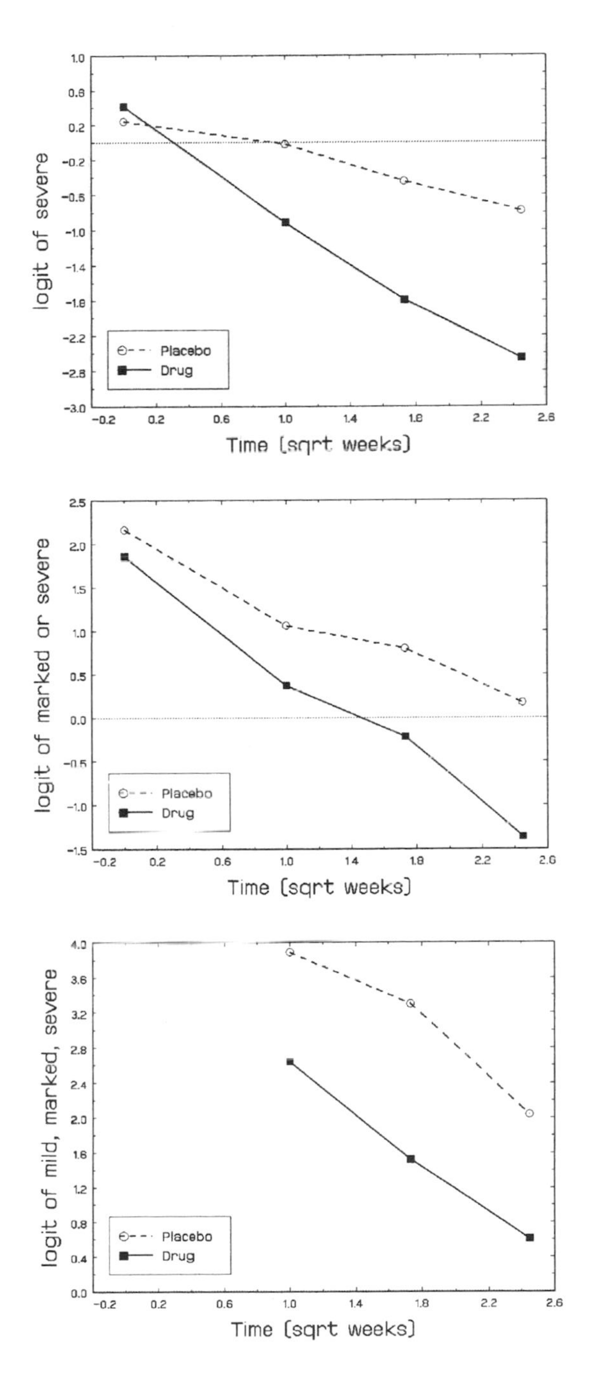

Figure 10.8. Observed cumulative logits across time by treatment group.

Table 10.2. NIMH Schizophrenia Collaborative Study: Severity of Illness Across Time (N = 437)—Random Intercept Ordinal Logistic Regression Results

	Estimate	SE	Z	$p <$
Intercept	5.858	0.343	17.08	.001
Threshold_2	3.033	0.132	22.91	.001
Threshold_3	5.152	0.179	28.74	.001
Drug (0 = placebo; 1 = drug)	−0.055	0.311	−0.18	.86
Time ($\sqrt{\text{week}}$)	−0.766	0.120	−6.39	.001
Drug by time	−1.206	0.133	−9.06	.001
Intercept sd	1.944	0.128		

$-2 \log L = 3402.72$

SE = standard error.

In terms of the fixed effects, the results indicate that the treatment groups do not significantly differ at baseline, the placebo group does improve over time, and that the drug group has greater improvement over time, relative to the placebo group. As one would expect, these conclusions agree with those from the analysis of the dichotomous version of this outcome in the last chapter.

In terms of the degree of subject heterogeneity, for this model the intraclass correlation (ICC), or in this case the intra-person correlation, is estimated as

$$r = \frac{1.944^2}{1.944^2 + \pi^2/3} = .53.$$

Thus, the responses are fairly highly correlated within subjects. For longitudinal data, this level of intra-person correlation would generally be expected, though the timing of the measurements plays a role. In general, measurement schedules with intervals closer in time (*e.g.*, weeks) yield higher ICCs than those with distal time intervals (*e.g.*, months).

Next, we consider a random intercept and trend model. For this, the within-subjects (level-1) model is unchanged, however, the between-subjects (level-2) model is

$$\begin{aligned} b_{0i} &= \beta_0 + \beta_2 Drug_i + v_{0i}, \\ b_{1i} &= \beta_1 + \beta_3 Drug_i + v_{1i}. \end{aligned} \tag{10.25}$$

This model posits subject heterogeneity in terms of both the intercept and time terms. Here the vector of random subject effects $\boldsymbol{v}_i$ is bivariate normal in the population with mean $\mathbf{0}$ and variance–covariance matrix $\boldsymbol{\Sigma}_v = \boldsymbol{T}\boldsymbol{T}'$, where $\boldsymbol{T}$ is the Cholesky factor. As noted, the above level-2 model is written in terms of the unstandardized random effects, namely $\boldsymbol{v}_i = \boldsymbol{T}\boldsymbol{\theta}_i$, for convenience and to agree with the representation in the continuous outcome models described in Chapters 4 and 5.

Putting the level-1 and level-2 models together yields (with D denoting $Drug$ and W denoting $Week$)

$$\log\left[\frac{P_{ijc}}{1-P_{ijc}}\right] =$$
$$\gamma_c - [\beta_0 + \beta_1\sqrt{W}_j + \beta_2 D_i + \beta_3(D_i \times \sqrt{W}_j) + \upsilon_{0i} + \upsilon_{1i}\sqrt{W}_j].$$

In this random intercept and trend model, the Cholesky factor equals

$$\boldsymbol{T} = \begin{bmatrix} \sigma_{\upsilon_0} & 0 \\ \frac{\sigma_{\upsilon_{01}}}{\sigma_{\upsilon_0}} & \left(\sigma^2_{\upsilon_1} - \frac{\sigma^2_{\upsilon_{01}}}{\sigma^2_{\upsilon_0}}\right)^{1/2} \end{bmatrix},$$

thus, the relationship between the unstandardized and standardized random effects is

$$\upsilon_{0i} = \sigma_{\upsilon_0}\theta_{0i} \qquad (10.26)$$

$$\upsilon_{1i} = \frac{\sigma_{\upsilon_{01}}}{\sigma_{\upsilon_0}}\theta_{0i} + \left(\sigma^2_{\upsilon_1} - \frac{\sigma^2_{\upsilon_{01}}}{\sigma^2_{\upsilon_0}}\right)^{1/2}\theta_{1i}. \qquad (10.27)$$

The results of this analysis are listed in Table 10.3. Comparing this model to the previous one via a likelihood ratio test to assess whether the random trends are significant (*i.e.*, $H_0 : \sigma^2_{\upsilon_1} = \sigma_{\upsilon_0\upsilon_1} = 0$) yields $X^2_2 = 77.90, p < .001$. Clearly there is strong evidence of subject heterogeneity in the the time trends.

Table 10.3. NIMH Schizophrenia Collaborative Study: Severity of Illness Across Time (N = 437)—Random Intercept and Trend Ordinal Logistic Regression Results

	Estimate	SE	Z	$p <$
Intercept	7.309	0.484	15.10	.001
Threshold$_2$	3.912	0.214	18.29	.001
Threshold$_3$	6.528	0.291	22.43	.001
Drug (0=placebo, 1 = drug)	0.111	0.402	0.28	.78
Time ($\sqrt{\text{week}}$)	−0.875	0.236	−3.71	.001
Drug by time	−1.724	0.270	−6.40	.001
Cholesky				
Intercept sd	2.669	0.256	$(\hat{\sigma}^2_{\upsilon_0} = 7.13)$	
Intercept–time covariance	−0.588	0.159	$(r_{\upsilon_0\upsilon_1} = -.41)$	
Time sd	1.308	0.126	$(\hat{\sigma}^2_{\upsilon_1} = 2.06)$	
$-2\log L = 3324.82$				

SE = standard error.

In terms of the fixed effects, the conclusions are the same as before (*i.e.*, from the random intercept model of Table 10.2). What is interesting here, however, is the additional degree of subject heterogeneity in responses across time that is indicated by this analysis. Notice that the estimated variance in the time trends is $\hat{\sigma}^2_{\upsilon_1} = 2.06$. Thus, while the average trends for the placebo and drug groups are estimated as $-.875$ and $-.875 - 1.724 = -2.599$, respectively, there is quite a bit of variance in these trends. For example, for the population of placebo subjects approximately 95% of subjects have slopes in the interval $-.875 \pm (1.96 \times \sqrt{2.06}) = -3.69$ to 1.94. Similarly, for the population of drug subjects, the interval is calculated as $-2.599 \pm (1.96 \times \sqrt{2.06}) = -5.42$ to $.21$. Notice that both intervals include negative and positive slopes reflecting the wide heterogeneity in trends for both subject groups. Finally, the covariance between the intercept and linear trend is negative; expressed as a correlation it equals $-.41$, which is moderately large. This suggests that patients who are initially more severely ill (*i.e.*, greater intercepts) improve at a greater rate (*i.e.*, more pronounced negative slopes). An alternative explanation, though, is that of a floor effect due to the IMPS79 rating scale. Namely, patients with lower initial scores have a more limited range for improving their scores than those with higher initial scores.

To get a further sense of the subject heterogeneity in time trends for this sample, Figures 10.9 and 10.10 depict the empirical Bayes estimated trends for the probability of being in the first category (normal or borderline ill) across time for each of the control and drug patients, respectively.

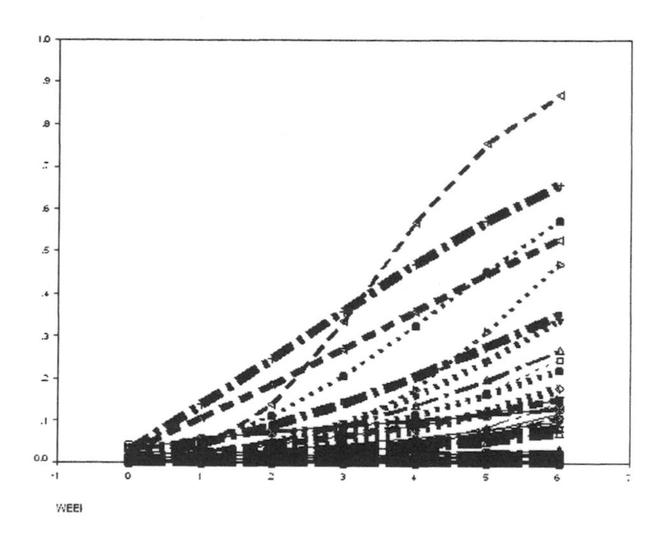

Figure 10.9. Estimated probabilities across time for placebo patients.

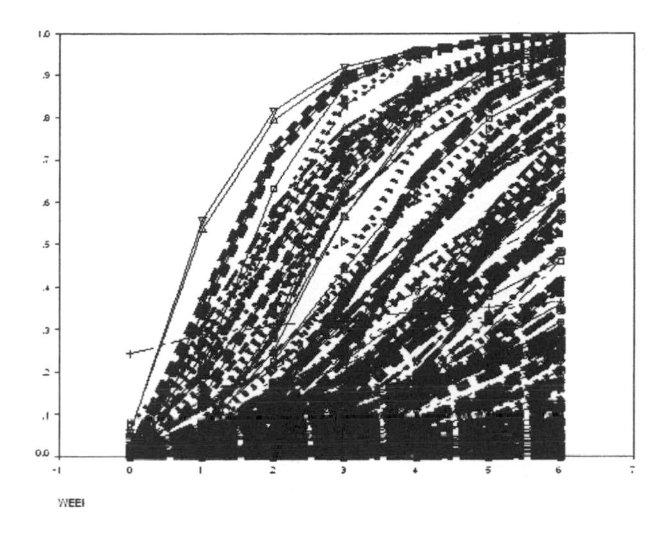

Figure 10.10. Estimated probabilities across time for drug patients.

These trends are generated using the estimated regression coefficients from Table 10.3 and the logistic cdf formula (noting that $\gamma_1 = 0$ by definition):

$$\begin{aligned} &Pr(Y_{ij} = 1) = \\ &\quad 1/\{1 + \exp[-(\gamma_1 - (\boldsymbol{x}'_{ij}\hat{\boldsymbol{\beta}} + \boldsymbol{z}'_{ij}\hat{\boldsymbol{v}}))]\} = \\ &\quad 1/\{1 + \exp[7.31 + .11\, D_i - .88\sqrt{W_j} - 1.72\, D_i \times \sqrt{W_j} + \hat{v}_{0i} + \hat{v}_{1i}\sqrt{W_j}]\}, \end{aligned}$$

with values for each subject's group, $Drug$ (0 or 1) denoted as D, the seven values of $Week$ (0 to 6) denoted as W, and the empirical Bayes estimates $\hat{v}_{0i}$ and $\hat{v}_{1i}$ for each subject. As these figures reveal, though some placebo patients do improve across time, not many have too great a probability of being in the normal category by the end of the study. For the drug group there is considerable heterogeneity, though many subjects have a good chance of being normal by the last timepoint. Similar plots can be generated for the other response categories using

$$\begin{aligned} Pr(Y_{ij} = 2) &= 1/\{1 + \exp[-(\hat{\gamma}_2 - (\boldsymbol{x}'_{ij}\hat{\boldsymbol{\beta}} + \boldsymbol{z}'_{ij}\hat{\boldsymbol{v}}))]\} - Pr(Y_{ij} = 1), \\ Pr(Y_{ij} = 3) &= 1/\{1 + \exp[-(\hat{\gamma}_3 - (\boldsymbol{x}'_{ij}\hat{\boldsymbol{\beta}} + \boldsymbol{z}'_{ij}\hat{\boldsymbol{v}}))]\} - Pr(Y_{ij} = 2), \\ Pr(Y_{ij} = 4) &= 1 - 1/\{1 + \exp[-(\hat{\gamma}_3 - (\boldsymbol{x}'_{ij}\hat{\boldsymbol{\beta}} + \boldsymbol{z}'_{ij}\hat{\boldsymbol{v}}))]\}. \end{aligned}$$

The above subject-specific plots are very useful for conveying the degree of differential response across time at the individual level. Ultimately, though, one might also be interested in how well the model fits the observed marginal proportions. For this, Figure 10.11 displays the marginalized model fit of the observed proportions for this ordinal outcome across time by treatment group. As can be seen, the fit appears quite reasonable.

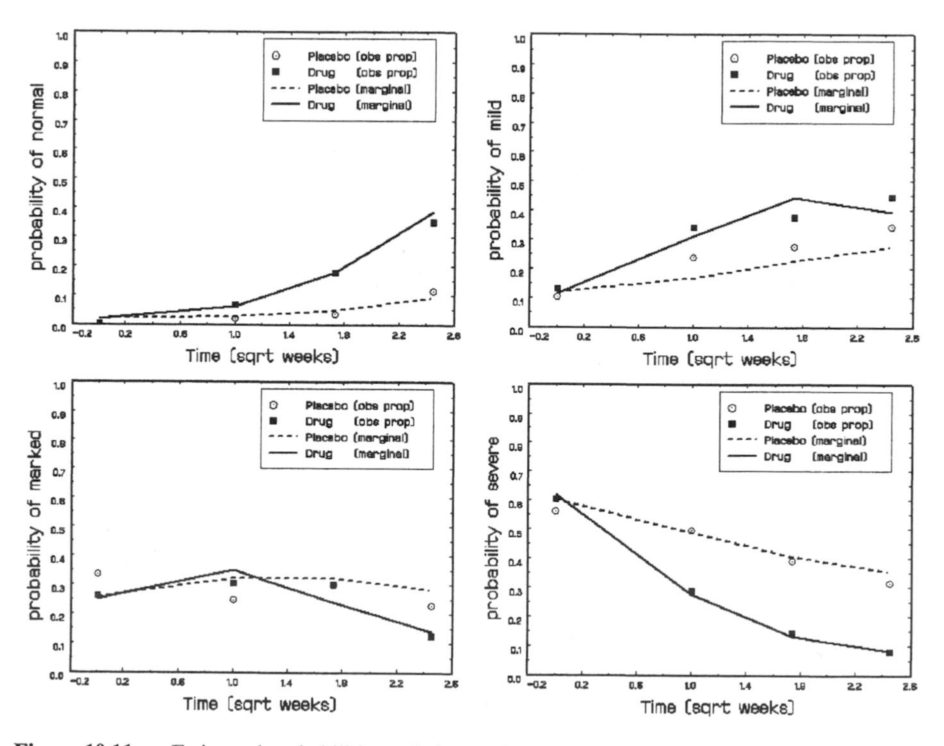

Figure 10.11. Estimated probabilities and observed proportions across time by treatment group.

To obtain these marginal probability estimates, Tables 10.4 and 10.5 presents SAS PROC IML code based on the parameter estimates in Table 10.3. In Table 10.4, the covariate matrices for the two treatment groups are defined as `x0` and `x1`, respectively, and the common random-effects design matrix as `zmat`. The estimated parameter values are given for the model intercept `int`, Cholesky factor of the random-effects variance covariance matrix `chol`, regression coefficients `beta`, and thresholds `thresh`. The marginalization vector `sdy` is derived following the description in Section 9.7.2. Table 10.5 then lists the code to obtain the marginalized cumulative logits for the two treatment groups: `za0`, `zb0`, and `zc0` for the placebo group; and `za1`, `zb1`, and `zc1` for the drug group. The remaining lines translate these cumulative logits to cumulative probabilities, using the logistic cdf formula. Finally, the estimated marginalized category probabilities for the two groups are printed out. As in previous IML examples, several SAS PROC IML functions are utilized including `J` (matrix initialization), `diag` (construction of a diagonal matrix), `T` (matrix transpose), and `vecdiag` (extraction of the diagonal of a matrix).

Finally, allowing for the covariate effects to vary across the three cumulative logits of the model provides a way to assess whether the proportional odds assumption is reasonable for these data. In the present case, there are three covariates: drug, time, and the drug by time interaction. A nonproportional odds model allows these three covariates to have separate effects across the three cumulative logits, estimating a total of nine covariate effects. Fitting this model (not shown) yields $-2 \log L = 3321.11$, and so a likelihood ratio test of the proportional odds assumption is $X^2 = 3324.82 - 3321.11 = 3.71$, which is not significant on 6 degrees of freedom. The degrees of freedom equal 6 because of the difference in estimated covariate parameters: 3 for the proportional odds model and 9 for the nonproportional odds model. Thus, for these data the proportional odds model is reasonable.

Table 10.4. SAS PROC IML Code: Computing Marginal Ordinal Probabilities

```
PROC IML;
/* Results from random intercept & trend model */;
x0     =  {   0 0.00000 0,
              0 1.00000 0,
              0 1.73205 0,
              0 2.44949 0};
x1     =  {   1 0.00000 0.00000,
              1 1.00000 1.00000,
              1 1.73205 1.73205,
              1 2.44949 2.44949};
int       =  {7.309};
chol      =  {2.669 0,
               -.588 1.308};
beta      =  { .111, -.875, -1.724};
thresh    =  {3.912, 6.528};
/* Approximate Marginalization Method */;
pi     =   3.141592654;
nt     =   4;
ivec   =   J(nt,1,1);
zmat   =   {1 0.00000,
            1 1.00000,
            1 1.73205,
            1 2.44949};
evec   =   (15/16)**2 * (pi**2)/3 * ivec;
/* nt by nt matrix with evec on the diagonal and zeros elsewhere */;
emat = diag(evec);
/* variance--covariance matrix of underlying latent variable */;
vary = zmat * chol * T(chol) * T(zmat) + emat;
sdy = sqrt(vecdiag(vary) / vecdiag(emat));
```

Table 10.5. SAS PROC IML Code: Computing Marginal Ordinal Probabilities (Continued)

```
za0 = (0 - (int + x0*beta)) / sdy ;
zb0 = (thresh[1] - (int + x0*beta)) / sdy;
zc0 = (thresh[2] - (int + x0*beta)) / sdy;
za1 = (0 - (int + x1*beta)) / sdy;
zb1 = (thresh[1] - (int + x1*beta)) / sdy;
zc1 = (thresh[2] - (int + x1*beta)) / sdy;
grp0a = 1 / ( 1 + EXP(0 - za0));
grp0b = 1 / ( 1 + EXP(0 - zb0));
grp0c = 1 / ( 1 + EXP(0 - zc0));
grp1a = 1 / ( 1 + EXP(0 - za1));
grp1b = 1 / ( 1 + EXP(0 - zb1));
grp1c = 1 / ( 1 + EXP(0 - zc1));
print 'Random intercept and trend model';
print 'Approximate Marginalization Method';
print 'marginal prob for group 0 - catg 1' grp0a [FORMAT=8.4];
print 'marginal prob for group 0 - catg 2' (grp0b-grp0a) [FORMAT=8.4];
print 'marginal prob for group 0 - catg 3' (grp0c-grp0b) [FORMAT=8.4];
print 'marginal prob for group 0 - catg 4' (1-grp0c) [FORMAT=8.4];
print 'marginal prob for group 1 - catg 1' grp1a [FORMAT=8.4];
print 'marginal prob for group 1 - catg 2' (grp1b-grp1a) [FORMAT=8.4];
print 'marginal prob for group 1 - catg 3' (grp1c-grp1b) [FORMAT=8.4];
print 'marginal prob for group 1 - catg 4' (1-grp1c) [FORMAT=8.4];
```

10.4 HEALTH SERVICES RESEARCH EXAMPLE

An aim of the McKinney Homeless Research Project (MHRP) study [Hough et al., 1997; Hurlburt et al., 1996] in San Diego, CA was to evaluate the effectiveness of using section 8 certificates as a means of providing independent housing to the severely mentally ill homeless. Section 8 housing certificates were provided from the Department of Housing and Urban Development (HUD) to local housing authorities in San Diego. These housing certificates, which require clients to pay 30% of their income toward rent, are designed to make it possible for low income individuals to choose and obtain independent housing in the community. Three hundred sixty-one clients took part in this longitudinal study employing a randomized factorial design. Clients were randomly assigned to one of two types of supportive case management (comprehensive vs. traditional) and to one of two levels of access to independent housing (using section 8 certificates). Eligibility for the project was restricted to individuals diagnosed with a severe and persistent mental illness who were either homeless or at high risk of becoming homeless at the start of the study. Individuals' housing status was classified at baseline and at 6, 12, and 24 month follow-ups.

In this illustration, focus will be on examining the effect of access to section 8 certificates on repeated housing outcomes across time. At each timepoint the housing status of each subject was classified as either streets/shelters, community housing, or independent housing. This outcome can be thought of as ordinal with increasing categories indicating improved housing outcomes. The observed sample sizes and response proportions for these three outcome categories by group are presented in Table 10.6.

Table 10.6. Housing Status Across Time by Group: Response Proportions and Sample Sizes

Group	Status	Timepoint			
		Baseline	6 months	12 months	24 months
Control	Street	.555	.186	.089	.124
	Community	.339	.578	.582	.455
	Independent	.106	.236	.329	.421
	N	180	161	146	145
Section 8	Street	.442	.093	.121	.120
	Community	.414	.280	.146	.228
	Independent	.144	.627	.732	.652
	N	181	161	157	158

These observed proportions suggest a general decrease in street living and an increase in independent living across time for both groups. The increase in independent housing, however, appears to occur sooner for the section 8 group relative to the control group. Regarding community living, across time this increases for the control group and decreases for the section 8 group.

There are some missing data across time; attrition rates of 19.4% and 12.7% are observed at the final timepoint for the control and section 8 groups, respectively. Since estimation is based on a full-likelihood approach, the missing data are assumed to be "ignorable" conditional on both the model covariates and the observed housing outcomes [Laird, 1988]. As previously noted, ignorable nonresponse falls under the "missing at random" (MAR) assumption introduced by Rubin [1976], in which the missingness depends only on observed data. In what follows, because the focus is on describing application of the model, we will make the MAR assumption. Further approaches, however, which do not rely on the MAR assumption are described in detail in Chapter 14, which describes missing data models.

To prepare for the ordinal analyses, the observed cumulative logits across time for the two groups are plotted in Figures 10.12 and 10.13. The first cumulative logit compares independent and community housing versus street living (*i.e.*, categories 2 & 3 combined versus 1), while the second cumulative logit compares independent housing versus community housing and street living (*i.e.*, category 3 versus 2 and 1 combined).

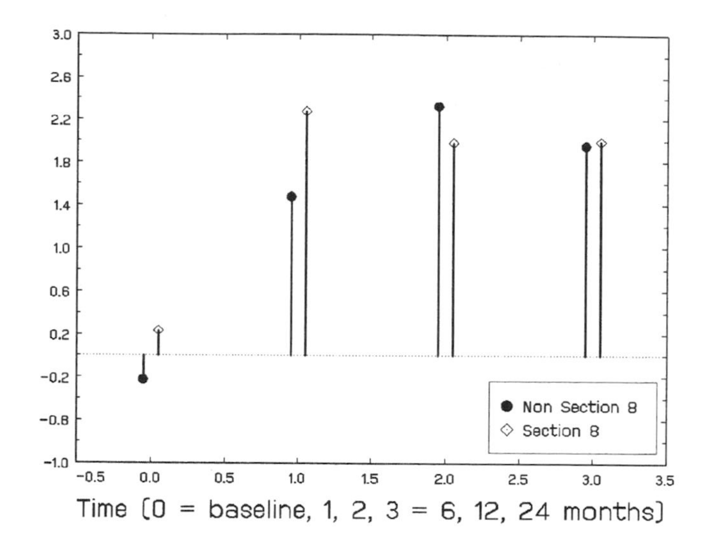

Figure 10.12. First cumulative logit values across time by group.

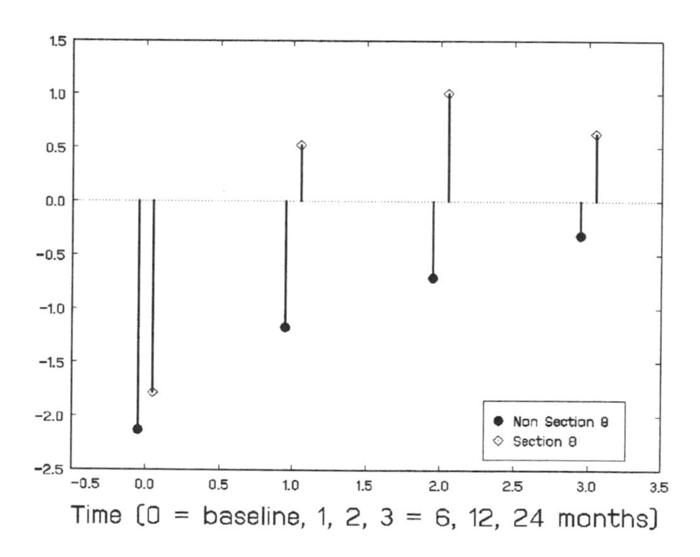

Figure 10.13. Second cumulative logit values across time by group.

For the proportional odds model to hold, these two plots should look approximately the same, with the only difference being the scale difference on the y axis. As can be seen, these plots do not look that similar. For example, the post-baseline group differences do not appear to be the same for the two cumulative logits. In particular, it appears that the section 8 group does better more consistently in terms of the second cumulative logit (*i.e.*, independent versus community and street housing). This suggests that the proportional odds model may not be reasonable for these data.

To assess this more rigorously, two mixed-effects ordinal logistic models were fit to these data, the first assuming a proportional odds model and the second relaxing this assumption. For both analyses, the repeated housing status classifications were modeled in terms of time effects (6, 12, and 24 month follow-ups compared to baseline), a group effect (section 8 versus control), and group by time interaction terms. The first analysis assumes these effects are the same across the two cumulative logits of the model, whereas the second analysis estimates effects for each explanatory variable on each of the two cumulative logits. For both, only a random subject effect was included to account for the longitudinal responses. Results from these analyses are given in Table 10.7.

Table 10.7. Housing Status Across Time: Random Intercept Ordinal Logistic Regression Results

	Proportional Odds		Nonproportional Odds			
			Non-street[a]		Independent[b]	
Term	Estimate	SE	Estimate	SE	Estimate	SE
Intercept	−.220	.203	−.322	.218		
Threshold	**2.744**	.110			**2.377**	.279
t1 (6 month vs. base)	**1.736**	.233	**2.297**	.298	**1.079**	.358
t2 (12 month vs. base)	**2.315**	.268	**3.345**	.450	**1.645**	.336
t3 (24 month vs. base)	**2.499**	.247	**2.821**	.369	**2.145**	.339
Section 8 (yes = 1, no = 0)	*.497*	.280	*.592*	.305	.323	.401
Section 8 by t1	**1.408**	.334	.566	.478	**2.023**	.478
Section 8 by t2	**1.173**	.360	−.958	.582	**2.016**	.466
Section 8 by t3	*.638*	.331	−.366	.506	**1.073**	.472
Subject sd	**1.459**	.106	**1.457**[c]	.112	**1.457**[c]	.112
$-2 \log L$	2274.39		2222.25			

SE = standard error, **bold** indicates $p < .05$, *italic* indicates $.05 < p < .10$.
[a] logit comparing independent and community housing vs. street.
[b] logit comparing independent housing vs. community housing and street.
[c] these are constrained to be equal in the model.

The proportional odds model indicates significant time effects for all timepoints relative to baseline, but only significant group by time interactions for the 6 and 12 month follow-ups. Marginally significant effects are obtained for the section 8 effect and the section 8 by t3 (24-months) interaction. Thus, the analysis indicates that the control group moves away from street living to independent living across time, and that this improvement is more pronounced for section 8 subjects at the 6 and 12 month follow-up. Because the section 8 by t3 interaction is only marginally significant, the groups do not differ significantly in housing status at the 24-month follow-up as compared to baseline.

However, comparing log-likelihood values clearly rejects the proportional odds assumption (likelihood ratio $X^2_7 = 52.14, p < .001$) indicating that the effects of the explanatory variables cannot be assumed identical across the two cumulative logits. Interestingly, none of the section 8 by time interaction terms are significant in terms of the non-street logit (*i.e.*, comparing independent and community housing combined versus street), while all of these interactions are significant in terms of the independent logit (*i.e.*, comparing independent housing versus community housing and street combined). Thus, as compared to baseline, section 8 subjects are more likely to be in independent housing at all follow-up timepoints, relative to the control group.

In terms of the random subject effect, it is clear that the data are correlated within subjects. Expressed as an intraclass correlation, the attributable variance at the subject-level equals .39 for both models. Specifically, for the proportional odds model it is calculated as $1.459^2/(1.459^2 + \pi^2/3)$, while for the nonproportional odds model it is $1.457^2/(1.457^2 + \pi^2/3)$. It should be noted that the latter model does not allow the random-effect variance term (*i.e.*, the subject sd) to vary across the cumulative logits. Thus, the estimated subject sd of 1.457 pertains to both cumulative logits in this latter model. In terms of significance, the Wald test easily rejects the null hypothesis that the (subject) population standard deviation equals zero. Strictly speaking, as noted by Bryk and Raudenbush [1992] and others, this test is not to be relied upon, especially as the population variance is close to zero. In the present case, the actual significance test is not critical because it is more or less assumed that the population distribution of subject effects will not have zero variance.

10.5 SUMMARY

Mixed-effects ordinal logistic regression models are useful for analysis of longitudinal outcomes with more than two response categories, when the categories have a natural ordering. By and large, the models are seen as extensions of the mixed-effects logistic regression model for binary response data. The ordinal model uses cumulative dichotomizations of the categorical outcome.

Mixed-effects models are useful for describing both population and individual trends for longitudinal data. The analysis of the schizophrenia dataset illustrated how both types of trends could be generated and what each indicates. In particular, the individual trends give a sense of the degree of heterogeneity of response that exists, and thus can point to the influence of subjects on their trends. Aggregating over individuals to produce marginal results is also possible, though additional work is necessary, as has been shown.

For ordinal data, both proportional odds and nonproportional odds models were considered. Since, as noted by Peterson and Harrell [1990], examples of nonproportional odds are not difficult to find, the latter models are especially attractive for analyzing ordinal outcomes. In the homelessness example presented in this chapter, the nonproportional odds model provided more specific information about the effect of section 8 certificates. Namely, as compared to baseline, these certificates were effective in increasing independent housing (versus community housing and street living combined) at all follow-up timepoints. Interestingly, the same could not be said when comparing independent and community housing combined versus street living. Thus, the use of the nonproportional odds model was helpful in elucidating a more focused analysis of the effect of the section 8 program.

As there is more-readily available software for continuous data, it is important to note advantages of using the ordinal model presented in this chapter, rather than simply analyzing ordinal responses as continuous outcomes. As is well known, the probit or logistic specifications take into account the ceiling and floor effects of the dependent variable, whereas linear models for measurement data clearly do not. As McKelvey and Zavoina [1975] point out, due to the ceiling and floor effects of the dependent variable, values of the residuals and regressors will be correlated when linear models for measurement data are applied to ordinal outcomes, which can result in biased estimates of the regression coefficients. Furthermore, as Winship and Marc [1984] note, the advantage of ordinal regression models in accounting for ceiling and floor effects of the dependent variable is most critical when the dependent variable is highly skewed, or when groups, defined by different covariate values, are compared which have widely varying skewness in the dependent variable.

CHAPTER 11

MIXED-EFFECTS REGRESSION MODELS FOR NOMINAL DATA

When the response measure of interest has more than two categories that have no particular natural ordering (*e.g.*, republican, democrat, undecided; or inpatient, outpatient, emergency room health service utilization), the outcome represents a "nominal" response variable. In some cases, a plausible ordering of the categories may be hypothesized, and ordinal and nominal regression models are compared to determine which provides a better fit to the observed proportions. There are two primary differences between ordinal and nominal regression models. First, for the proportional odds ordinal regression model, a single set of coefficients relate the predictors to all categories of the ordinal response variable, whereas in the nominal model $C-1$ sets of predictors are estimated that describe unique relationships between the predictors and $C-1$ contrasts between the nominal response categories. Note that for nonproportional odds ordinal regression models, this distinction is blurred to the extent that those predictors that interact with the thresholds also have category-specific effects. Second, in the case of mixed-effects ordinal and nominal regression models, each category has its own random-effect variance under the nominal model, but a common random-effect variance under the ordinal model. As such, the nominal model is a more general model than the ordinal model which imposes order restrictions on the categorical outcome measure of interest.

For nominal responses, there have been some developments in terms of multilevel models as well. An early example is the item response theory model for nominal educational test data described by Bock [1972]. This model includes a random effect for the level-2 subjects and fixed item parameters for the level-1 item responses nested within subjects. While Bock's model is a full-information maximum likelihood approach, using Gauss–Hermite

Longitudinal Data Analysis. By Donald Hedeker and Robert D. Gibbons

quadrature to integrate over the random-effects distribution, it doesn't include covariates or multiple random effects. More general regression models for multilevel nominal data have been considered by Goldstein [1995, Chapter 7], Daniels and Gatsonis [1997], and Revelt and Train [1998], though these approaches use either more approximate or Bayesian methods to handle the integration over the random effects. Also, these models generally adopt a reference cell approach for modeling the nominal response variable in which one of the categories is chosen as the reference cell and parameters are characterized in terms of the remaining $C-1$ comparisons to this reference cell. Bock's model, alternatively, was written in terms of any set of $C-1$ comparisons across the nominal response categories. This approach for the category comparisons was utilized in an article by Hedeker [2003]. Finally, Hartzel et al. [2001] synthesizes much of the work in this area, describing a general mixed-effects model for both clustered ordinal and nominal responses.

This chapter describes multilevel models for nominal data that accommodate multiple random effects and allow for a general form for model covariates. The model is referred to as a mixed-effects multinomial regression model, where it is assumed that the nominal response measure has a multinomial distribution in the population of subjects. This assumption is quite general in that any set of potential response proportions is permissible and no order restrictions are imposed. For nominal response data, models using both reference cell and more general category comparisons are described. To illustrate application of the various mixed models for nominal responses, analysis of a longitudinal psychiatric dataset is described. Finally, we discuss use of the model in competing risk survival analysis and illustrate the application with an example from solid organ transplantation.

11.1 MIXED-EFFECTS MULTINOMIAL REGRESSION MODEL

Let Y_{ij} denote the value of the nominal variable associated with level-2 unit i and level-1 unit j. Adding random effects to the fixed-effects multinomial logistic regression model [Agresti, 2002; Long, 1997], the probability that $Y_{ij} = c$ (a response occurs in category c) for a given level-2 unit i, conditional on the random effects $\boldsymbol{\theta}$ is given by

$$p_{ijc} = \Pr(Y_{ij} = c \mid \boldsymbol{\theta}) = \frac{\exp(z_{ijc})}{1 + \sum_{h=2}^{C} \exp(z_{ijh})} \quad \text{for } c = 2, 3, \ldots, C, \quad (11.1)$$

$$p_{ij1} = \Pr(Y_{ij} = 1 \mid \boldsymbol{\theta}) = \frac{1}{1 + \sum_{h=2}^{C} \exp(z_{ijh})}, \quad (11.2)$$

where the multinomial logit $z_{ijc} = \boldsymbol{x}'_{ij}\boldsymbol{\beta}_c + \boldsymbol{z}'_{ij}\boldsymbol{T}_c\boldsymbol{\theta}_j$. Comparing this to the logit for ordered responses, we see that the covariate effects $\boldsymbol{\beta}_c$ vary across categories ($c = 2, 3, \ldots, C$), as do the random-effect variance terms $\boldsymbol{T}_c$. As written above, an important distinction between the model for ordinal and nominal responses is that the former uses cumulative comparisons of the categories whereas the latter uses comparisons to a reference category.

This model generalizes Bock's model for educational test data [Bock, 1972] by including covariates $\boldsymbol{x}_{ij}$, and by allowing a general random-effects design vector $\boldsymbol{z}_{ij}$ including the possibility of multiple random effects $\boldsymbol{\theta}_j$. As discussed by Bock [1972], the model has a plausible interpretation. Namely, each nominal category is assumed to be related to an underlying latent "response tendency" for that category. The category c associated with the response variable Y_{ij} is then the category for which the response tendency is maximal.

These latent response tendencies are assumed to be independently distributed following approximately normal distributions (*i.e.*, logistic distributions due to the logistic regression formulation). Notice that this assumption of C latent variables differs from the ordinal model where only one underlying latent variable is assumed. Bock [1975] refers to the former as the extremal concept and the latter as the aforementioned threshold concept, and he notes that both were introduced into psychophysics by Thurstone [1927]. The two are equivalent only for the dichotomous case (*i.e.*, when there are only two response categories).

The model as written above allows estimation of any pairwise comparisons among the C response categories. As characterized in Bock [1972], it is beneficial to write the nominal model to allow for any possible set of $C - 1$ contrasts. For this, the category probabilities are written as

$$p_{ijc} = \frac{\exp(z_{ijc})}{\sum_{h=1}^{C} \exp(z_{ijh})} \quad \text{for } c = 1, 2, \ldots, C, \tag{11.3}$$

where now

$$z_{ijc} = \boldsymbol{x}'_{ij}\boldsymbol{\Gamma}\boldsymbol{d}_c + (\boldsymbol{z}'_{ij} \otimes \boldsymbol{\theta}'_i)\mathrm{J}'_{r^*}\boldsymbol{\Lambda}\boldsymbol{d}_c. \tag{11.4}$$

Here, $\boldsymbol{D}$ is the $(C-1) \times C$ matrix containing the contrast coefficients for the $C-1$ contrasts between the C logits and $\boldsymbol{d}_c$ is the cth column vector of this matrix. The $p \times (C - 1)$ parameter matrix $\boldsymbol{\Gamma}$ contains the regression coefficients associated with the p covariates for each of the $C - 1$ contrasts. Similarly, $\boldsymbol{\Lambda}$ contains the random-effect variance parameters for each of the $C - 1$ contrasts. Specifically,

$$\boldsymbol{\Lambda} = [\ \mathrm{v}(\boldsymbol{T}_1) \quad \mathrm{v}(\boldsymbol{T}_2) \quad \ldots \quad \mathrm{v}(\boldsymbol{T}_{C-1})\ ],$$

where $\mathrm{v}(\boldsymbol{T}_c)$ is the $r^* \times 1$ vector ($r^* = r[r+1]\,/\,2$) of elements below and on the diagonal of the Cholesky (lower-triangular) factor $\boldsymbol{T}_c$, and J_{r^*} is the transformation matrix of Magnus [1988], which eliminates the elements above the main diagonal. This latter matrix is necessary to ensure that the appropriate terms from the $1 \times r^2$ vector resulting from the Kronecker product $(\boldsymbol{z}'_{ij} \otimes \boldsymbol{\theta}'_i)$ are multiplied with the $r^* \times 1$ vector resulting from $\boldsymbol{\Lambda}\boldsymbol{d}_c$. For the case of a random intercept model, the model simplifies to

$$z_{ijc} = \boldsymbol{x}'_{ij}\boldsymbol{\Gamma}\boldsymbol{d}_c + \boldsymbol{\Lambda}\boldsymbol{d}_c\theta_i, \tag{11.5}$$

with $\boldsymbol{\Lambda}$ as the $1 \times (C - 1)$ vector, namely, $\boldsymbol{\Lambda} = [\ \sigma_1 \quad \sigma_2 \quad \ldots \quad \sigma_{C-1}\ ]$.

Notice that if $\boldsymbol{D}$ equals

$$\boldsymbol{D} = \begin{bmatrix} 0 & 1 & 0 & \ldots & 0 \\ 0 & 0 & 1 & \ldots & 0 \\ \cdot & \cdot & \cdot & \cdots & \cdot \\ 0 & 0 & 0 & \ldots & 1 \end{bmatrix}$$

the model simplifies to the earlier representation in (11.1) and (11.2). The current formulation, however, allows for a great deal of flexibility in the types of comparisons across the C response categories. For example, if the categories are ordered, an alternative to the cumulative logit model of the previous chapter is to employ Helmert contrasts [Bock, 1975] within the nominal model. For this, with $C = 4$, the following contrast matrix would be used:

$$\boldsymbol{D} = \begin{bmatrix} -\frac{3}{4} & \frac{1}{4} & \frac{1}{4} & \frac{1}{4} \\ 0 & -\frac{2}{3} & \frac{1}{3} & \frac{1}{3} \\ 0 & 0 & -\frac{1}{2} & \frac{1}{2} \end{bmatrix}$$

with the scale of each contrast set to equal unity in terms of the difference of contrast coefficients. Helmert contrasts of the logits are similar to the comparisons within continuation-ratio logit models, as described within a mixed model formulation by Ten Have and Uttal [1994]. The difference is that the Helmert contrasts above are applied to the category logits rather than the category probabilities as in continuation-ratio models.

11.1.1 Intraclass Correlation

For a random intercept model, it is often of interest to express the between-subjects variance in terms of the intraclass correlation. One way to obtain this expression utilizes the underlying latent response tendencies, denoted as y_{ijc}. Also, for simplicity, this will be done for the reference-cell formulation, though it applies to the more general contrast situation as well. The random intercept regression model for the latent variable y_{ijc}, including level-1 residuals ε_{ijc}, is written as

$$y_{ijc} = \boldsymbol{x}'_{ij}\boldsymbol{\beta}_c + \sigma_c\theta_i + \varepsilon_{ijc} \quad \text{for } c = 1, 2, \ldots, C. \tag{11.6}$$

As mentioned, for a particular ijth unit, the category c associated with the observed nominal response Y_{ij} is the one for which y_{ijc} is maximal. Since, in the present formulation, $c = 1$ is the reference category, we have $\boldsymbol{\beta}_1 = \sigma_1 = 0$, and so the model can be rewritten as

$$y_{ijc} = \boldsymbol{x}'_{ij}\boldsymbol{\beta}_c + \sigma_c\theta_i + (\varepsilon_{ijc} - \varepsilon_{ij1}) \quad \text{for } c = 2, \ldots, C, \tag{11.7}$$

for the latent response tendency of category c relative to the reference category. It can be shown that the level-1 residuals ε_{ijc} for each category are distributed according to a type I extreme-value distribution (see Maddala [1983, page 60]). It can further be shown that the standard logistic distribution is obtained as the difference of two independent type I extreme-value variates (see McCullagh and Nelder [1989, pages 20 and 142]). As a result, the level-1 variance is given by $\pi^2/3$, which is the variance for a standard logistic distribution. The estimated intraclass correlations are thus calculated as $r_c = \hat{\sigma}_c^2/(\hat{\sigma}_c^2 + \pi^2/3)$, where $\hat{\sigma}_c^2$ is the estimated between-subjects variance assuming normally distributed random intercepts. Notice that $C - 1$ intraclass correlations are estimated. As such, the subject influence on the longitudinal responses is allowed to vary across the nominal response categories.

11.1.2 Parameter Estimation

Estimation follows the procedure described for binary and ordinal outcomes. Specifically, letting $\boldsymbol{Y}_i$ denote the vector of nominal responses from level-2 unit i (for the n_i level-1 units nested within), the probability of any $\boldsymbol{Y}_i$ conditional on the random effects $\boldsymbol{\theta}$ is equal to the product of the probabilities of the level-1 responses:

$$\ell(\boldsymbol{Y}_i \mid \boldsymbol{\theta}) = \prod_{j=1}^{n_i} \prod_{c=1}^{C} (p_{ijc})^{\mathrm{y_{ijc}}} \tag{11.8}$$

where $\mathrm{y}_{ijc} = 1$ if $Y_{ij} = c$, and 0 otherwise. The marginal log-likelihood from the N level-2 units,

$$\log L = \sum_{i}^{N} \log h(\boldsymbol{Y}_i) = \sum_{i}^{N} \int_{\boldsymbol{\theta}} \ell(\boldsymbol{Y}_i \mid \boldsymbol{\theta})\; g(\boldsymbol{\theta})\; d\boldsymbol{\theta}, \tag{11.9}$$

is maximized to obtain maximum marginal likelihood estimates of $\boldsymbol{\Gamma}$ and $\boldsymbol{\Lambda}$. Specifically, using $\boldsymbol{\Delta}$ to represent either parameter matrix, and denoting the conditional likelihood as ℓ_i and the marginal density as h_i,

$$\frac{\partial \log L}{\partial \boldsymbol{\Delta}'} = \sum_{i=1}^{N} h_i^{-1} \int_{\boldsymbol{\theta}} \left[\sum_{j=1}^{n_i} \boldsymbol{D}\,(\mathbf{y}_{ij} - \boldsymbol{P}_{ij}) \otimes \partial \boldsymbol{\Delta} \right] \ell_i\; g(\boldsymbol{\theta})\; d\boldsymbol{\theta}, \tag{11.10}$$

where

$$\partial \boldsymbol{\Gamma} = \boldsymbol{x}'_{ij}, \quad \partial \boldsymbol{\Lambda} = \left[\mathrm{J}_{r^*}(\boldsymbol{\theta} \otimes \boldsymbol{z}_{ij}) \right]', \tag{11.11}$$

$\mathbf{y}_{ij}$ is the $C \times 1$ indicator vector, and $\boldsymbol{P}_{ij}$ is the $C \times 1$ vector obtained by applying (11.3) for each category. As in the binary and ordinal cases, Fisher's method of scoring can be used to provide the solution to these likelihood equations. Also, as in those cases, integration over the random effects must be performed; the same techniques and issues described in Chapter 9 pertain here as well.

11.2 HEALTH SERVICES RESEARCH EXAMPLE

Returning to the McKinney Homeless Research Project (MHRP) study data [Hough et al., 1997; Hurlburt et al., 1996], described in Chapter 10, we now illustrate an analysis using a mixed-effects multinomial regression model. For the initial set of analyses, reference category contrasts were used and street/shelter was chosen as the reference category. Thus, the first comparison compares community to street responses, and the second compares independent to street responses. A second analysis using Helmert contrasts is also described.

Corresponding observed logits for the reference-cell comparisons by group and time are given in Figures 11.1 and 11.2. Comparing these plots, different patterns for the post-baseline group differences are suggested. It seems that the non-section 8 group does better in terms of the community versus street comparison, whereas the section 8 group is improved for the independent versus street comparison. Further, the group differences appear to vary across time. The subsequent analyses examine these visual impressions of the data.

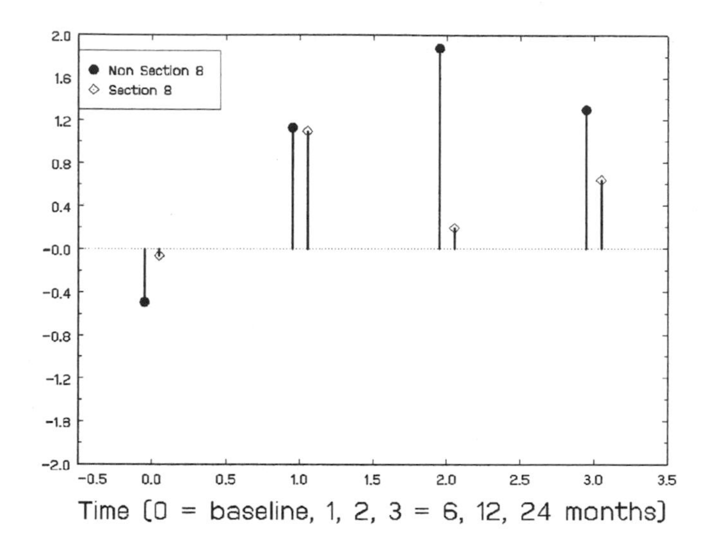

Figure 11.1. First reference-cell logit values (community versus street) across time by group.

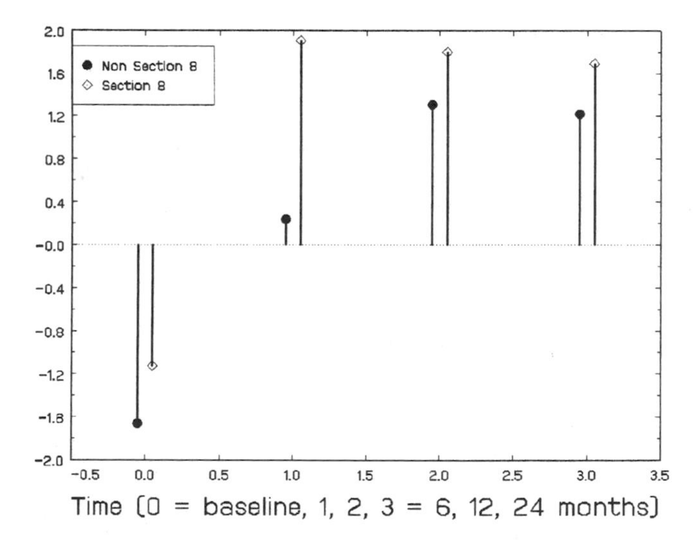

Figure 11.2. Second reference-cell logit values (independent versus street) across time by group.

To examine the sensitivity of the results to the normality assumption for the random effects, two models were fit to these data assuming the random effects were normally and uniformly distributed, respectively. Table 11.1 lists results for the two models, and for the two response category comparisons of community versus street (upper half of table) and independent versus street (lower half of table), respectively. The time and group effects are the same as in the previous ordinal analyses of the last chapter.

Table 11.1. Housing Status Across Time: Nominal Model Estimates and Standard Errors (SE)

	Normal Prior		Uniform Prior	
Term	Estimate	SE	Estimate	SE
Community versus Street				
Intercept	**−.452**	.192	**−.473**	.184
t1 (6 month vs. base)	**1.942**	.312	**1.850**	.309
t2 (12 month vs. base)	**2.820**	.466	**2.686**	.457
t3 (24 month vs. base)	**2.259**	.378	**2.143**	.375
Section 8 (yes = 1, no = 0)	*.521*	.268	*.471*	.258
Section 8 by t1	−.135	.490	−.220	.484
Section 8 by t2	**−1.917**	.611	**−1.938**	.600
Section 8 by t3	*−.952*	.535	*−.987*	.527
Subject sd	**.871**	.138	**.153**	.031
Independent versus Street				
Intercept	**−2.675**	.367	**−2.727**	.351
t1 (6 month vs. base)	**2.682**	.425	**2.540**	.422
t2 (12 month vs. base)	**4.088**	.559	**3.916**	.551
t3 (24 month vs. base)	**4.099**	.469	**3.973**	.462
Section 8 (yes = 1, no = 0)	.781	.491	.675	.460
Section 8 by t1	**2.003**	.614	**2.016**	.605
Section 8 by t2	.548	.694	.645	.676
Section 8 by t3	.304	.615	.334	.600
Subject sd	**2.334**	.196	**.490**	.040
$-2 \log L$	2218.73		2224.74	

Bold indicates $p < .05$, *italic* indicates $.05 < p < .10$.

The results are very similar for the two models. Thus, the random-effects distributional form does not seem to play an important role for these data. Subjects in the control group increase both independent and community housing relative to street housing at all three follow-ups, as compared to baseline. Compared to controls, the increase in community versus street housing is less pronounced for section 8 subjects at 12 months, but not statistically different at 6 months and only marginally different at 24 months. Conversely, as compared to controls, the increase in independent versus street housing is more pronounced for section 8 subjects at 6 months, but not statistically different at 12 or 24 months.

As in the ordinal case (see Chapter 10), the Wald tests are all significant for the inclusion of the random effects variance terms. A likelihood ratio test also clearly supports inclusion of the random subject effect (likelihood ratio $X_2^2 = 134.3$ and 128.3 for the normal and uniform distribution, respectively, as compared to the fixed-effects model, not shown). This use of the likelihood ratio test has also been called into question, with some advocating halved p-values for such testing of variance parameters (see Snijders and Bosker [1999], pages 90–91). In the present case, the difference in log-likelihood values between the fixed and random-effects models is so large, relative to the degrees of freedom, that the preference for the random-effects model is clear. Expressed as intraclass correlations, $r_1 = .19$ and $r_2 = .62$ for community versus street and independent versus street, respectively. Thus, the subject influence is much more pronounced in terms of distinguishing independent versus street living, relative to community versus street living. This is borne out by contrasting models with separate versus a common random-effect variance across the two category contrasts (not shown), which yields a highly significant likelihood ratio test statistic ($X_1^2 = 49.2$) favoring the model with separate variance terms.

For the nominal model assuming normally distributed random effects, model fit to the actual data is depicted in Figures 11.3 and 11.4. These figures compare the observed marginal logits (which were also plotted in Figures 11.1 and 11.2) with the "marginalized" logits of the mixed-effects model obtained using the quadrature method described in Section 9.6.3. In terms of community versus street housing, Figure 11.1 shows the dramatic increase for the control group relative to the section 8 group. As the analysis indicated, these groups differ most at 12 months. Figure 11.2, which illustrates the logits of independent versus street housing clearly, depicts the beneficial effect of section 8 certificates at 6 months. Thereafter, it is seen that the control group catches up to some degree. Considering these plots along with the results of the multilevel analysis, it is seen that both groups reduce the degree of street housing, but do so in somewhat different ways. The control group subjects are shifted more towards community housing, whereas section 8 subjects are more quickly shifted towards independent housing.

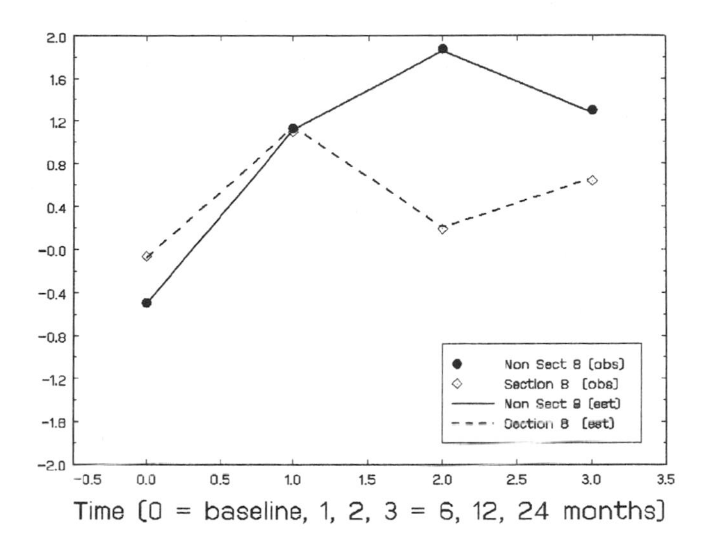

Figure 11.3. Model fit of first reference-cell logits (community versus street) across time by group.

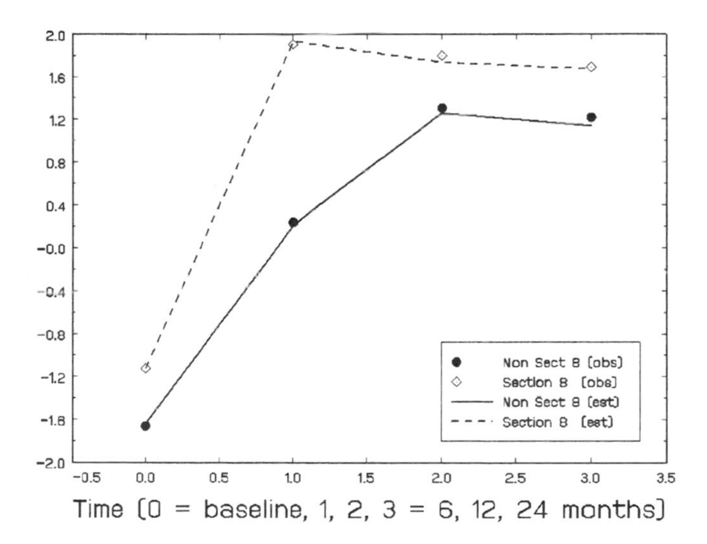

Figure 11.4. Model fit of second reference-cell logits (independent versus street) across time by group.

An analysis was also done to examine if the random-effect variance terms varied significantly by treatment group. The deviance ($-2\log L$) for this model, assuming normally distributed random effects, equaled 2218.43, which was nearly identical to the value of 2218.73 (from Tables 11.1 and 11.2) for the model assuming homogeneous variances across groups. The control group and section 8 group estimates of the subject standard deviations were respectively .771 (SE = .182) and .966 (SE = .214) for the community versus street comparison, and 2.228 (SE = .299) and 2.432 (SE = .266) for the independent versus street comparison. Thus, the homogeneity of variance assumption across treatment groups is reasonable.

Finally, Table 11.2 lists results obtained for an analysis assuming normally-distributed random effects and using Helmert contrasts for the three response categories. From this analysis, it is interesting that none of the section 8 by time interaction terms are observed to be statistically significant for the first Helmert contrast (*i.e.*, comparing non-street to street housing). Thus, group assignment is not significantly related to housing when considering simply non-street versus street housing outcomes. However, the second Helmert contrast that contrasts the two types of non-street housing (*i.e.*, independent versus community) does reveal the beneficial effect of the section 8 certificate in terms of the positive group by time interaction terms. Again, the section 8 group is more associated with independent housing, relative to community housing, than the non-section 8 group. In many ways, the Helmert contrasts, with their intuitive interpretations, represent the best choice for the analysis of these data.

Table 11.2. Housing Status Across Time: Nominal Model Estimates and Standard Errors (SE) using Helmert Contrasts

	Independent & Community vs. Street		Independent vs. Community	
Term	Estimate	SE	Estimate	SE
Intercept	**–1.564**	.244	**–2.224**	.326
t1 (6 month vs. base)	**2.312**	.322	**.741**	.375
t2 (12 month vs. base)	**3.454**	.484	**1.268**	.352
t3 (24 month vs. base)	**3.179**	.387	**1.839**	.358
Section 8 (yes = 1, no = 0)	*.651*	.334	.260	.425
Section 8 by t1	*.934*	.495	**2.138**	.505
Section 8 by t2	–.684	.601	**2.465**	.512
Section 8 by t3	–.324	.517	**1.256**	.509
Subject sd	**1.602**	.148	**1.463**	.166

$-2\log L = 2218.73$

Bold indicates $p < .05$, *italic* indicates $.05 < p < .10$.

11.3 COMPETING RISK SURVIVAL MODELS

Efron [1988] proposed an alternative parameterization of the Cox proportional hazards model termed a "partial logistic regression" model or "person-time logistic regression" model [Ingram and Kleinman, 1989] (see also Section 10.2.3). This clever idea allows the fitting of the proportional hazard survival model using commonly available logistic regression software. The method works as follows. First, we obtain a series of sequential records from each subject for the period of time that they were observed in the study. For example, a patient who was followed for 4 months and then died would have 4 records with an outcome of 0 for months 1–3 and an outcome of 1 (transplant) on month 4, assumming a monthly measurement interval. By contrast, a person who remained in the study for 2 months and was then lost to follow-up, would have two monthly records, each with an outcome of zero, indicating that they were censored at the end of month 2. Finally, a subject who survived for the entire 12 month study would have 12 monthly records all with an outcome of zero. Efron [1988] and Ingram and Kleinman [1989] have shown that this approach to modeling time to event data provides excellent agreement with the traditional proportional hazards survival model and becomes identical as the time intervals go to zero (*i.e.*, approach continuous time).

Using this idea, the Institute of Medicine [IOM, 1999] committee on organ procurement and transplantation further generalized the approach to incorporate competing risks of wait-list mortality and transplantation, by substituting a multinomial logistic regression for the usual logistic regression for binary response. To examine issues of geographic equity, they further generalized the approach to include random effects that accounted for the clustering within geographic allocation regions known as organ procurement organizations. In the following, we provide an overview of this application of mixed-effects multinomial regression models. The interested reader is referred to the original IOM report, and the subsequent papers by Gibbons and co-workers [Gibbons et al., 2000, 2003].

11.3.1 Waiting for Organ Transplantation

Since the enactment of the National Organ Transplant Act of 1984, the number of people receiving organs has increased steadily over time. In 1998, more than 21,000 Americans—about 57 people a day—were transplanted with a kidney, liver, heart, lung, or other organ. On any given day, approximately 62,000 people are waiting for an organ and every 16 minutes a new name is added to the national waiting list [UNOS, 1999]. Moreover, although the number of donors has increased steadily since 1988, donation rates are not growing as quickly as the demand for organs [USGAO, 1997]. As a result, approximately 4,000 Americans die each year (11 people per day) while waiting for a solid organ transplant [UNOS, 1999].

One of the most visible and contentious issues regarding the fairness of the original system of organ procurement and transplantation was the argument that it resulted in great disparities in the amount of time potential liver transplant patients wait for a transplant, depending on where the patient lived. (The term "waiting time" is used to refer to the time from registration at a transplantation center to transplant, death, or removal from the waiting list for other reasons.) An additional concern was that minorities and the poor may have had less access to organ transplants than did whites of higher socioeconomic status.

In response to concerns expressed about possible inequities in the original system of organ procurement and transplantation, the U.S. Department of Health and Human Services (DHHS) published a new regulation (Final Rule) in April 1998 (42 CFR Part 121) to "assure that allocation of scarce organs will be based on common medical criteria, not accidents of geography" [DHHS, 1998]. The stated principles underlying the Final Rule included increasing federal oversight, increasing public access to data, implementing consistent medical listing criteria, placing emphasis on medical need, and reducing disparities in waiting times for transplants among different areas of the country.

Issuance of the Final Rule generated considerable controversy in the transplant community. Concerns were expressed that its implementation would increase the cost of transplantation, force the closure of small transplant centers, adversely affect access to transplantation on the part of minorities and low-income patients, discourage organ donation, and result in fewer lives saved.

In October 1998, the U.S. Congress suspended implementation of the Final Rule for 1 year to allow further study of its potential impact. During that time, Congress asked the Institute of Medicine (IOM) of the National Academy of Sciences (NAS) to conduct a study to review current Organ Procurement Transplantation Network (OPTN) policies and the potential impact of the Final Rule. The IOM study was completed in July of 1999 [IOM, 1999; Gibbons et al., 2000]. The report was based on analysis of approximately 68,000 U.S. liver transplant waiting list records that describe every transition made by every patient on the waiting list from 1995 through the first quarter of 1999. The committee focused on liver transplants because (1) disparities in median waiting times for liver transplants was a primary factor in DHHS's rationale for developing the final rule, (2) liver allocation policies have been especially contentious with the OPTN making several changes in the recent past, (3) the time that a liver can viably survive outside of the body (cold ischemic time) is much longer than for hearts and lungs, making changes to the allocation system possible, and (4) the medical urgency is greater than some other organs, such as kidneys, where medical alternatives such as dialysis are available.

A fundamental issue in the liver allocation system is the classification of patients into "status levels" based on the current medical severity of their illness. These status levels reflect the life expectancy of patients in the absence of transplantation, which vary from a few days for status 1 (*i.e.*, the most severely ill patients with an average life expectancy of one week), from days to months in status 2 (divided into severe (2A) and less severe (2B) chronic conditions), to potentially years in status 3 (patients in need of transplantation but not at serious risk at this stage of their illness). Moreover, more than 50% of the patients on the list are in status 3.

Model Specification. Stratified analyses were performed separately for the time spent in each status level, with the exception of 2A for which there were too few subjects. For status 1, time refers to days, whereas for status 2B and 3, time refers to months. In all analyses, the outcome measure was the nominal measure of transplant, death or other. Other can be shifting to another status level and never returning to the status level in question, being too sick to transplant, being delisted, being transplanted at another center, or still waiting. Covariates included age (0–5, 6–17, 18 and over), sex (female = 0, male = 1), race (black = 1, else = 0), blood type (O or B = 1 else = 0) and OPO transplant volume (small, medium, and large based on number of transplanted patients in 1995–1999; OPO denotes Organ Procurement Organization). For blood type, contrasts between types O and B versus A and AB were selected because the former two can only receive donation from a subset of donors whereas the latter can receive donation from almost all potential donors.

Results. A summary of several statistics of interest that help characterize the sample and waiting time distributions is presented in Table 11.3. Maximum (marginal) likelihood

(ML) estimates, standard errors (SE) and corresponding probabilities for Wald test statistics are presented in Table 11.4, separately for status levels 1, 2B and 3 respectively. In the following we provide an overview of the most important findings.

Table 11.3. Characteristics of Liver Transplant Patients by Status, 1995–1999

	Totals	Status 1	Status 2	Status 3
Total patients, 1995–1999	33,286	5,294	14,264	26,907
Percentage receiving a transplant	47.1	52.4	50.2	21.3
Percentage dying prior to transplantation	8.3	9.2	6.1	5.2
Percentage post-transplant mortality	5.4	11.1	5.0	1.9
Percentage male	58.7	54.1	59.9	58.7
Percentage with A or AB blood type	16.0	15.3	15.4	15.8
Percentage African American	7.7	11.2	8.3	6.9
Mean age (years)	45.0	36.3	44.9	46.1
Mean waiting time (days)	255.6	4.8	56.8	285.1

The "Totals" column reflects number of unique listings and therefore does not equal the sum of the other three columns which count patients within status levels (a given patient may be counted in up to three status levels for a particular listing [IOM, 1999]).

Geographic Inequity. Systematic OPO-specific effects accounted for less than 5% of the total variance (*i.e.*, intra-cluster correlation of 0.045) in transplantation rates for status 1 patients (see Table 11.4). The geographic distribution for the most severely ill patients is therefore reasonably equitable with mean waiting time of 4.8 days (see Table 11.4). In contrast, OPO-specific effects accounted for 13% of the variability in transplantation rates for status 2B patients and 35% of the variability for status 3 patients (see Table 11.5). This implies that the systematic variation in waiting time across OPOs is almost completely determined by variations in waiting times for the less severely ill patients, with little variation for the most severely ill patients. This finding is further illustrated in Figure 11.5 where the empirical Bayes (EB) estimates of the OPO-specific adjusted transplantation effects are displayed.

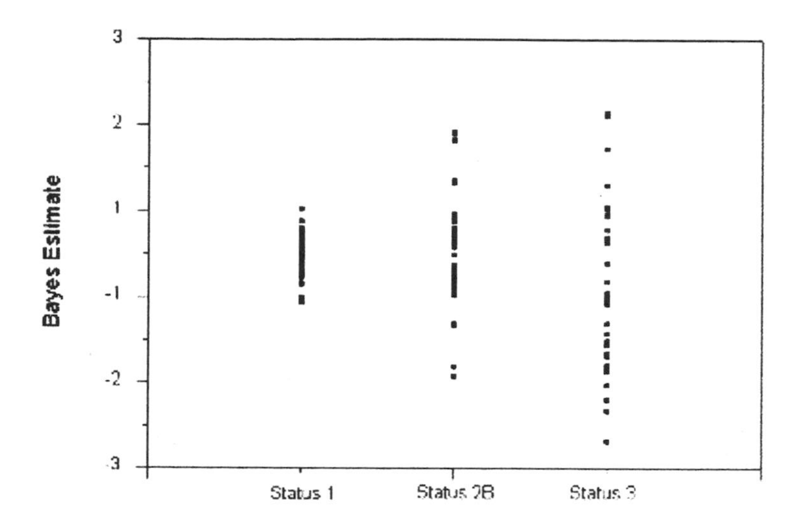

Figure 11.5. OPO-specific Bayes estimates $(\sigma_1\theta_i)$ of transplantation rates by status category adjusted for competing risk of mortality and model covariates (*e.g.*, sex, race, blood type).

The interpretation of the EB estimate is OPO i's deviation from the population rate, adjusted for covariates. As such, the y axis in Figure 11.5 is in a log-odds scale, where values of 1, 2, and 3 represents increases in the likelihood of transplantation by factors of 2.7, 7.4 and 20.1 respectively. Negative values represent corresponding decreases in probability of transplantation relative to the overall population rate.

Inspection of Figure 11.5 clearly illustrates that the greatest geographic variation in adjusted transplantation rates is for the least severely ill patients. Moreover, while transplantation rates of these less severely ill patients vary significantly, they have little relationship to mortality. In all cases, systematic OPO differences in pre-transplantation mortality rates accounted for less than 1% of the variation in overall adjusted mortality rates. No significant effects of race or gender were observed indicating that the system is equitable for women and minorities once they are listed.

OPO Volume and Size. Smaller OPOs are more likely than larger OPOs to transplant status 2B and 3 patients (see Table 11.5). For status 1 patients, OPO size played no role in transplantation or mortality rates. In contrast, for status 2B and 3 patients, OPO size was significantly related to transplantation rates. For large OPOs (9+ million) the initial one month transplantation rates were 5% for status 2B patients and 3% for status 3 patients. By contrast in the smaller OPOs (4 million or less), initial one-month transplantation rates were as high as 17% for status 2B patients and 9% for status 3 patients. Based on these results the IOM recommended that at least 9 million people be included in an organ allocation region to maximize the chance of transplantation for the most severely ill patients (for a detailed analysis of OPO size, see the original IOM report).

Table 11.4. Mixed-Effects Competing Risk Survival Models for Patient Time in Status Levels 1, 2B, and 3—Maximum Likelihood Estimates (Standard Errors)

	Status 1		Status 2B		Status 3	
Transplant Versus Other						
Intercept	−1.829 (0.276)		−2.077 (0.129)		−3.593 (0.210)	
Day (1), month (2B, 3)	0.016 (0.015)		−0.092 (0.016)	***	−0.220 (0.030)	***
Age 0–5 vs. ≥ 18	−0.907 (0.188)	***	0.470 (0.103)	***	1.156 (0.154)	***
Age 6–17 vs. ≥ 18	−0.362 (0.234)		0.135 (0.243)		0.844 (0.268)	**
Gender (1 = male)	−0.098 (0.198)		0.126 (0.087)		0.054 (0.186)	
Race (1 = black)	−0.275 (0.268)		0.134 (0.222)		0.158 (0.304)	
Blood type (1 = B or O)	−0.076 (0.196)		−0.577 (0.062)	***	−0.477 (0.098)	***
OPO volume (M vs. L)	−0.054 (0.319)		0.590 (0.157)	***	1.179 (0.149)	***
OPO volume (S vs. L)	0.261 (0.336)		0.560 (0.187)	**	0.757 (0.228)	***
Random OPO effect SD	0.393 (0.144)	**	0.689 (0.064)	***	1.335 (0.162)	***
Mortality versus Other						
Intercept	−3.685 (0.482)		−3.313 (0.227)		−3.654 (0.172)	
Day (1), month (2B, 3)	0.023 (0.047)		−0.213 (0.039)	***	−0.216 (0.041)	***
Age 0–5 vs. Adult	−0.968 (0.378)	**	−0.195 (0.381)		−2.119 (2.099)	
Age 6-17 vs. Adult	−1.001 (0.551)		−0.516 (0.641)		−1.193 (2.000)	
Gender (1 = male)	0.077 (0.371)		0.014 (0.191)		−0.063 (0.268)	
Race (1 = black)	0.162 (0.448)		−0.082 (0.359)		0.027 (0.544)	
Blood type (1 = B or O)	0.003 (0.433)		−0.005 (0.164)		−0.017 (0.231)	
OPO volume (M vs. L)	0.203 (0.491)		0.202 (0.126)		−0.526 (0.300)	
OPO volume (S vs. L)	−0.230 (0.930)		0.355 (0.151)	**	−0.658 (0.358)	
Random OPO effect SD	0.042 (0.298)		0.116 (0.049)	**	0.137 (0.157)	

*$p < .05$, **$p < .01$, ***$p < .001$.

The Effect of Sharing. The IOM further examined this finding by analyzing the results of several regional and state-wide sharing arrangements among two or more OPOs, most typically for status 1 patients. their analysis of these "natural experiments" revealed that sharing significantly increased the status 1 transplantation rate from 42% without sharing to 52% with sharing, lowered average status 1 waiting times from 4 to 3 days, and decreased status 1 pre-transplantation mortality from 9% to 7%. Not surprisingly, sharing significantly decreased the rate of transplantation for less severely ill patients. For example, among small OPOs that served a population of 2 million or less, the status 3 transplantation rate decreased from 31% for those OPOs that did not share to 6% for those that did share, making more organs available for more severely ill patients. Though sharing decreased status 3 transplantation rates, there was no increase in pre-transplantation mortality of status 3 patients.

Waiting Times and Need for Transplant. Other interesting results of the analysis concerned the relationship of transplantation and pre-transplantation mortality to waiting times. For status 1 patients, the rates were constant over the first 12 days of listing at approximately 15% for transplantation per day and 3% for mortality (see Figure 11.6 panel A), but for status 2B and 3 patients both rates decreased rapidly over time (see Figure 11.6 panels B and C).

For status 2B, transplantation rates decreased from 12% to 5% per month over a 12-month period while pre-transplant mortality rates decreased from 3% to 0.3% per month. For status 3 patients, transplantation rates decreased from 4% to 0.05% per month over a 12-month period and pre-transplant mortality rates decreased from 2% to 0.2% per month. These findings indicate that waiting time is inversely related to medical need in the less severely ill patients and should therefore not be used as a criterion for transplantation (as it had traditionally been) in status 2B and 3. Figure 11.7 displays estimated cumulative time-to-event distributions for status 1 (panel A), status 2B (panel B), and status 3 (panel C). Inspection of Figure 11.7 reveals that after 12 days 80% of the status 1 patients at risk are transplanted whereas 10% die while waiting. For status 2B, 60% of patients at risk through 12 months are transplanted and 7% die while waiting. For status 3, 20% of patients at risk through 12 months are transplanted and 8% die while waiting.

Specification of the Random-Effect Distribution. In examining the robustness of the model to the choice of random-effect distribution, the IOM compared the ML estimates and SEs for the status 1 patient data for models with Gaussian and uniform (*i.e.*, rectangular) random-effect distributions. The results of this comparison are presented in Table 11.5. Table 11.5 reveals that there is virtually no effect of the specification of the random-effect distribution on the estimates and standard errors for the fixed effects in the model. As expected, the random-effect variance for the uniform distribution is different from that estimated for Gaussian random-effects. In either case (*i.e.*, Gaussian or uniform random-effect distributions) the random-effects are given by $\sigma_k \theta_i$, however, in the Gaussian case the variance of θ_i is unity, whereas in the uniform case the variance of θ_i is $(a-b)^2/12$, where a and b are the maximum and minimum quadrature nodes respectively. In the current example, $a = 4.859$ and $b = -4.859$ (*i.e.*, the extreme Gauss–Hermite quadrature nodes with 10 points). As such, the adjusted random-effect standard deviation for transplantation is $\sigma_1 = \sqrt{.131^2(9.718^2/12)} = 0.368$ which is similar to the value of $\sigma_1 = .393$ estimated for the case of Gaussian random effects (see Table 11.5).

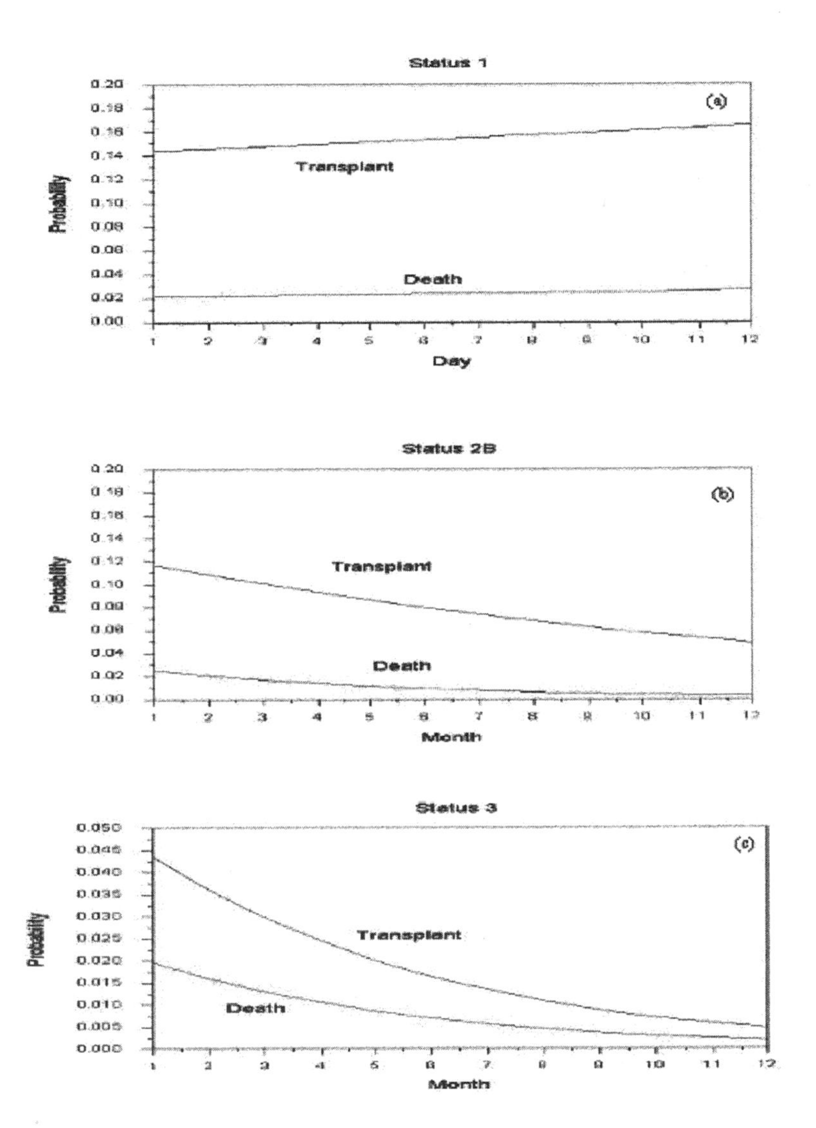

Figure 11.6. Estimated hazard rates for (a) status 1, (b) status 2B, and (c) status 3 patients awaiting liver transplantation. The hazard rate describes the likelihood of transplantation or mortality at a given point in time (using one whole day (status 1) or one whole month (status 2B and 3) as the unit) adjusted for the competing risks (*i.e.*, transplantation or mortality) and the model covariates (*e.g.*, sex, race, blood type).

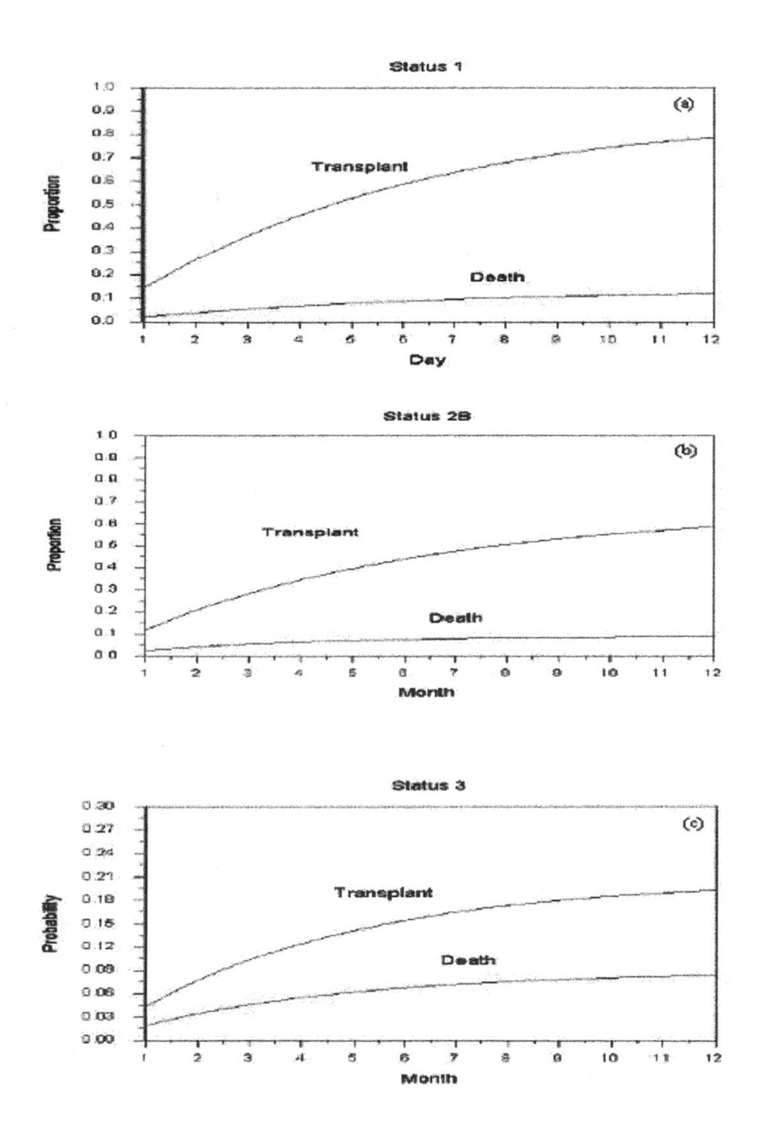

Figure 11.7. Estimated cumulative time-to-event distributions for (a) status 1, (b) status 2B, and (c) status 3 patients awaiting liver transplantation. The cumulative time-to-event distribution describes the overall adjusted likelihood of transplantation or mortality up to a particular point in time.

Table 11.5. Comparison of Normal and Rectangular Random-Effects Distributions for Patient Time in Status Levels 1—Maximum Likelihood Estimates (Standard Errors)

	Normal	Rectangular
Transplant Versus Other		
Intercept	−1.829 (0.276)	−1.837 (0.261)
Day (1), month (2B, 3)	0.016 (0.015)	0.016 (0.015)
Age 0–5 vs. ≥ 18	−0.907 (0.188)	−0.894 (0.188)
Age 6–17 vs. ≥ 18	−0.362 (0.234)	−0.364 (0.236)
Gender (1 = male)	−0.098 (0.198)	−0.101 (0.196)
Race (1 = black)	−0.275 (0.268)	−0.279 (0.282)
Blood Type (1 = B or O)	−0.076 (0.196)	−0.070 (0.195)
OPO volume (M vs. L)	−0.054 (0.319)	−0.070 (0.317)
OPO volume (S vs. L)	0.261 (0.336)	0.263 (0.336)
Random OPO effect SD	0.393 (0.144)	0.131 (0.040)
Mortality versus Other		
Intercept	−3.685 (0.482)	−3.693 (0.458)
Day (1), month (2B, 3)	0.023 (0.047)	0.025 (0.046)
Age 0–5 vs. Adult	−0.968 (0.378)	−0.970 (0.409)
Age 6–17 vs. Adult	−1.001 (0.551)	−1.001 (0.552)
Gender (1 = male)	0.077 (0.371)	0.081 (0.344)
Race (1 = black)	0.162 (0.448)	0.173 (0.493)
Blood Type (1 = B or O)	0.003 (0.433)	−0.010 (0.441)
OPO volume (M vs. L)	0.203 (0.491)	0.208 (0.470)
OPO volume (S vs. L)	−0.230 (0.930)	−0.217 (0.908)
Random OPO effect SD	0.042 (0.298)	0.014 (0.101)

11.4 SUMMARY

Mixed-effects multinomial logistic regression models are described for the analysis of nominal categorical data. These models are useful for analysis of outcomes with more than two unordered response categories. By and large, the models are seen as extensions of the mixed logistic regression model. However, the nominal model typically uses dichotomizations that are based on selecting one category as the reference that the others are each compared to. This chapter has also described how other comparisons can be imbedded within the nominal model.

For the nominal model, both reference cell and Helmert contrasts were applied in the analysis of the homelessness data. The former indicated an increase for community relative to street housing for the non-section 8 group, as well as an increase for independent relative to street housing for the section 8 group. Alternatively, the Helmert contrasts indicated that the groups did not differ in terms of non-street versus street housing, but did differ in terms of the type of non-street housing (*i.e.*, the section 8 group was more associated with independent housing). In either case, the nominal model makes an assumption that has been referred to as "independence of irrelevant alternatives" in the econometric literature [Maddala, 1983]. This is because the effect of an explanatory variable comparing two categories is the same regardless of the total number of categories considered. This assumption is generally reasonable when the categories are distinct and dissimilar, and unreasonable as the nominal categories are seen as substitutes for one another [McFadden, 1973; Amemiya, 1981]. McFadden [1980] notes that the multinomial logistic regression model is relatively robust in many cases in which this assumption is implausible. In our examples, the outcome categories are fairly distinct and so the assumption would seem to be reasonable for these data. A model that relaxes this assumption for multilevel nominal data is described in Hartzel et al. [2001].

The organ transplantation example illustrates how the mixed-effects model for nominal data can be used to solve a fundamental problem in survival analysis; competing risk survival analysis with random effects. The illustration is also useful in that it shows how this model was used to address a critical problem in public health policy.

CHAPTER 12

MIXED-EFFECTS REGRESSION MODELS FOR COUNTS

The focus of many scientific investigations involves the measurement of the frequency of a specific behavior or activity. For example, in health services research, the number of hospitalizations within a fixed period of time is often the focus of study. When the data are clustered either by obtaining repeated observation periods from the same subjects or counts from individuals clustered within units (*e.g.*, suicide rates in counties), multilevel approaches are required as well. There are several approaches to the analysis of count data in the context of fixed-effects and mixed-effects models. Alternatives include: (1) Ignore the quantitative nature of the data and analyze the presence or absence of the behavior (*e.g.*, service use) as a binary outcome; (2) create an ordinal response variable with categories of, for example, 0 visits, 1 visit, 2 visits, 3 visits, 4 or more visits; (3) model the counts as a Poisson distribution in a Poisson fixed-effects or mixed-effects regression model. All of these approaches have strengths and limitations. The binary approach is simple to use (see Chapter 9), but discards the quantitative information that the investigator went to the trouble to collect. The ordinal approach (see Chapter 10) relies on often unrealistic or arbitrary cut-points and typically assumes that the covariates have a proportional effect over the categories. Fixed-effects and mixed-effects Poisson regression models are more sophisticated and appropriate for the distributional form of frequency data, but can often underestimate the number of subjects who do not exhibit the behavior (*e.g.*, no service use). Generalizations of the Poisson model, known as zero-inflated Poisson (ZIP) regression models [Lambert, 1992], often remedy this situation by providing a mixture of regression models, one to estimate effects of predictors on the presence or absence of the behavior or activity, and one to estimate the effects of predictors on the frequency of the behavior or activity conditional on its' use. Recently, the ZIP model has been generalized to the mixed-effects case [Hall, 2000; Hur et al., 2002] so that analysis of longitudinal and/or clustered frequency data is now possible.

In this chapter, we explore the development and application of mixed-effects Poisson regression models for the analysis of multilevel frequency data, and we describe generalizations to the case of ZIP models and their application.

12.1 POISSON REGRESSION MODEL

Poisson regression has been increasingly used in a variety of settings where the dependent variable is a nonnegative count consisting of the number of events of some type occurring over a given period of time. Examples of applications include the number of patents issued to firms [Hausman et al., 1984], the number of trips to the doctor [Cameron and Trivedi, 1986], and the occurrence of homicides per year [Grogger, 1990].

In Poisson regression modeling, the data are modeled as Poisson counts whose means are expressed as a function of covariates. Let y_i be a response variable and $\boldsymbol{x_i} = (1, x_{i1}, \ldots, x_{ip})'$ be a $(p+1) \times 1$ vector of covariates for the ith individual. Then the Poisson regression model y_i is

$$f(y_i; \lambda_i) = \frac{\exp(-\lambda_i)(\lambda_i^{y_i})}{y_i!}, \quad y_i = 0, 1, 2, \ldots, N \tag{12.1}$$

where $\lambda_i = \exp(\boldsymbol{x_i}'\boldsymbol{\beta})$ and $\boldsymbol{\beta} = (\beta_0, \beta_1, \ldots, \beta_p)'$ is a $(p+1)$-dimensional vector of unknown parameters corresponding to $\boldsymbol{x_i}$. An important property of the Poisson distribution is the equality of the mean and variance:

$$\mathrm{E}(y_i) = \mathrm{Var}(y_i) = \lambda_i. \tag{12.2}$$

The likelihood for N independent observations from (12.1) is

$$L = \prod_{i=1}^{N} \frac{\exp(-\lambda_i)(\lambda_i^{y_i})}{y_i!} \tag{12.3}$$

and the corresponding log-likelihood function is

$$\log L = -\sum_{i}^{N} \left[\lambda_i + y_i \log \lambda_i - \log(y_i!)\right]. \tag{12.4}$$

The first and second partial derivatives with respect to unknown parameter $\boldsymbol{\beta}$ are given by

$$\frac{\partial \log L}{\partial \boldsymbol{\beta}} = \sum_{i}^{N} (y_i - \lambda_i)\boldsymbol{x_i} \tag{12.5}$$

and

$$\frac{\partial^2 \log L}{\partial \boldsymbol{\beta} \partial \boldsymbol{\beta}'} = -\sum_{i}^{N} \lambda_i \boldsymbol{x_i} \boldsymbol{x_i}', \tag{12.6}$$

respectively. Since the above equations are not linear, iterative procedures such as the Newton–Raphson algorithm or Fisher's scoring algorithm are used to obtain the maximum likelihood estimator of β, denoted by $\hat{\beta}$. A consistent estimator of the variance–covariance matrix of $\hat{\beta}$, $V(\hat{\beta})$, is the inverse of the information matrix, $I(\hat{\beta})$, which is the inverse of the matrix of negative second partial derivatives evaluated at the MLE. Inferences on the regression parameter β can be made using $V(\hat{\beta})$.

12.2 MODIFIED POISSON MODELS

In analyzing count data, we often observe an excess of zeros compared with the Poisson distribution with the same mean. The excess of zeros often results from a mixture distribution with two components, one of which generates binary outcomes (with values 0 or 1 as in the logistic regression model), while the other generates the count outcome which may be equal to zero as well (as in the Poisson regression model). As an example, consider the following survey question: "How many times have you gone to a mental health professional this year?" There would be two types of zero responses. The first set of zero responses would be from a group of people who are not in need of mental health treatment and would not plan on going to a mental health clinic this year. The second set of zero responses would be from a group of people who are in need of mental health treatment but have not yet visited a mental health clinic. Interpretation of this model would be that in one population (a group of people who are not in need of treatment) one observes only zeros while in the other population (a group of people who are in need of treatment but may or may not obtain it in the current year) one observes counts including zero from a discrete distribution such as Poisson. Hence, fitting a classical Poisson model to these data would understate the theoretical probability of zero in the Poisson model.

Various modifications of the Poisson model have been proposed to accommodate these extra zeros. For example, Cohen [1954] discussed a model without any covariates which is further described by Johnson et al. [1992]. For models with covariates, Mullahy [1986] and King [1989] proposed Poisson "hurdle" regression models. Heilbron [1989, 1994] introduced a zero-altered Poisson or ZAP model. Lambert [1992] proposed a zero-inflated Poisson (ZIP) model and ZIP(τ) model. Both the ZIP and ZIP(τ) models are mixtures of logistic and Poisson regression models where the logistic portion contributes to the probability of a count of zero (*e.g.*, a person who does not need mental health service and does not use them), and the Poisson portion contributes to the frequency of utilization conditional on use. Note that for the Poisson portion of the model a zero is a valid count, for example, a person who is in need of mental health services but does not use them. The ZIP model permits different covariates and coefficient values between the logistic and Poisson portions of the model, whereas the ZIP(τ) model does not. Note that in both models, the variance functions for the logistic and Poisson portions have unique estimates. Greene [1994] extended this approach to a zero-inflated negative binomial model, which does not impose the restriction that the mean and variance of the Poisson portion of the model are the same. Jovanovic et al. [1994] discussed use of existing software for fitting extra zero Poisson data. Zorn [1996] showed that the hurdle event count model of King [1989] and Lambert's and Greene's ZIP models are special cases of a more general dual regime data generating process which results in extra Poisson zero counts. Some other examples of zero-inflated Poisson distributions can be found in Singh [1963], Goraski [1977], and Martin and Katti [1965].

12.3 THE ZIP MODEL

Following Lambert [1992], a ZIP model for the response y_i is given by

$$\begin{array}{lll} y_i \sim 0 & \text{with probability} & \pi_i, \\ y_i \sim \text{Poisson}(\lambda_i) & \text{with probability} & (1-\pi_i), \qquad i=1,\ldots,N, \end{array} \tag{12.7}$$

where

$$\text{logit}(\pi_i) = \boldsymbol{w_i}'\boldsymbol{\gamma} \tag{12.8}$$

and

$$\log(\lambda_i) = \boldsymbol{x_i}'\boldsymbol{\beta}. \tag{12.9}$$

Note that the equations in (12.8) and (12.9) are the canonical link functions of the Poisson model and the logistic model in the generalized linear model framework described in McCullagh and Nelder [1989]. Lambert referred to $y_i = 0$ as a perfect state or zero state, and $y_i \sim \text{Poisson}(\lambda_i)$ as an imperfect state or Poisson state due to the fact that her primary interest was on the number of defects in manufacturing. Zorn [1996] named those the transition and event stages, respectively, in his dual regime model.

The covariates that are associated with the Poisson part of the ZIP model may or may not be the same covariates that are associated with the logistic part of the ZIP model. If the same covariates effect λ_i and π_i, and if π_i can be written as a function of λ_i such that

$$\text{logit}(\pi_i) = -\tau\boldsymbol{x_i}'\boldsymbol{\beta} \tag{12.10}$$

and

$$\log(\lambda_i) = \boldsymbol{x_i}'\boldsymbol{\beta} \tag{12.11}$$

for an unknown parameter τ, then the ZIP model in (12.7)–(12.9) with logit link for π_i and log link for λ_i, and shape parameter τ is called a ZIP(τ) model [Lambert, 1992]. Lambert also suggested the log-log link defined by $\log(-\log(\pi_i)) = \tau\boldsymbol{x_i}'\boldsymbol{\beta}$, and the complementary log-log link defined by $\log(-\log(1-\pi_i)) = -\tau\boldsymbol{x_i}'\boldsymbol{\beta}$ as alternatives for the mixing parameter π_i in the ZIP(τ) model.

It can be seen that the zero outcome is a result of the perfect state as well as the Poisson state by rewriting the above ZIP model (12.7)–(12.9) as

$$\begin{aligned} \Pr(y_i = 0) &= \pi_i + (1-\pi_i)\exp(-\lambda_i), \\ \Pr(y_i = k) &= (1-\pi_i)\frac{\exp(-\lambda_i)(\lambda_i^k)}{k!}, \quad k = 1, 2, \ldots, \end{aligned} \tag{12.12}$$

where λ_i and π_i are defined as in (12.8)–(12.11). This model is termed WZ (with zeros) by Johnson et al. [1992] and has been studied by Heilbron [1989] and termed a ZAP model. Interpretation of this model is different from the Poisson "hurdle" model in that the Poisson "hurdle" model does not allow a zero count after the hurdle is crossed (truncated Poisson). However, in the ZIP model, zero counts can be observed from both the perfect and Poisson states. When $\pi_i = 0$, the Poisson distribution is obtained.

Estimation of the parameters of the ZIP or ZIP(τ) models is straightforward. Lambert [1992] introduced two separate log-likelihoods, one each for the perfect and Poisson states, and used the EM algorithm [Dempster et al., 1977] for the ZIP model and the Newton–Raphson algorithm for the ZIP(τ). For faster convergence, a straightforward gradient method was suggested by Greene [1994]. To compute the estimate of the asymptotic covariance matrix, Greene used the Berndt et al. [1974] (BHHH) estimator, which is the inverse of the sum of the outer products of the first derivatives.

Let $\pi_i = \Psi(\boldsymbol{w_i}'\gamma)$, where $\Psi(\cdot)$ is the cumulative logistic probability such that $\pi_i = \exp(\boldsymbol{w_i}'\gamma)/(1+\exp(\boldsymbol{w_i}'\gamma))$, and let $f(\cdot)$ Poisson density function with mean λ_i. Greene formulated the probability density function for the random variable, y_i as

$$\Pr(y_i) = \mathrm{p}(y_i) = (1-\pi_i)f(y_i) + \mathrm{I}(y_i)\pi_i, \tag{12.13}$$

where $\mathrm{I}(y_i)$ is an indicator function taking a value of 1 if $y_i = 0$ and a value of 0 otherwise, and λ_i and π_i are previously defined in (12.8)–(12.11). Then the log likelihood based on n independent individuals is

$$\begin{aligned}\log L &= \sum_{i=1}^{N} \log \mathrm{p}(y_i) \\ &= \sum_{i} \log[(1-\pi_i)f(y_i) + \mathrm{I}(y_i)\pi_i].\end{aligned} \tag{12.14}$$

Thus, the first derivatives with respect to the Poisson parameter vector β and the logistic parameter vector γ are obtained as

$$\begin{aligned}\frac{\partial \log L}{\partial \beta} &= \sum_{i=1}^{N} \frac{\partial \log \mathrm{p}(y_i)}{\partial \beta} \\ &= \sum_{i} \frac{1}{\mathrm{p}(y_i)} \frac{\partial \mathrm{p}(y_i)}{\partial \beta} \\ &= \sum_{i} \frac{1}{\mathrm{p}(y_i)} \left[(1-\pi_i)f(y_i)\frac{\partial \log f(y_i)}{\partial \beta} + (\mathrm{I}(y_i) - f(y_i))\frac{\partial \pi_i}{\partial \beta}\right]\end{aligned} \tag{12.15}$$

and

$$\begin{aligned}\frac{\partial \log L}{\partial \gamma} &= \sum_{i=1}^{N} \frac{\partial \log \mathrm{p}(y_i)}{\partial \gamma} \\ &= \sum_{i} \frac{1}{\mathrm{p}(y_i)} \frac{\partial \mathrm{p}(y_i)}{\partial \gamma} \\ &= \sum_{i} \frac{1}{\mathrm{p}(y_i)} [\mathrm{I}(y_i) - f(y_i)] \frac{\partial \pi_i}{\partial \gamma},\end{aligned} \tag{12.16}$$

where $\partial \log f(y_i)/\partial \boldsymbol{\beta}$ is given in (12.5) and $\partial \pi_i/\partial \boldsymbol{\gamma} = \partial \Psi(\boldsymbol{w_i}'\boldsymbol{\gamma})/\partial \boldsymbol{\gamma} = \pi_i' \boldsymbol{w_i}$ with $\pi_i' = \pi_i(1-\pi_i)$ for the ZIP model. For the ZIP(τ) model, $\boldsymbol{w_i}$ is replaced with $\boldsymbol{x_i}'\boldsymbol{\beta}$. When λ_i and π_i are not related, $\partial \pi_i/\partial \boldsymbol{\beta} = 0$ and when λ_i and π_i are functionally related $\partial \pi_i/\partial \boldsymbol{\beta} = -\tau \pi_i' \boldsymbol{x_i}$. The corresponding estimated information matrices for the ZIP and the ZIP(τ) models, respectively, are

$$\hat{\mathrm{I}}(\hat{\boldsymbol{\beta}}, \hat{\boldsymbol{\gamma}}) = -\begin{bmatrix} \frac{\partial^2 \log \mathrm{p}(y_i)}{\partial \boldsymbol{\beta} \partial \boldsymbol{\beta}'} & \frac{\partial^2 \log \mathrm{p}(y_i)}{\partial \boldsymbol{\beta} \partial \boldsymbol{\gamma}} \\ \frac{\partial^2 \log \mathrm{p}(y_i)}{\partial \boldsymbol{\beta} \partial \boldsymbol{\gamma}} & \frac{\partial^2 \log \mathrm{p}(y_i)}{\partial \boldsymbol{\gamma} \partial \boldsymbol{\gamma}'} \end{bmatrix} \tag{12.17}$$

and

$$\hat{\mathrm{I}}(\hat{\boldsymbol{\beta}}, \hat{\tau}) = \begin{bmatrix} \frac{\partial^2 \log \mathrm{p}(y_i)}{\partial \boldsymbol{\beta} \partial \boldsymbol{\beta}'} & \frac{\partial^2 \log \mathrm{p}(y_i)}{\partial \boldsymbol{\beta} \partial \tau} \\ \frac{\partial^2 \log \mathrm{p}(y_i)}{\partial \boldsymbol{\beta} \partial \tau} & \frac{\partial^2 \log \mathrm{p}(y_i)}{\partial \tau \partial \tau'} \end{bmatrix}. \tag{12.18}$$

A major limitation of these models is that they do not accommodate random effects. For example, we may wish to determine if rates of some behavior (*e.g.*, service utilization) are increasing over time in individuals who receive a particular intervention relative to those who do not.

12.4 MIXED-EFFECTS MODELS FOR COUNTS

The traditional Poisson regression model for count responses and the logistic regression model for binary or dichotomous responses assume statistical independence of observations. However, in many cases the frequency data are clustered or longitudinal and the assumption of independence is not reasonable. For example, in studies of mental health service utilization, changes in utilization frequency over time is often the primary focus of the investigation. Likewise, in cross-sectional studies where observations are clustered within geographic units (*e.g.*, suicide rates in counties), observations within the geographic units are more likely to be correlated than observations between geographic units.

For count response data, Goldstein [1991] proposed a multilevel log-linear model. Breslow [1984] described a Poisson regression model with normally distributed random effects, and Lawless and Willmot [1989] studied a Poisson regression model with random effects having an inverse Gaussian distribution. Siddiqui [1996] developed a Poisson mixed-effects regression model for clustered count data, and compared models with normally distributed and gamma-distributed random effects. Other pertinent references include Thall [1988], Stukel [1993], and Albert [1992]. Note that these models have a single set of regression coefficients and do not incorporate zero inflation.

12.4.1 Mixed-Effects Poisson Regression Model

In this section, we describe parameter estimation of a mixed-effects Poisson regression model. The difference between this model and the ZIP model is the increased probability mass at zero and the ability to estimate separate regression models for the logistic (*i.e.*, binary) and Poisson (*i.e.*, count) components of the ZIP model.

Suppose that there are $n = \sum_i n_i$ nonnegative observations y_{ij} for $i = 1, \ldots, N$ subjects and $j = 1, \ldots, n_i$ observations for subject i and a $(p+1)$ dimensional unknown parameter vector, $\boldsymbol{\beta}$, associated with a covariate vector $\boldsymbol{x_{ij}} = (1, x_{ij1}, \cdots, x_{ijp})'$. For simplicity, consider a model with a single random effect v_i, and assume that v_i is normally distributed with mean 0 and variance σ^2 and independent of the covariate vector $\boldsymbol{x_{ij}}$. Then the conditional density function of the n_i individual responses for subject i is written as

$$\begin{aligned} f(\boldsymbol{y_i}|\theta) &= \prod_{j=1}^{n_i} \mathrm{f}(y_{ij}; \lambda_{ij}) \\ &= \prod_{j} \frac{\exp(-\lambda_{ij})\lambda_{ij}^{y_{ij}}}{y_{ij}!} \end{aligned} \tag{12.19}$$

with

$$\begin{aligned} \lambda_{ij} &= \exp(\boldsymbol{x'_{ij}}\boldsymbol{\beta} + v_i) \\ &= \exp(\boldsymbol{x'_{ij}}\boldsymbol{\beta} + \sigma\theta_i), \end{aligned} \tag{12.20}$$

where $\theta_i = v_i/\sigma$ such that $\theta_i \sim N(0,1)$. The log-likelihood function corresponding to (12.19) becomes

$$\log l(\boldsymbol{y_i}|\theta) = -\sum_{j=1}^{n_i} \left[\exp(\boldsymbol{x'_{ij}}\boldsymbol{\beta} + \sigma\theta_i) + y_{ij}(\boldsymbol{x'_{ij}}\boldsymbol{\beta} + \sigma\theta_i) - \log(y_{ij}!) \right]. \tag{12.21}$$

The first and second derivatives with respect to $\boldsymbol{\beta}$ and σ are

$$\begin{bmatrix} \frac{\partial \log l(\boldsymbol{y_i}|\theta)}{\partial \boldsymbol{\beta}} \\ \frac{\partial \log l(\boldsymbol{y_i}|\theta)}{\partial \sigma} \end{bmatrix} = \begin{bmatrix} \sum_{j=1}^{n_i} (y_{ij} - \exp(\boldsymbol{x'_{ij}}\boldsymbol{\beta} + \sigma\theta_i))\boldsymbol{x_{ij}} \\ \sum_{j=1}^{n_i} (y_{ij} - \exp(\boldsymbol{x'_{ij}}\boldsymbol{\beta} + \sigma\theta_i))\theta_i \end{bmatrix} \tag{12.22}$$

and

$$\begin{bmatrix} \frac{\partial^2 \log l(\boldsymbol{y_i}|\theta)}{\partial \boldsymbol{\beta}\partial \boldsymbol{\beta}'} \\ \frac{\partial^2 \log l(\boldsymbol{y_i}|\theta)}{\partial \sigma \partial \sigma'} \end{bmatrix} = \begin{bmatrix} -\sum_{j=1}^{n_i} \exp(\boldsymbol{x'_{ij}}\boldsymbol{\beta} + \sigma\theta_i)\boldsymbol{x_{ij}}\boldsymbol{x'_{ij}} \\ -\sum_{j=1}^{n_i} \exp(\boldsymbol{x'_{ij}}\boldsymbol{\beta} + \sigma\theta_i)\theta_i^2 \end{bmatrix}. \tag{12.23}$$

The marginal density of $\boldsymbol{y_i}$ is approximated as

$$\begin{aligned} h(\boldsymbol{\beta}, \sigma, \theta; \boldsymbol{y_i}) &= h(\boldsymbol{y_i}) \\ &= \int_\theta f(\boldsymbol{y_i}|\theta)g(\theta)d\theta \\ &\approx \sum_{q=1}^{Q} \left[\prod_{j=1}^{n_i} \frac{\exp(-\lambda_{ijq})\lambda_{ijq}^{y_{ij}}}{y_{ij}!} \right] A(B_q), \end{aligned} \tag{12.24}$$

where

$$\lambda_{ijq} = \exp(\boldsymbol{x}'_{ij}\boldsymbol{\beta} + \sigma B_q), \tag{12.25}$$

and B_q and $A(B_q)$ are the values of the Gauss–Hermite quadrature node and weight at q, $q = 1, \ldots, Q$ (see Section 9.6.3). The first derivatives of the full log-likelihood are

$$\frac{\partial \log L}{\partial \boldsymbol{\beta}} = \sum_{i=1}^{N} \frac{1}{h(y_i)} \sum_{q}^{Q} \sum_{j=1}^{n_i} (y_{ij} - \lambda_{ijq}) \boldsymbol{x_{ij}} \frac{\exp(-\lambda_{ijq}) \lambda_{ijq}{}^{y_{ij}}}{y_{ij}} A(B_q) \tag{12.26}$$

and

$$\frac{\partial \log L}{\partial \sigma} = \sum_{i=1}^{N} \frac{1}{h(y_i)} \sum_{q}^{Q} \sum_{j=1}^{n_i} (y_{ij} - \lambda_{ijq}) \theta_i \frac{\exp(-\lambda_{ijq}) \lambda_{ijq}{}^{y_{ij}}}{y_{ij}} A(B_q). \tag{12.27}$$

Parameters of the model can be estimated by setting the above first derivatives to zero and iteratively solving using the Fisher scoring or Newton–Raphson method described in the previous section.

12.4.2 Estimation of Random Effects

As previously discussed for several of the other random-effect models it is often of interest to estimate values of the random effects θ_i within a sample. For this, the (EAP) estimator $\bar{\theta}_i$ has been suggested by Bock and Aitkin [1981]. The estimator $\bar{\theta}_i$ given $\boldsymbol{y_i}$ for subject i can be obtained as

$$\bar{\theta}_i = \mathrm{E}(\theta_i | \boldsymbol{y_i}) = \frac{1}{h(\boldsymbol{y_i})} \int_{\theta} \theta_i l(\boldsymbol{y_i} | \theta_i) \, g(\theta) \, d\theta. \tag{12.28}$$

Similarly, the variance of the posterior distribution of $\bar{\theta}_i$, which may be used to make inferences regarding the EAP estimator (*i.e.*, the posterior variance is an estimate of precision of the estimate of $\bar{\theta}_i$), is given by

$$\mathrm{Var}(\bar{\theta}_i | \boldsymbol{y_i}) = \frac{1}{h(\boldsymbol{y_i})} \int_{\theta} (\theta_i - \bar{\theta}_i)^2 l(\boldsymbol{y_i} | \theta_i) \, g(\theta) \, d\theta. \tag{12.29}$$

Again, the integrations can be approximated numerically using Gauss–Hermite quadrature by summing over the specified number of quadrature nodes and corresponding quadrature weights.

In clustered data, the empirical Bayes estimator ($\bar{\theta}_i$) of cluster-specific effects has been used by Thomas et al. [1992] in hospital mortality rate analysis. In their presentation, these cluster-specific effects or hospital-specific effects represent how much the death rates for patients at hospital i differ (on the logit scale) from the national rates for patients with the same covariate values. In this context, cluster-specific effects may be linked to service utilization rates for individuals (*i.e.*, longitudinal case) or clinics, hospitals, units, families etc. (*i.e.*, the clustered case). Longford [1994] gives extensive references to applications involving empirical Bayes estimates of random effects.

12.4.3 Mixed-Effects ZIP Regression Model

Mixed-effects ZIP models have been described by Hur et al. [2002], who developed a mixed-effects ZIP model for a single random effect, and Hall [2000], who incorporated random effects into a ZIP model for analysis of repeated measures data. However, in Hall's model, the random effects were only added into the Poisson component of the ZIP model (*i.e.*, not for the binary component) and he used the EM algorithm to estimate parameters (which is slow and does not guarantee convergence).

It should be noted that it is possible to have different distributions for the random effects, such as a uniform (or rectangular), a gamma distribution for the Poisson model [Hausman et al., 1984; Lawless, 1987; Dean and Lawless, 1989] or a beta distribution for the logistic model [Prentice, 1986]. However, as noted by Longford [1994], "the normal is the only well-established multivariate distribution with a full range of correlation structures" for models with multiple random effects. Preisler [1989] also noted that interpretation of parameters is more straightforward because the random and fixed effects are on the same scale. Thus, while other distributions may have computational advantages (*i.e.*, the gamma and beta), the normal provides a more natural choice for interpretation of results in the general case. Note that one possible alternative to the multivariate normal distribution is the multivariate gamma distribution. The multivariate gamma is attractive because it can accommodate similar correlation structures to the multivariate normal.

The ZIP model results from a mixture distribution where the data generating process may be separated into two parts: a binomial data generating process governed by the mixing parameter q and a Poisson count data generating process governed by the Poisson parameter λ. Depending on the relationships between the Poisson parameter (λ) and the logistic parameter (q), a mixed-effects ZIP model or a mixed-effects ZIP(τ) model is used. When covariate effects from the Poisson and logistic parts are different or functionally not related to each other, a mixed-effects ZIP model is presented. Alternatively, when covariates are the same and the parameters λ and q are functionally related, a mixed-effect ZIP(τ) model is used.

Let y_{ij} represent the response or dependent variable, and let $\boldsymbol{x_{ij}} = (1, x_{ij1}, \ldots, x_{ijp})'$ and $\boldsymbol{w_{ij}} = (1, w_{ij1}, \ldots, w_{ijr})'$ be the $(p+1) \times 1$ and $(s+1) \times 1$ covariate vectors for the Poisson and logistic parts, respectively. Here, $i = 1, \ldots, N$ subjects and $j = 1, \ldots, n_i$ repeated observations for subject i, so that the total number of observations is $n = \sum_{i=1}^{N} n_i$. Denote $\boldsymbol{\beta} = (\beta_0, \beta_1, \ldots, \beta_p)'$ and $\boldsymbol{\gamma} = (\gamma_0, \gamma_1, \ldots, \gamma_s)'$ as the Poisson and logistic regression parameter vectors associated with covariates $\boldsymbol{x_{ij}}$ and $\boldsymbol{w_{ij}}$. Then the mixed-effects ZIP model for the normally distributed random effects $v_i \sim N(0, {\sigma_1}^2)$ and $\nu_i \sim N(0, {\sigma_2}^2)$ is defined as follows:

$$\Pr(y_{ij}) = \mathrm{p}(y_{ij}) = (1 - \pi_{ij}) f(y_{ij}) + \mathrm{I}(y_{ij})\pi_{ij}, \tag{12.30}$$

where

$$f(y_{ij}) = \frac{\exp(-\lambda_{ij})\lambda_{ij}^{y_{ij}}}{y_{ij}!}, \tag{12.31}$$

$$\begin{aligned}\text{logit}(\pi_{ij}) &= \boldsymbol{w_{ij}}'\boldsymbol{\gamma} + \nu_i \\ &= \boldsymbol{w_{ij}}'\boldsymbol{\gamma} + \sigma_2\theta_{2i},\end{aligned} \tag{12.32}$$

and

$$\begin{aligned}\log(\lambda_{ij}) &= \boldsymbol{x}'_{\boldsymbol{ij}}\boldsymbol{\beta} + \upsilon_i \\ &= \boldsymbol{x}'_{\boldsymbol{ij}}\boldsymbol{\beta} + \sigma_1\theta_{1i}.\end{aligned} \tag{12.33}$$

The function $\text{I}(y_{ij})$ is an indicator function taking a value of 1 if the observed response is zero ($y_{ij} = 0$) and a value of 0 if the observed response is positive ($y_{ij} > 0$). Note that since the covariates that effect the Poisson and logistic parts of the model are not necessarily the same, two different sets of covariate vectors $\boldsymbol{x_{ij}}$ and $\boldsymbol{w_{ij}}$ are in this model. The two separate random effects ν_i and υ_i for the logistic and Poisson parts are standardized such that $\theta_{1i} = \upsilon_i/\sigma_1$ and $\theta_{2i} = \nu_i/\sigma_2$. This allows for the variation of the random effects to be different corresponding to the Poisson and logistic components of the model.

The above mixed-effects model is termed a subject-specific model, as opposed to a population-averaged model [Neuhaus and Jewell, 1990; Neuhaus et al., 1991; Ten Have et al., 1993, 1996], and in this approach the probability distribution of y_{ij} is modeled as a function of the covariates $\boldsymbol{x_{ij}}$ and $\boldsymbol{w_{ij}}$ and parameters ν_i and υ_i specific to the ith subject. Here, $\boldsymbol{\beta}$ measures the change in the conditional log of the mean response with the covariates $\boldsymbol{x_{ij}}$ for subject i described by υ_i. Likewise, $\boldsymbol{\gamma}$ measures the change in the conditional logit of the probability of response with the covariates $\boldsymbol{w_{ij}}$ for subject i described by ν_i. The parameters σ_1 and σ_2 represent subject variation for the Poisson and logistic components, respectively.

By rewriting the model as functions of covariates $\boldsymbol{x_{ij}}$ and $\boldsymbol{w_{ij}}$, we obtain

$$\pi_{ij} = \frac{\exp(\boldsymbol{w_{ij}}'\boldsymbol{\gamma} + \sigma_2\theta_{2i})}{1 + \exp(\boldsymbol{w_{ij}}'\boldsymbol{\gamma} + \sigma_2\theta_{2i})} \tag{12.34}$$

and

$$\lambda_{ij} = \exp(\boldsymbol{x}'_{\boldsymbol{ij}}\boldsymbol{\beta} + \sigma_1\theta_{1i}). \tag{12.35}$$

For estimation, define $\boldsymbol{\eta} = (\boldsymbol{\eta}_1, \boldsymbol{\eta}_2)' = (\boldsymbol{\beta}, \sigma_1, \boldsymbol{\gamma}, \sigma_2)'$. Although the computations involving the derivatives are complex, estimation of the parameters of the mixed-effects ZIP model is a straightforward application of the previously described maximum likelihood methods. Assuming that there are N subjects and n_i observations for subject i, then the likelihood function of the mixed-effects ZIP model for a given subject i can be expressed as the product of the individual likelihoods within a subject i:

$$l(\boldsymbol{y_i}|\theta) = \prod_{j=1}^{n_i} \left[(1 - \pi_{ij})f(y_{ij}) + \text{I}(y_{ij})\pi_{ij}\right], \tag{12.36}$$

where the log-likelihood of (12.36) is

$$\log l(\boldsymbol{y_i}|\theta) = \sum_{j=1}^{n_i} \log[(1-\pi_{ij})f(y_{ij}) + \mathrm{I}(y_{ij})\pi_{ij}] \tag{12.37}$$

and the first derivatives of the marginal log-likelihood for $\boldsymbol{\eta}_1 = (\boldsymbol{\beta}, \sigma_1)$ and $\boldsymbol{\eta}_2 = (\boldsymbol{\gamma}, \sigma_2)$ are

$$\frac{\partial \log l(\boldsymbol{y_i}|\theta)}{\partial \boldsymbol{\eta}_1} = \begin{bmatrix} \sum_{j=1}^{n_i} \frac{1}{\mathrm{p}(y_{ij})} \left[(1-\pi_{ij})f(y_{ij})\frac{\partial \log f(y_{ij})}{\partial \beta_0}\right] \\ \sum_{j=1}^{n_i} \frac{1}{\mathrm{p}(y_{ij})} \left[(1-\pi_{ij})f(y_{ij})\frac{\partial \log f(y_{ij})}{\partial \beta_1}\right] \\ \vdots \\ \sum_{j=1}^{n_i} \frac{1}{\mathrm{p}(y_{ij})} \left[(1-\pi_{ij})f(y_{ij})\frac{\partial \log f(y_{ij})}{\partial \beta_p}\right] \\ \sum_{j=1}^{n_i} \frac{1}{\mathrm{p}(y_{ij})} \left[(1-\pi_{ij})f(y_{ij})\frac{\partial \log f(y_{ij})}{\partial \sigma_1}\right] \end{bmatrix} \tag{12.38}$$

and

$$\frac{\partial \log l(\boldsymbol{y_i}|\theta)}{\partial \boldsymbol{\eta}_2} = \begin{bmatrix} \sum_{j=1}^{n_i} \frac{1}{\mathrm{p}(y_{ij})} [(\mathrm{I}(y_{ij}) - f(y_{ij})] \frac{\partial \pi_{ij}}{\partial \gamma_0}] \\ \sum_{j=1}^{n_i} \frac{1}{\mathrm{p}(y_{ij})} [(\mathrm{I}(y_{ij}) - f(y_{ij})] \frac{\partial \pi_{ij}}{\partial \gamma_1}] \\ \vdots \\ \sum_{j=1}^{n_i} \frac{1}{\mathrm{p}(y_{ij})} [(\mathrm{I}(y_{ij}) - f(y_{ij})] \frac{\partial \pi_{ij}}{\partial \gamma_s}] \\ \sum_{j=1}^{n_i} \frac{1}{\mathrm{p}(y_{ij})} [(\mathrm{I}(y_{ij}) - f(y_{ij})] \frac{\partial \pi_{ij}}{\partial \sigma_2}] \end{bmatrix}, \tag{12.39}$$

where

$$\begin{bmatrix} \frac{\partial \log f(y_{ij})}{\partial \beta_0} \\ \frac{\partial \log f(y_{ij})}{\partial \beta_1} \\ \vdots \\ \frac{\partial \log f(y_{ij})}{\partial \beta_p} \\ \frac{\partial \log f(y_{ij})}{\partial \sigma_1} \end{bmatrix} = \begin{bmatrix} (y_{ij} - \lambda_{ij})1 \\ (y_{ij} - \lambda_{ij})x_{ij1} \\ \vdots \\ (y_{ij} - \lambda_{ij})x_{ijp} \\ (y_{ij} - \lambda_{ij})\theta_{1i} \end{bmatrix} \tag{12.40}$$

and

$$\begin{bmatrix} \frac{\partial \pi_{ij}}{\partial \gamma_0} \\ \frac{\partial \pi_{ij}}{\partial \gamma_1} \\ \vdots \\ \frac{\partial \pi_{ij}}{\partial \gamma_s} \\ \frac{\partial \pi_{ij}}{\partial \sigma_2} \end{bmatrix} = \begin{bmatrix} \pi_{ij}(1-\pi_{ij})1 \\ \pi_{ij}(1-\pi_{ij})w_{ij1} \\ \vdots \\ \pi_{ij}(1-\pi_{ij})w_{ijs} \\ \pi_{ij}(1-\pi_{ij})\theta_{2i} \end{bmatrix}. \tag{12.41}$$

The second derivatives of the above log-likelihood are complex, but can be approximated using the BHHH estimator (the sum of the outer product of the first derivatives) to yield the observed information matrix. The solutions of the log-likelihood function can now be obtained iteratively using the Newton–Raphson algorithm as previously described.

Mixed-Effects ZIP(τ) Model. When the covariates for the Poisson model are the same as the covariates for the logistic model and the Poisson parameter vector λ_{ij} and the logistic parameter vector π_{ij} are functionally related (as previously described in the ZIP(τ) model), the normally distributed mixed-effects ZIP(τ) model is given by

$$\Pr(y_{ij}) = \mathrm{p}(y_{ij}) = (1 - \pi_{ij})f(y_{ij}) + \mathrm{I}(y_{ij})\pi_{ij}, \tag{12.42}$$

where $f(y_{ij})$ is defined in (12.31) and

$$\begin{aligned} \mathrm{logit}(\pi_{ij}) &= -\tau \boldsymbol{x}'_{ij}\boldsymbol{\beta} + \nu_i \\ &= -\tau \boldsymbol{x}'_{ij}\boldsymbol{\beta} + \sigma_2\theta_{2i} \end{aligned} \tag{12.43}$$

and $\log(\lambda_{ij})$ is defined in (60). The unknown parameter τ becomes

$$\tau = -\frac{\mathrm{E}\left[\log \frac{\pi_{ij}}{(1-\pi_{ij})}\right]}{\mathrm{E}\left[\log \lambda_{ij}\right]}, \tag{12.44}$$

where $\mathrm{E}(x)$ means the expected value of the random variable x.

Unlike a mixed-effects ZIP model, the covariates of the logistic part are functionally related to the Poisson covariates, and so only one set of covariate vectors $\boldsymbol{x}_{ij}$ is necessary in this model. However, the variations of the random effects are allowed to be different for the Poisson and logistic as was the case with the mixed-effects ZIP model. The vector $\boldsymbol{\beta}$ has the same interpretation as before.

Rewriting λ_{ij} and π_{ij} as functions of the covariates yields

$$\pi_{ij} = \frac{\exp(-\tau \boldsymbol{x}'_{ij}\boldsymbol{\beta} + \sigma_2\theta_{2i})}{1 + \exp(-\tau \boldsymbol{x}'_{ij}\boldsymbol{\beta} + \sigma_2\theta_{2i})} \tag{12.45}$$

and

$$\lambda_{ij} = \exp(\boldsymbol{x}'_{ij}\boldsymbol{\beta} + \sigma_1\theta_{1i}). \tag{12.46}$$

The first derivatives of the marginal log-likelihood function of the mixed-effects ZIP(τ) model with respect to $\boldsymbol{\eta}_1 = (\boldsymbol{\beta}, \sigma_1)$ and $\boldsymbol{\eta}_2 = (\tau, \sigma_2)$ are

$$\frac{\partial \log l(\boldsymbol{y}_i|\theta)}{\partial \boldsymbol{\eta}_1} = \begin{bmatrix} \sum_{j=1}^{n_i} \frac{1}{\mathrm{p}(y_{ij})} \left[(1-\pi_{ij})f(y_{ij})\frac{\partial \log f(y_{ij})}{\partial \beta_0} + [\mathrm{I}(y_{ij}) - f(y_{ij})]\frac{\partial \pi_{ij}}{\partial \beta_0}\right] \\ \sum_{j=1}^{n_i} \frac{1}{\mathrm{p}(y_{ij})} \left[(1-\pi_{ij})f(y_{ij})\frac{\partial \log f(y_{ij})}{\partial \beta_1} + [\mathrm{I}(y_{ij}) - f(y_{ij})]\frac{\partial \pi_{ij}}{\partial \beta_1}\right] \\ \vdots \\ \sum_{j=1}^{n_i} \frac{1}{\mathrm{p}(y_{ij})} \left[(1-\pi_{ij})f(y_{ij})\frac{\partial \log f(y_{ij})}{\partial \beta_p} + [\mathrm{I}(y_{ij}) - f(y_{ij})]\frac{\partial \pi_{ij}}{\partial \beta_p}\right] \\ \sum_{j=1}^{n_i} \frac{1}{\mathrm{p}(y_{ij})} \left[(1-\pi_{ij})f(y_{ij})\frac{\partial \log f(y_{ij})}{\partial \sigma_1} + [\mathrm{I}(y_{ij}) - f(y_{ij})]\frac{\partial \pi_{ij}}{\partial \sigma_1}\right] \end{bmatrix} \tag{12.47}$$

and

$$\frac{\partial \log l(\boldsymbol{y_i}|\theta)}{\partial \boldsymbol{\eta}_2} = \begin{bmatrix} \sum_{j=1}^{n_i} \frac{1}{\mathrm{p}(y_{ij})}[\mathrm{I}(y_{ij}) - f(y_{ij})]\frac{\partial \pi_{ij}}{\partial \tau} \\ \sum_{j=1}^{n_i} \frac{1}{\mathrm{p}(y_{ij})}[\mathrm{I}(y_{ij}) - f(y_{ij})]\frac{\partial \pi_{ij}}{\partial \sigma_2} \end{bmatrix}, \tag{12.48}$$

where

$$\begin{bmatrix} \frac{\partial \log f(y_{ij})}{\partial \beta_0} \\ \frac{\partial \log f(y_{ij})}{\partial \beta_1} \\ \vdots \\ \frac{\partial \log f(y_{ij})}{\partial \beta_p} \\ \frac{\partial \log f(y_{ij})}{\partial \sigma_1} \end{bmatrix} = \begin{bmatrix} (y_{ij} - \lambda_{ij})1 \\ (y_{ij} - \lambda_{ij})x_{ij1} \\ \vdots \\ (y_{ij} \quad \lambda_{ij})x_{ijp} \\ (y_{ij} - \lambda_{ij})\theta_{1i} \end{bmatrix}, \tag{12.49}$$

$$\begin{bmatrix} \frac{\partial \pi_{ij}}{\partial \beta_0} \\ \frac{\partial \pi_{ij}}{\partial \beta_1} \\ \vdots \\ \frac{\partial \pi_{ij}}{\partial \beta_p} \\ \frac{\partial \pi_{ij}}{\partial \sigma_1} \end{bmatrix} = \begin{bmatrix} -\tau\pi_{ij}(1 - \pi_{ij})1 \\ -\tau\pi_{ij}(1 - \pi_{ij})x_{ij1} \\ \vdots \\ -\tau\pi_{ij}(1 - \pi_{ij})x_{ijp} \\ 0 \end{bmatrix}, \tag{12.50}$$

and

$$\begin{bmatrix} \frac{\partial \pi_{ij}}{\partial \tau} \\ \frac{\partial \pi_{ij}}{\partial \sigma_2} \end{bmatrix} = \begin{bmatrix} -\pi_{ij}(1 - \pi_{ij})\boldsymbol{x}'_{ij}\boldsymbol{\beta} \\ \pi_{ij}(1 - \pi_{ij})\theta_{2i} \end{bmatrix}. \tag{12.51}$$

The derivatives of the full log-likelihood of the ZIP(τ) model can be obtained as previously described for the mixed-effects ZIP model. The second derivatives of the above log-likelihood can be approximated using the BHHH estimator, and so the solutions of the log-likelihood function can be obtained iteratively using the Newton–Raphson algorithm.

12.5 ILLUSTRATION

The enormous human cost of suicide makes research and prevention a national priority. Worldwide, there are about one million suicides annually. In the last 25 years about 750,000 people suicided in the United States, and suicides outnumber homicides by at least 3 to 2 [Goldsmith et al., 2002]. Deaths from suicide exceeded deaths from AIDS by 200,000 in the past 20 years. The estimated cost to the nation in lost income alone is 11.8 billion dollars [Goldsmith et al., 2002]. Biological, psychological, social, and cultural factors all have a significant impact on the risk of suicide; however, over 90% of suicides in the United States are associated with psychiatric illness [Goldsmith et al., 2002]. There is a paucity of randomized, controlled clinical trials evaluating the safety and efficacy of any antidepressant medication in depressed subjects at risk for suicide such as based on a history of suicidal behavior. In the absence of such data, and the huge number of subjects such studies would require to have adequate statistical power due to the low base rate of suicide, alternative approaches are needed to provide an indication of the relationship of antidepressant treatment to suicide.

As an alternative, Gibbons et al. [2005] obtained data from the National Center for Health Statistics on suicide rates for 1996–1998 for each U.S. county broken down by sex, race (African American versus other), and age (5–14, 15–24, 25–44, 45–64, and 65 and over). County-level antidepressant medication prescription rate data designed to provide estimates of national and local prescription volumes came from a random sample of 20,000 pharmacies from the 36,000 pharmacies in the IMS database, representing over half of all retail pharmacies in the continental United States. The data do not include hospital prescriptions. For each county, prescription rates (number of pills per county from 1996 to 1998) were obtained for three antidepressant subclasses: tricyclic antidepressants (TCAs), selective serotonin reuptake inhibitors (SSRIs: citalopram, paroxetine, fluoxetine, fluvoxamine and sertraline), and other non-SSRI antidepressants (nefazodone, mirtazepine, buproprion and venlafaxine). To relate antidepressant use to suicide rate adjusting for county-specific case-mix (age, sex, and race), they used a mixed-effects Poisson regression model. The model estimates overall suicide rate conditional on age, sex, race, and antidepressant use and can also be used to estimate covariate adjusted county-specific estimates of suicide rates. In terms of antidepressant drug use, they used the natural logarithm of number of pills per person per year to: (a) adjust for differential population size of counties and (b) eliminate excessive influence of counties with extremely high or low antidepressant utilization. In the model, age, sex, and race were considered fixed-effects and the intercept and antidepressant drug effects were treated as random effects. This model specification allows the suicide rate to vary from county to county, and allows the relationship between antidepressant drug use and suicide to vary from county to county. As such, they estimated county-specific changes in suicide rates attributable to changes in antidepressant drug use, adjusted for the age, sex, and race composition of each county. The effect of policy changes (*e.g.*, adding or eliminating a particular type of antidepressant medication) was then estimated by accumulating the county-specific estimates over all counties. To test the possibility that the observed associations were simply due to access to quality health care, they included the effects of median county-level income as a covariate in a second model. To decompose the overall relationship between suicide rate and antidepressant drug use into intra-county and inter-county components, a third model was fitted using the county mean drug use and yearly deviation from the mean for each class of drugs as covariates in the model (see Chapter 4). The estimated coefficient for the mean drug use corresponds to between-county effect, and the estimated coefficient for the yearly deviations from the county mean corresponds to the within-county effect.

To establish goodness of fit of the mixed-effects Poisson regression model to the observed data, observed and estimated suicide rates (broken down by age, sex, and race) are displayed in Table 12.1. In general, suicide rate increases with age, is higher in males, and is lower in African Americans. Black females have the lowest suicide rates across the age range. In white males the suicide rate is increasing with age, whereas in all other groups the suicide rate either is constant or decreases after age 65 years. Comparison of the observed and expected frequencies reveals that the mixed-effects Poisson regression model fits the observed data extremely well. The observed number of suicides from 1996 to 1998 was 91,673 and the estimated rate (based on actual drug use) was 90,973, a difference of 233 suicides per year (0.76%). Estimated marginal suicide rates for a GEE model are also presented in Table 12.1, and the GEE model also fits the data extremely well. The primary difference between the GEE and mixed-effects regression models in this context is that the mixed-effects models provide county-specific covariate adjusted suicide rate estimates whereas the GEE model only provides overall population suicide rates.

The overall relationship between all antidepressant drug use and suicide rate was not statistically significant ($Z = -1.46, p < .14$). To determine whether antidepressant class-specific associations exist, a model with TCA and SSRI+non-SSRI was fitted to the data. The combination of SSRI and non-SSRI was used to avoid multicolinearity because these two drug classes were highly correlated, $r = 0.98$.

Table 12.1. Observed and Expected Suicide Rates by Age, Race, and Sex, 1996–1998

Age Group	Race	Sex	Number of Suicides	Population	Observed Rate	Expected Rate[a]	Expected Rate[b]
5–14	Black	Male	79	9,256,227	0.000009	0.000010	0.000009
5–14	Black	Female	28	8,978,221	0.000003	0.000003	0.000002
5–14	Other	Male	620	50,356,003	0.000012	0.000014	0.000012
5–14	Other	Female	206	47,847,778	0.000004	0.000005	0.000004
15–24	Black	Male	1,333	8,389,386	0.000159	0.000177	0.000160
15–24	Black	Female	191	8,352,196	0.000023	0.000024	0.000021
15–24	Other	Male	9,482	47,906,710	0.000198	0.000222	0.000198
15–24	Other	Female	1,673	45,396,608	0.000037	0.000042	0.000037
25–44	Black	Male	2,546	15,274,935	0.000167	0.000184	0.000164
25–44	Black	Female	474	17,191,095	0.000028	0.000033	0.000030
25–44	Other	Male	27,209	109,106,670	0.000249	0.000283	0.000250
25–44	Other	Female	6,977	108,864,081	0.000064	0.000072	0.000064
45–64	Black	Male	861	7,741,680	0.000111	0.000124	0.000111
45–64	Black	Female	224	9,633,227	0.000023	0.000026	0.000023
45–64	Other	Male	17,358	72,740,945	0.000239	0.000267	0.000239
45–64	Other	Female	5,307	76,289,629	0.000070	0.000078	0.000070
65+	Black	Male	415	3,295,133	0.000126	0.000142	0.000131
65+	Black	Female	83	5,140,632	0.000016	0.000014	0.000013
65+	Other	Male	14,074	38,889,596	0.000362	0.000398	0.000361
65+	Other	Female	2,814	55,229,051	0.000051	0.000057	0.000051

[a] Mixed-effects regression model.

[2] GEE model.

Maximum likelihood estimates (MLEs) indicated that the combination of SSRIs and non-SSRIs/non-TCAs had a significant negative association with suicide rate (MLE = $-.15$, $p < .001$). TCA use had a significant positive association with suicide rate (MLE = .20, $p < .001$), meaning increases in TCA use are associated with increases in the suicide rate, whereas increases in SSRI and non-SSRI/non-TCA use are associated with decreases in the suicide rate. The ratio of SSRIs + non-SSRIs to TCAs use is related to suicide rate. When this ratio is large (greater SSRI and non-SSRI/non-TCA use), the suicide rate is lower, and when it is smaller (greater TCA use) the suicide rate is higher.

Regarding magnitude of the case-mix adjusted effects, the average population, 1996–1998, was 248,060,988, the number of suicides over this three-year period was 91,673, making an annual rate of 12.32 suicides per 100,000. Going from actual TCA use at that time versus hypothetically no TCA use at that time (adjusted for age, sex, race, and SSRIs + non-SSRIs versus non-TCA use) we estimate 10,237 fewer suicides per year or a 4.12 per 100,000 reduction in the annual suicide rate (*i.e.*, a 33% decrease). By contrast, if SSRIs and non-SSRIs/nonTCAs use was hypothetically eliminated, the estimated number of suicides would increase by 15,202 suicides per year or the rate would increase 6.13 suicides per 100,000 (*i.e.*, a 50% increase). These effects are apparent in the raw data (*i.e.*, unadjusted for the effects of sex, age, and race—see Figure 12.1). For the lowest decile (high TCA use) the overall observed suicide rate is approximately 15 per 100,000, whereas for the highest decile (low TCA use) the suicide rate is approximately 10 per 100,000. Note that the finding of an association between TCA use and suicide, does not necessarily imply that it is a causal association in which TCA use leads to suicide. TCA use may be an indicator of one of a number of possible problems in the health care delivery system, such as limited access to quality mental health care.

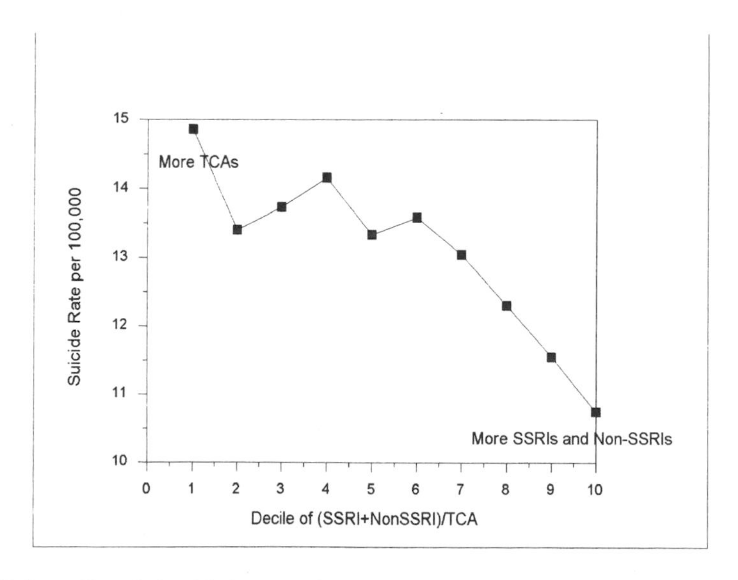

Figure 12.1. The relationship between the ratio of SSRIs and non-SSRIs to TCAs and observed suicide rate per 100,000.

Despite relatively little geographic clustering of antidepressant drug use ratios, there appears to be an association with population size (see Figure 12.2). Counties with larger populations use relatively less TCAs than the counties with smaller populations. This suggests there may be an urban-rural distinction in prescription practice. Rural areas may be poorer, have less access to psychiatrists and under treatment of depressive illness perhaps through use of subtherapeutic doses of side effect prone older tricyclic antidepressant medications that are more lethal on overdose, elevating suicide rates. To shed light on this hypothesis, median income for each county from the 2000 U.S. Census was included as a predictor in the model (in addition, to age, sex, race, and antidepressant medications). Income is related to suicide rate and had a negative coefficient (MLE = $-.01$, $p < .001$). Counties with higher income have lower suicide rates (adjusting for age, sex, and race composition). When income was included in the model, the overall magnitude of the antidepressant medication effects was reduced, but both remained statistically significant and in the same direction as without income adjustment (TCA MLE = 0.07, $p < .001$; SSRIs and non-SSRIs versus non-TCA MLE = -0.04, $p < .001$).

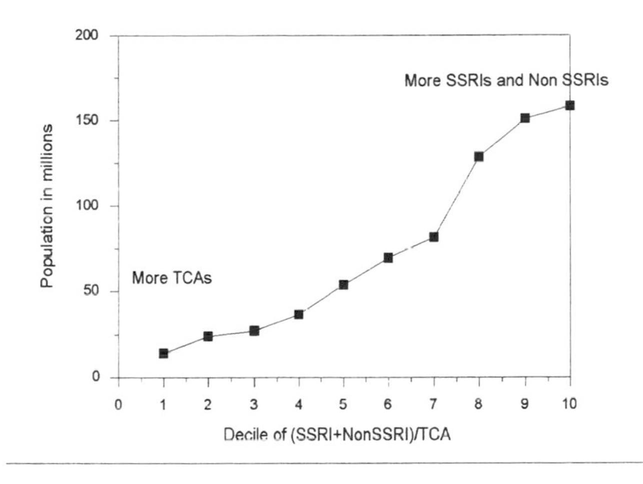

Figure 12.2. The relationship between the ratio of SSRIs and non-SSRIs to TCAs and population size.

Finally, drug effects were decomposed into between and within-county effects. Significant between-county effects were observed as before: positive association between TCA use and suicide ($p < .0001$) and a negative association between non-TCA use and suicide ($p < .0001$). Within-county effects were also significant for non-TCA use and suicide ($p < .0001$), where increase in non-TCA use over the three year time interval within counties was associated with lower suicide rates. By contrast, no significant within-county association between suicide rate and TCA use was observed, although the estimated coefficient was positive ($p < .39$). This finding adds further support to the notion that high TCA use in a county may be a marker of poorer access to high quality mental health care and/or poor detection or recognition of mental disorders, which in turn may be associated with increased suicide rates. As such, elimination of TCA prescriptions in and of itself may not produce the decreases in suicide rate predicted by the statistical model.

12.6 SUMMARY

In this chapter we have described mixed-effects Poisson and mixed-effects ZIP models for the analysis of longitudinal count data. The latter is especially appropriate when there are an excess of zeros in a dataset. In terms of the random effects, we have assumed that they come from a normal or multivariate normal distribution. Alternatively, several authors have proposed mixed Poisson-gamma regression models for analyzing longitudinal interval count data (*e.g.*, Thall [1988]). The advantage of using the gamma distribution for the (univariate) random effects is that the integration over the random effects distribution can be expressed in closed form. The disadvantage is that it specifies a distribution that is different from the normal distribution, and clearly, the latter is generally preferred for many reasons. As described, in order to allow for normally distributed random effects, numerical integration (*i.e.*, Gaussian quadrature) can be used. This provides us with a common technique for handling all types of non-linear mixed models (*i.e.*, dichotomous, ordinal, nominal), and is the approach used in several software packages (*e.g.*, SAS PROC NLMIXED, MIXPREG, LIMDEP).

Our example highlighted the use of mixed-effects Poisson regression to address a very important research question, namely, modeling of county-specific suicide rates as a function of antidepressant use, among other variables. For more information on this application, the interested reader is referred to Gibbons et al. [2005].

CHAPTER 13

MIXED-EFFECTS REGRESSION MODELS FOR THREE-LEVEL DATA

To this point, we have considered two-level data structures in which experimental units (*e.g.*, subjects) are repeatedly measured over time, or possibly nested within some social, cultural, or ecological strata (*e.g.*, school, classroom, clinic, hospital, county, etc.). In many longitudinal studies, the data are collected not at one site, but at multiple centers. This may occur when the number of subjects at any one center is inadequate, and so by obtaining a sample of centers a reasonable number of subjects are recruited for the study. Often, data of this sort are analyzed ignoring the fact that subjects are nested within different centers, or clusters. This type of analysis treats all subjects as independent observations. It is likely though, that the data from subjects within a cluster are not independent, but are correlated to some degree, and so an analysis which ignores this association may be misleading.

The amount of dependency in the data that is observable due to the clustering of the data is measured by the intraclass correlation. When the intraclass correlation equals 0, there is no association among subjects from the same cluster and analysis which ignores the clustering of the data is valid. For certain variables, however, intraclass correlation levels have been observed between 5% to 12% for data from spouse pairs and 0.05% to 0.85% for data clustered by counties. As the intraclass correlation increases, the amount of independent information from the data decreases, inflating the Type I error rate of an analysis which ignores this correlation. Thus, statistical analysis which treats all subjects as independent observations may yield tests of significance that are generally too liberal.

In previous papers [Hedeker et al., 1991, 1994], we have shown how mixed-effects regression models are useful in the analysis of clustered data which are not longitudinal. In that case, outcomes at the subject-level are modeled in terms of both subject (*i.e.*, sex and age) and cluster (*i.e.*, center size) level variables, while concurrently estimating and adjusting for the amount of intraclass correlation present in the data. Since mixed-effects regression models make no assumption regarding cluster sample size, a varying number of subjects within each cluster can be accommodated.

In this chapter, we describe how mixed-effects regression models can be used to analyze data which are both longitudinal and clustered. Data of this sort are oftentimes referred to as three-level data and the models as three-level models [Raudenbush and Bryk, 2002; Goldstein, 1995; Longford, 1987], the levels referring to the repeated observations (level 1) being nested within the subjects (level 2) who are nested within clusters (level 3).

13.1 THREE-LEVEL MIXED-EFFECTS LINEAR REGRESSION MODEL

In what follows, for simplicity, we will describe the 3-level model in terms of repeated observations (level 1) nested within subjects (level 2) who are nested within clusters (level 3). However, the model is also applicable to all types of three-level designs (*e.g.*, subjects nested within classrooms, and classrooms nested within schools). In terms of notation, suppose that we have $i = 1, \ldots, N$ clusters, $j = 1, \ldots, n_i$ subjects in cluster i, and $k = 1, \ldots, n_{ij}$ observations for subject j in cluster i. Further denote $N_i = \sum_{j=1}^{n_i} n_{ij}$, which represents the total number of observations within cluster i. Suppose that we wanted to fit a model with a single random cluster effect (*i.e.*, intercept) and multiple random subject effects (*i.e.*, intercept and trend). In terms of the random effects, this model could be written as follows

$$\begin{bmatrix} \boldsymbol{y}_{i1} \\ \boldsymbol{y}_{i2} \\ \boldsymbol{y}_{i3} \\ \cdots \\ \cdots \\ \cdots \\ \boldsymbol{y}_{in_i} \end{bmatrix} = \begin{bmatrix} 1 & \boldsymbol{Z}_{i1} & 0 & 0 & \ldots & 0 \\ 1 & 0 & \boldsymbol{Z}_{i2} & 0 & \ldots & 0 \\ 1 & 0 & 0 & \boldsymbol{Z}_{i3} & \ldots & 0 \\ \cdots & \cdots & \cdots & \cdots & \cdots & \cdots \\ \cdots & \cdots & \cdots & \cdots & \cdots & \cdots \\ \cdots & \cdots & \cdots & \cdots & \cdots & \cdots \\ 1 & 0 & 0 & 0 & \ldots & \boldsymbol{Z}_{in_i} \end{bmatrix} \begin{bmatrix} \gamma_i \\ \boldsymbol{v}_{i1} \\ \boldsymbol{v}_{i2} \\ \boldsymbol{v}_{i3} \\ \cdots \\ \boldsymbol{v}_{in_i} \end{bmatrix} + \begin{bmatrix} \boldsymbol{\varepsilon}_{i1} \\ \boldsymbol{\varepsilon}_{i2} \\ \boldsymbol{\varepsilon}_{i3} \\ \cdots \\ \cdots \\ \cdots \\ \boldsymbol{\varepsilon}_{in_i} \end{bmatrix},$$

or

$$\boldsymbol{y}_i = \boldsymbol{Z}_i \boldsymbol{v}_i^* + \boldsymbol{\varepsilon}_i,$$

where $\gamma_i \sim \mathcal{N}(0, \sigma_\gamma^2)$, $\boldsymbol{v}_{ij} \sim \mathcal{N}(0, \boldsymbol{\Sigma}_v)$, and $\varepsilon_{ijk} \sim \mathcal{N}(0, \sigma_\varepsilon^2)$. Here, in this model with a single random cluster effect, the number of random effects equals $r + 1$, where r is the number of random subject effects. However, for a given cluster i the dimension of $\boldsymbol{v}_i^*$ equals $n_i \times r + 1$. Let us denote this as R_i, where the subscript i indicates that this value varies depending on the size of the cluster. Adding covariates to this model yields

$$\boldsymbol{y}_i = \boldsymbol{X}_i \boldsymbol{\beta} + \boldsymbol{Z}_i \boldsymbol{v}_i^* + \boldsymbol{\varepsilon}_i, \tag{13.1}$$

where $\boldsymbol{X}_i$ is a known $N_i \times p$ design matrix for the fixed effects, $\boldsymbol{\beta}$ is the $p \times 1$ vector of unknown fixed regression parameters, $\boldsymbol{Z}_i$ is a known $N_i \times R_i$ design matrix for the random effects, $\boldsymbol{v}_i^*$ is the $R_i \times 1$ vector of unknown random effects, and $\boldsymbol{\varepsilon}_i$ is the $N_i \times 1$ error vector.

As a result, the observations $\boldsymbol{y}$ and random coefficients $\boldsymbol{v}^*$ have the joint multivariate normal distribution:

$$\begin{bmatrix} \boldsymbol{y}_i \\ \boldsymbol{v}_i^* \end{bmatrix} \sim \mathcal{N}\left(\begin{bmatrix} \boldsymbol{X}_i\boldsymbol{\beta} \\ 0 \end{bmatrix}, \begin{bmatrix} \boldsymbol{Z}_i\boldsymbol{\Sigma}_i\boldsymbol{Z}_i' + \sigma_\varepsilon^2\boldsymbol{I}_i & \boldsymbol{Z}_i\boldsymbol{\Sigma}_i \\ \boldsymbol{\Sigma}_i\boldsymbol{Z}_i' & \boldsymbol{\Sigma}_i \end{bmatrix} \right),$$

where the random coefficients $\boldsymbol{v}_i^*$ and variance-covariance matrix $\boldsymbol{\Sigma}_i$ are given as

$$\begin{bmatrix} \gamma_i \\ \boldsymbol{v}_{i1} \\ \boldsymbol{v}_{i2} \\ \cdots \\ \boldsymbol{v}_{in_i} \end{bmatrix} \sim \mathcal{N}\left(\begin{bmatrix} 0 \\ 0 \\ 0 \\ \cdots \\ 0 \end{bmatrix} \begin{bmatrix} \sigma_\gamma^2 & 0 & 0 & \cdots & 0 \\ 0 & \boldsymbol{\Sigma}_v & 0 & \cdots & 0 \\ 0 & 0 & \boldsymbol{\Sigma}_v & \cdots & 0 \\ \cdots & \cdots & \cdots & \cdots & \cdots \\ 0 & 0 & 0 & \cdots & \boldsymbol{\Sigma}_v \end{bmatrix} \right).$$

The subscript i for the variance–covariance matrix $\boldsymbol{\Sigma}_i$ merely reflects the number of observations within the cluster, not the number of parameters which is the same for all clusters.

The mean of the posterior distribution of $\boldsymbol{v}^*$, given $\boldsymbol{y}_i$, yields the EAP ("expected *a posteriori*"), or empirical Bayes, estimator of the cluster and individual trend parameters:

$$\bar{\boldsymbol{v}}^* = \left[\boldsymbol{Z}_i'(\boldsymbol{\sigma}_\varepsilon^2\boldsymbol{I}_i)^{-1}\boldsymbol{Z}_i + \boldsymbol{\Sigma}_i^{-1}\right]^{-1} \boldsymbol{Z}_i'(\boldsymbol{\sigma}_\varepsilon^2\boldsymbol{I}_i)^{-1}(\boldsymbol{y}_i - \boldsymbol{X}_i\boldsymbol{\beta})$$

with covariance matrix

$$\boldsymbol{\Sigma}_{v^*|y_i} = \left[\boldsymbol{Z}_i'(\boldsymbol{\sigma}_\varepsilon^2\boldsymbol{I}_i)^{-1}\boldsymbol{Z}_i + \boldsymbol{\Sigma}_i^{-1}\right]^{-1}.$$

Differentiating the log-likelihood, $\log L = \sum_{i=1}^N \log h(\boldsymbol{y}_i)$, yields

$$\begin{aligned}
\frac{\partial \log L}{\partial \boldsymbol{\beta}} &= \sigma_\varepsilon^2 \sum_{i=1}^N \boldsymbol{X}_i'\boldsymbol{\mu}_i, \\
\frac{\partial \log L}{\partial \boldsymbol{\Sigma}^*} &= \frac{1}{2}\sum_{i=1}^N \boldsymbol{D}_i'\boldsymbol{G}_i'\text{vec}\left[\boldsymbol{\Sigma}_i^{-1}(\boldsymbol{\Sigma}_{v^*|y_j} + \bar{\boldsymbol{v}}^*(\bar{\boldsymbol{v}}^*)' - \boldsymbol{\Sigma}_i)\boldsymbol{\Sigma}_i^{-1}\right], \\
\frac{\partial \log L}{\partial \sigma_\varepsilon^2} &= \frac{1}{2}\sigma_\varepsilon^{-4}\sum_{i=1}^N -n_i\sigma^2 + \boldsymbol{\mu}_i'\boldsymbol{\mu}_i + \text{tr}\left[\boldsymbol{\Sigma}_{v^*|y_i}\boldsymbol{Z}_i'\boldsymbol{Z}_i\right].
\end{aligned}$$

where $\boldsymbol{\mu}_i = \boldsymbol{y}_i - \boldsymbol{Z}_i\bar{\boldsymbol{v}}^* - \boldsymbol{X}_i\boldsymbol{\beta}$, and $\text{vech}\boldsymbol{\Sigma}_i = \boldsymbol{D}_i\boldsymbol{\Sigma}^*$. In this notation, $\boldsymbol{\Sigma}^*$ represents the vector with only the unique variance–covariance parameters, and $\boldsymbol{D}_i$ is a matrix of ones and zeros that is necessary to pick off only the correct terms. Note that these derivatives are functionally the same as those for the 2-level model. Thus, the same approach using the EM algorithm and Fisher scoring solution can be used.

13.1.1 Illustration

The following data were collected as part of the National Institute of Mental Health schizophrenia collaborative study on treatment-related changes in overall severity using the Inpatient Multidimensional Psychiatric Scale (IMPS) [Lorr and Klett, 1966]. Item 79, "severity of illness," was scored in the following way:

1. Normal, not at all ill
2. Borderline mentally ill
3. Mildly ill
4. Moderately ill
5. Markedly ill
6. Severely ill
7. Among the most extremely ill

Nine centers participated in this study, and within each center subjects were randomly assigned to a placebo condition or one of three drug conditions (chlorpromazine, fluphenazine, or thioridazine). Previous analysis of these data [Gibbons et al., 1988] revealed very similar effects for the three drug groups, so for this example we will concentrate only on subjects assigned to either the placebo or chlorpromazine condition. The sample sizes for these two groups across the timeframe of the study are displayed in Table 13.1. As can be seen from Table 13.1, the sample sizes across time are quite unbalanced, there being fairly large differences in the number of observations made in the six weeks of treatment.

Table 13.1. Sample Sizes Across Time

Treatment Group	Sample Sizes at Week						
	0	1	2	3	4	5	6
Placebo	110	108	5	89	2	2	72
Chlorpromazine	110	108	3	96	4	5	87

In the first analysis, the model included random subject intercepts and linear week trends. Additionally, fixed effects for treatment group, week, and an interaction of group by week were also included. In terms of the week effect, this was expressed as the square root of week, which is used here to linearize the relationship between the outcome variable and time. Results for this model are displayed in Table 13.2.

Table 13.2. Estimates, Standard Errors, and Probability Values for the Two-Level Mixed-Effects Regression Model

Parameter	Estimate	SE	$p <$
Placebo intercept (β_0)	5.345	0.085	.001
Placebo vs. Chlorpromazine (β_2)	0.029	0.120	.812
Placebo slope β_1	−0.343	0.066	.001
Placebo vs. Chlorpromazine (β_3)	−0.564	0.091	.001
Intercept variance $\sigma^2_{\upsilon_0}$	0.342	0.082	
Slope variance $\sigma^2_{\upsilon_1}$	0.224	0.044	
Intercept, slope covariance $\sigma_{\upsilon_0 \upsilon_1}$	0.007	0.047	
Error variance σ^2	0.571	0.042	

Inspection of Table 13.2, reveals the following. First, the average severity at baseline for the placebo group was a rating of 5.35, which is between "severely" and "markedly" ill. The precision of this estimate is approximately one decimal point (*i.e.*, SE = .09) and it is clearly differentiable from zero ($p < .001$). The chlorpromazine group did not differ significantly from the placebo group in their initial severity, its intercept was estimated at 5.38. The rate of change for the placebo group was $\hat{\beta}_1 = -.343$ which reflects an improvement of .343 units per week on the square root scale, or a total 6 week improvement of .839 units in the original metric. The chlorpromazine group exhibited a significantly increased improvement rates relative to the placebo group. The rate of change was .907 units per (square root) week for the chlorpromazine group, corresponding to an overall 6 week improvement rating of 2.22 units in the original metric. Taking the intercepts and slopes together, we obtain estimated 6 week average ratings of $5.345 - 0.343\sqrt{6} = 4.50$ for placebo (moderately to markedly ill), $5.373 - 0.907\sqrt{6} = 3.15$ for chlorpromazine (mildly ill).

In terms of the individual intercept and slope terms, there was considerable variability for both effects as evidenced by the appreciable $\sigma^2_{\upsilon_0}$ and $\sigma^2_{\upsilon_1}$ estimates. Taken together, these indicate that individuals within both treatment groups varied significantly in terms of their initial level of severity and their rate of improvement over time, as compared to the estimated overall trends for these two groups. Interestingly, the correlation between initial severity and rate of change was negligible ($r_{\upsilon_0 \upsilon_1} = .024$), indicating that initial severity had no significant influence on the rate of a patient's recovery.

Longitudinal Data From Multiple Centers

As was noted, longitudinal data are often obtained from a sample of centers and not just from one center. For example, the NIMH dataset analyzed above consisted of data collected from nine centers. Statistical analysis which ignores this clustering of the data, treats all individuals as independent observations. However, data from individuals within a cluster are likely to be correlated to some degree, and so any analysis which ignores this dependency in the data can be misleading. The amount of dependency in the data that is observable due to the clustering of the data is measured by the intraclass correlation. As the intraclass correlation increases, the amount of independent information from the data decreases, inflating the Type I error rate of an analysis which ignores this correlation [Blair et al., 1983]. As a result, significance tests are generally too liberal if intraclass correlation is present, but ignored.

The mixed model can be augmented to handle the clustering of data within centers. For this, we will consider a 3-level model for the measurement y made on occasion k for subject j within center i. Relative to the previous analysis, an additional random center effect, denoted γ_i, will be included. This term indicates the influence that the center is having on the response of the individual. We assume that the distribution of the center effects is normal with mean 0 and variance σ_γ^2. To the degree that clustering of individuals within centers influences the individual outcomes, the center effects γ_i deviate from zero and the population variance associated with these effects σ_γ^2 increases in value. Conversely, when the clustering of individuals within centers is having little influence on the individual outcomes, the center effects γ_i will all be near zero and the population variance σ_γ^2 will approach zero. Again, empirical Bayes estimation can be used for the center effects γ_i, with marginal maximum likelihood estimation of the population variance term σ_γ^2.

Treatment of the center effects as a random term in the model means that the specific centers used in the study are considered to be a representative sample from a larger population of potential centers. Conversely, if interest is only in making inferences about the specific centers of a dataset (*e.g.*, is there a difference between center A and center B?), the center could then be regarded as a "fixed" and not a "random" effect in the model. The random-effects approach is advised when there is interest in assessing the overall effect that any potential center may have on the data, and thus, in determining the degree of variability that the center accounts for in the data. This coincides with the manner in which the center is often conceptualized, that is, the center was drawn from a population of potential centers, and the 8 or 12 centers used in the study are not the population itself.

In addition to the random center effect, center-level covariates can be included in the model to assess the influence of, say, center size on the individual responses. Interactions between center-level, individual-level, and occasion-level covariates can also be included in the model. For example, the interaction of center size (center-level) by treatment group (individual-level) by time (occasion-level) can be included into the model to determine whether, say, treated subjects from small centers improve over time more dramatically than treated subjects from large centers. In this way, the model provides a useful method for examining and teasing out potential center-related effects.

As mentioned earlier, nine centers participated in the NIMH schizophrenia collaborative study. The sample sizes at baseline for the placebo and chlorpromazine groups at these nine centers are displayed in Table 13.3. As can be seen from Table 13.3, the sample sizes across the centers are fairly unbalanced.

Table 13.3. Baseline Sample Sizes Across Center

Treatment	Baseline Sample Sizes at Center								
Group	1	2	3	4	5	6	7	8	9
Placebo *(n = 107)*	13	20	13	15	13	7	10	10	6
Chlorpromazine *(n = 110)*	9	22	8	18	15	9	10	12	7

Table 13.4. Parameter Estimates, Standard Errors, and Probability Values for the Three-Level Mixed-Effects Regression Model

Parameter	Estimate	SE	$p <$
Placebo intercept (β_0)	5.339	0.106	.001
Placebo vs. Chlorpromazine (β_2)	0.038	0.116	.741
Placebo slope β_1	−0.341	0.066	.001
Placebo vs. Chlorpromazine (β_3)	−0.566	0.091	.001
Subject intercept variance $\sigma^2_{\upsilon_0}$	0.285	0.078	
Subject slope variance $\sigma^2_{\upsilon_1}$	0.225	0.044	
Subject intercept, slope covariance $\sigma_{\upsilon_0\,\upsilon_1}$	0.025	0.045	
Center variance σ^2_γ	0.039	0.031	
Error variance σ^2	0.570	0.040	

Results for the 3-level model are displayed in Table 13.4. Inspection of Table 13.4 reveals very similar results to those obtained earlier ignoring the center effect. The variance attributable to centers is not quite statistically significant, the standard deviation of the center effects represents approximately 1/5 of a rating unit ($\hat{\sigma}_\gamma = 0.198$). The variability associated with an individual's initial level is reduced relative to the previous 2-level model, though still very appreciable. The correlation between initial severity and rate of change is slightly larger, though still fairly negligible ($r_{\upsilon_0\,\upsilon_1} = .099$).

Table 13.5 lists the contributions of the three estimated sources of variability (residual, individual, and center), in terms of percentages of total variance, across the timepoints of the study. While the residual variance and center variance terms are constant over time, the amount of variance attributable to the individual varies with time; as a result, when expressed as percentages of total variance, all variance components change over time. The variance component attributable to the individual is a function of the estimated variance terms for the two random terms estimated per individual: the intercept and slope.

At baseline, the individual slope parameter does not influence the fitting of the data, that is, the baseline data are being fit by the individual intercept and the center effect. However, as time increases, the influence of the slope increases in fitting the data, and so as a result, the amount of variability attributable to the individual component increases appreciably over time.

While the amount of variability attributable to the center is relatively modest, between 1.66 to 4.39 percent of the total variance, this amount is consistent with what has been observed with other types of clustered data [Donner, 1982].

This example illustrates the application of mixed-effects regression models to the problem of unbalanced longitudinal data from multiple centers. Although in the present example the center effects were small, this cannot be assumed generally. The 3-level model described provides a way to empirically assess the influence of the center on the observations of the individual. If the center effect is observed to be negligible, then analysis by the 2-level model for longitudinal data is appropriate, otherwise, the results from the 2-level model may be misleading.

Regarding parameter estimates, the fixed effects are typically not greatly different between the 2- and 3-level models. However, the estimates of the standard errors, which determine the significance of these parameters, are influenced by the choice of model. In general, when a source of variability is present but ignored by the statistical model, the

Table 13.5. Estimated Variance Components expressed as Percent of Total Variance for the Three-Level Mixed-Effects Regression Model

Timepoint	Error	Subject	Center
Baseline	63.72	31.89	4.39
Week 1	48.73	47.91	3.36
Week 2	40.27	56.96	2.77
Week 3	34.41	63.22	2.37
Week 4	30.08	67.85	2.07
Week 5	26.74	71.42	1.84
Week 6	24.08	74.26	1.66

standard errors will be underestimated. Underestimation of standard errors results since the statistical model assumes that, conditional on the terms in the model, the observations are independent. However, when systematic variance is present, but ignored by the model, the observations are not independent, and the amount of independent information available in parameter estimation is erroneously inflated.

In the current context, there was very little difference in either the estimates or the standard errors between the 2- and 3-level models, since the amount of dependency present in the data as a result of the clustering of individuals within centers was small. The level of intraclass correlation observed in these data is consistent with reported levels from the literature [Donner, 1982; Donner and Klar, 2000; Jacobs et al., 1989; Murray, 1998; Siddiqui et al., 1996]. It is important to realize the level of intraclass correlation can vary depending on which dependent variable is being considered. When cluster and error variance terms, which determine the intraclass correlation, are being estimated in addition to other model parameters, the terms included in the model can affect the estimated value of intraclass correlation. Also, the sample size at various levels can influence precision of the variance terms, and so affect precision of intraclass correlation as well.

Finally, when dealing with multilevel data, the number of levels of data must be considered. Often, pooling higher-order levels is determined prior to the analysis for pragmatic or conceptual reasons; at other times, the decision can be empirically tested. From the example, one could empirically argue that 3-level analysis is unnecessary since the variance attributable to nesting of individuals within centers is not significant, so there is justification for pooling this additional level of the data. From a design perspective, on the other hand, there may be reason for including the center effect regardless of its statistical significance; for example, if centers were the unit of assignment in the randomization of treatment levels, one could argue that the random center term must remain in the model regardless of significance. When the variance attributable to a higher-order level is observed to be small and nonsignificant and the sample is of moderate size, the parameter estimates and standard errors will not differ greatly whether or not the higher-order level is included in the model. In this example, since difference between the 3-level analysis and the 2-level analysis was fairly small; both models gave similar results.

13.2 THREE-LEVEL MIXED-EFFECTS NONLINEAR REGRESSION MODELS

In this section, we discuss three-level nonlinear mixed-effects regression models, suitable for analysis of binary, ordinal, nominal and count response data. The primary difference between linear and nonlinear mixed-effects models is that in the nonlinear case, evaluation of the likelihood requires an r-dimensional integration over the joint distribution of the level-two and level-three random effects. This integration can either be performed numerically using fixed-point or adaptive quadrature (*e.g.*, Gauss–Hermite quadrature) or using some form of Monte Carlo simulation method (*e.g.*, Markov Chain Monte Carlo—MCMC). In the following sections we present the statistical foundation for several nonlinear mixed-effects regression models and an illustrative example.

13.2.1 Three-Level Mixed-Effects Probit Regression

To express the 3-level model in a general way, it is useful to use the following matrix representation. Stacking the unobservable latent response vectors of each subject within a cluster ($\boldsymbol{y}_{ij}$), the 3-level model for the resulting N_i response vector for the ith 3-level unit (classroom, clinic, etc.), $i = 1, 2, \ldots, N$, can be written as follows:

$$\underset{N_i\times 1}{\underbrace{\begin{bmatrix} \boldsymbol{y}_{i1} \\ \boldsymbol{y}_{i2} \\ \boldsymbol{y}_{i3} \\ \cdots \\ \cdots \\ \boldsymbol{y}_{in_i} \end{bmatrix}}_{\boldsymbol{y}_i}} = \underset{N_i\times((n_i\times r)+1)}{\underbrace{\begin{bmatrix} \mathbf{1}_{i1} & \boldsymbol{Z}_{i1} & 0 & 0 & \ldots & 0 \\ \mathbf{1}_{i2} & 0 & \boldsymbol{Z}_{i2} & 0 & \ldots & 0 \\ \mathbf{1}_{i3} & 0 & 0 & \boldsymbol{Z}_{i3} & \ldots & 0 \\ \cdots & \cdots & \cdots & \cdots & \cdots & \cdots \\ \cdots & \cdots & \cdots & \cdots & \cdots & \cdots \\ \mathbf{1}_{in_i} & 0 & 0 & 0 & \ldots & \boldsymbol{Z}_{in_i} \end{bmatrix}}_{\boldsymbol{Z}_i}} \underset{((n_i\times r)+1)\times 1}{\underbrace{\begin{bmatrix} \upsilon_{0i} \\ \upsilon_{i1} \\ \upsilon_{i2} \\ \upsilon_{i3} \\ \cdots \\ \upsilon_{in_i} \end{bmatrix}}_{\upsilon^*}} +$$

$$\underset{N_i\times(p+1)}{\underbrace{\begin{bmatrix} \mathbf{1}_{i1} & \boldsymbol{X}_{i1} \\ \mathbf{1}_{i2} & \boldsymbol{X}_{i2} \\ \mathbf{1}_{i3} & \boldsymbol{X}_{i3} \\ \cdots & \cdots \\ \cdots & \cdots \\ \mathbf{1}_{in_i} & \boldsymbol{X}_{in_i} \end{bmatrix}}_{\boldsymbol{X}_i}} \underset{(p+1)\times 1}{\underbrace{\begin{bmatrix} \beta_0 \\ \beta_1 \\ \cdots \\ \beta_p \end{bmatrix}}_{\boldsymbol{\beta}}} + \underset{N_i\times 1}{\underbrace{\begin{bmatrix} \varepsilon_{i1} \\ \varepsilon_{i2} \\ \varepsilon_{i3} \\ \cdots \\ \cdots \\ \varepsilon_{in_i} \end{bmatrix}}_{\varepsilon_i}}, \tag{13.2}$$

where the following independent components are distributed, $\upsilon_{0i} \sim \mathcal{N}(0, \sigma^2_{(3)})$, $\boldsymbol{\upsilon}_{ij} \sim \mathcal{N}(0, \boldsymbol{\Sigma}_{(2)})$, and $\varepsilon_i \sim \mathcal{N}(0, \sigma^2_\varepsilon \boldsymbol{I})$. Notice, there are n_i subjects within cluster i and N_i total observations within cluster i (the sum of all repeated observations for all subjects within the cluster). The number of random subject-level effects is r and the number of fixed covariates in the model (excluding the intercept) is p. In the case of binary responses, each person has a $n_{ij} \times 1$ vector $\boldsymbol{y}_{ij}$ of underlying response strengths, a $n_{ij} \times r$ design matrix $\boldsymbol{X}_{ij}$ for their r random effects $\boldsymbol{\upsilon}_{ij}$, and a $n_{ij} \times p$ matrix of covariates $\boldsymbol{X}_{ij}$. The covariate matrix usually includes the random effect design matrix so that the overall intercept, linear term, etc., is estimated and thus the random effects represent deviations from these overall terms.

A characteristic of the probit model is the assumption that there is an unobservable latent variable (y_{ijk}) related to the actual binary response through a "threshold concept" [Bock, 1975]. We assume the underlying latent variable y_{ijk} is continuous and that the binary response $\mathrm{y}_{ijk} = 1$ occurs when y_{ijk} exceeds a threshold γ (*i.e.*, $P(\mathrm{y}_{ijk} = 1) = P(y_{ijk} > \gamma)$). In terms of the latent response strength for subject j in cluster i on occasion k (y_{ijk}) we can rewrite (1) as

$$y_{ijk} = v_{0i} + \boldsymbol{z}'_{ijk}\boldsymbol{v}_{ij} + \boldsymbol{x}'_{ijk}\boldsymbol{\beta} + \varepsilon_{ijk}. \tag{13.3}$$

With the above mixed regression model for the latent variable y_{ijk}, the probability that $\mathrm{y}_{ijk} = 1$ (a positive response occurs), conditional on the random effects $\boldsymbol{v}^*$, is given by

$$\begin{aligned} \mathrm{P}(\mathrm{y}_{ijk} = 1 \mid \boldsymbol{v}^*) &= \frac{1}{\sqrt{(2\pi\sigma_\varepsilon^2)}} \int_\gamma^\infty \exp\left[-\frac{1}{2\sigma_\varepsilon^2}(y_{ijk} - v_{0i} - \boldsymbol{Z}'_{ijk}\boldsymbol{v}_{ij} - \boldsymbol{X}'_{ijk}\boldsymbol{\beta})^2\right] dy \\ &= \Phi[-(\gamma - \mathrm{z}_{ijk})/\sigma_\varepsilon], \end{aligned} \tag{13.4}$$

where $\mathrm{z}_{ijk} = v_{0i} + \boldsymbol{z}'_{ijk}\boldsymbol{v}_{ij} + \boldsymbol{x}'_{ijk}\boldsymbol{\beta}$ and $\Phi(\cdot)$ represents the cumulative standard normal density function. Without loss of generality, the origin and unit of z may be chosen arbitrarily. For convenience, let $\gamma = 0$ and to ensure identifiability let $\sigma_\varepsilon = 1$.

Let $\mathbf{y}_i$ be the vector of binary responses from cluster i for the n_i individuals examined at the n_{ij} timepoints. Assuming independence of the responses conditional on the random effects, the probability of any pattern $\mathbf{y}_i$, given $\boldsymbol{v}^*$, is equal to the product of the probabilities of the individual responses (both between and within individuals in cluster i):

$$\ell(\mathbf{y}_i \mid \boldsymbol{v}^*) = \prod_{j=1}^{n_i} \prod_{k=1}^{n_{ij}} [\Phi(\mathrm{z}_{ijk})]^{\mathrm{y}_{ijk}} [1 - \Phi(\mathrm{z}_{ijk})]^{1-\mathrm{y}_{ijk}}. \tag{13.5}$$

Then the marginal probability of $\mathbf{y}_i$ is expressed as the following integral of the likelihood, $\ell(\cdot)$, weighted by the prior density $g(\cdot)$:

$$h(\mathbf{y}_i) = \int_{\boldsymbol{v}^*} \ell(\mathbf{y}_i \mid \boldsymbol{v}^*)\, g(\boldsymbol{v}^*)\, d\boldsymbol{v}^*, \tag{13.6}$$

where $g(\boldsymbol{v}^*)$ represents the distribution of $\boldsymbol{v}^*$ in the population.

Orthogonalization of the Model Parameters. For numerical solution of the likelihood equations, Gibbons and Bock [1987] orthogonally transform the response model using the Cholesky decomposition of Σ_{v^*} [Bock, 1975]. Specifically, let $\boldsymbol{v}^* = \boldsymbol{T}^*\boldsymbol{\theta}^*$, where $\boldsymbol{T}^*\boldsymbol{T}^{*'} = \Sigma_{v^*}$ is the Cholesky decomposition of Σ_{v^*}. Then $\boldsymbol{\theta}^* = \boldsymbol{T}^{*-1}\boldsymbol{v}^*$, and so, $\mathcal{E}(\boldsymbol{\theta}^*) = \mathbf{0}$ and $\mathcal{V}(\boldsymbol{\theta}^*) = \boldsymbol{T}^{*-1}\Sigma_{v^*}(\boldsymbol{T}^{*-1})' = \boldsymbol{I}$. The reparameterized model is then

$$\mathrm{z}_{ijk} = \sigma_{(3)}\theta_{0i} + \boldsymbol{z}'_{ijk}\boldsymbol{T}\boldsymbol{\theta}_{ij} + \boldsymbol{x}'_{ijk}\boldsymbol{\beta}, \tag{13.7}$$

where θ_{0i} and $\boldsymbol{\theta}_{ij}$ are the standardized random effects for cluster i and individual j in cluster i respectively. Notice that since only a single random cluster effect is assumed, $\sigma_{(3)}$

is a scalar while $\boldsymbol{T}$ is the Cholesky (*i.e.*, square root) of the $r \times r$ matrix $\boldsymbol{\Sigma}_{(2)}$. The marginal probability then becomes

$$h(\mathbf{y}_i) = \int_{\boldsymbol{\theta}^*} \ell(\mathbf{y}_i \mid \boldsymbol{\theta}^*)\, g(\boldsymbol{\theta}^*)\, d\boldsymbol{\theta}^*, \tag{13.8}$$

where $g(\boldsymbol{\theta}^*)$ is the multivariate standard normal density.

The major problem with this representation of the marginal probability is that the dimensionality of $\boldsymbol{\theta}^*$ is $(n_i \times r) + 1$ and numerical integration of equation (13.8) would be exceedingly slow and computationally intractable if $(n_i \times r) + 1$ is greater than 10. Note, however, that conditional on the cluster-effect $\theta_{(3)}$, the responses from the n_i subjects in cluster i are independent, therefore the marginal probability can be rewritten as

$$h(\mathbf{y}_i) = \int_{\theta_{(3)}} \left\{ \prod_{j=1}^{n_i} \int_{\boldsymbol{\theta}_{(2)}} \left(\prod_{k=1}^{n_{ij}} [\Phi(z_{ijk})]^{1-y_{ijk}} \, [1 - \Phi(z_{ijk})]^{y_{ijk}} \right) g(\boldsymbol{\theta}_{(2)}) d\boldsymbol{\theta}_{(2)} \right\} g(\theta_{(3)}) d\theta_{(3)}, \tag{13.9}$$

where $\boldsymbol{\theta}_{(2)}$ are the r subject-level random effects. Here the integration is of dimensionality $r + 1$ and is tractable as long as the number of level two random effects is no greater than three or four. In longitudinal studies, we typically have one or two random effects at level two (*e.g.*, a random intercept and/or trend for each individual) and one random effect at level three (*e.g.*, a random cluster effect).

Estimation. The estimation of the covariate coefficients $\boldsymbol{\beta}$ and the population parameters in $\boldsymbol{T}$ requires differentiation of the log likelihood function with respect to these parameters. The log likelihood for the patterns from the N clusters can be written as

$$\log L - \sum_{i}^{N} \log h(\mathbf{y}_i). \tag{13.10}$$

Let $\boldsymbol{\eta}$ represent an arbitrary parameter vector; then for $\boldsymbol{\beta}$, $\mathrm{v}(\boldsymbol{T})$ (which denotes the unique elements of the Cholesky factor $\boldsymbol{T}$), and $\sigma_{(3)}$, we get

$$\begin{aligned} \frac{\partial \log L}{\partial \boldsymbol{\eta}} &= \sum_{i=1}^{N} \frac{1}{h}(\mathbf{y}_i) \int_{\theta_{(3)}} \ell_i(\theta_{(3)}) \\ &\left\{ \sum_{j=1}^{n_i} \frac{1}{h}(\mathbf{y}_{ij}) \int_{\boldsymbol{\theta}_{(2)}} \sum_{k=1}^{n_{ij}} \left(\frac{y_{ijk} - \Phi(z_{ijk})}{\Phi(z_{ijk})[1 - \Phi(z_{ijk})]} \right) \ell_{ij}(\boldsymbol{\theta}) \phi(z_{ijk}) \frac{\partial z_{ijk}}{\partial \boldsymbol{\eta}} g(\boldsymbol{\theta}_{(2)}) d\boldsymbol{\theta}_{(2)} \right\} \\ &\times g(\theta_{(3)}) d\theta_{(3)}, \end{aligned} \tag{13.11}$$

where

$$\ell_{ij}(\boldsymbol{\theta}) = \ell_{ij}(\boldsymbol{\theta}_{(2)}, \theta_{(3)}) = \prod_{k=1}^{n_{ij}} [\Phi(z_{ijk})]^{1-y_{ijk}} \, [1 - \Phi(z_{ijk})]^{y_{ijk}}, \tag{13.12}$$

$$\ell_i(\theta_{(3)}) = \prod_{j=1}^{n_i} \int_{\boldsymbol{\theta}_{(2)}} \ell_{ij}(\boldsymbol{\theta})\, g(\boldsymbol{\theta}_{(2)})\, d\boldsymbol{\theta}_{(2)}$$
$$= \prod_{j=1}^{n_i} h(\mathbf{y}_{ij}), \tag{13.13}$$

and

$$\frac{\partial \mathrm{z}_{ijk}}{\partial \boldsymbol{\beta}} = \boldsymbol{x}_{ijk}, \qquad \frac{\partial \mathrm{z}_{ijk}}{\partial \mathrm{v}(\boldsymbol{T})}, = \mathsf{J}_r(\boldsymbol{\theta}_{(2)} \otimes \boldsymbol{z}_{ijk}) \qquad \frac{\partial \mathrm{z}_{ijk}}{\partial \sigma_{(3)}} = \theta_{(3)},$$

and J_r is the transformation matrix of Magnus [1988], which eliminates the elements above the main diagonal.

As in the 2-level case described by Gibbons and Bock [1987] and Gibbons et al. [1994] the method of scoring can be used to provide (marginal) MLEs and numerical integration on the transformed $\boldsymbol{\theta}$ space can be performed [Stroud and Sechrest, 1966]. An advantage of numerical integration is that alternative distributions for the random effects can be considered. Thus, for example, we can compare parameter estimates for a normal versus rectangular prior to determine the degree to which our estimates are robust to deviation from the assumed normality of the distribution for the random effects.

13.2.2 Three-Level Logistic Regression Model for Dichotomous Outcomes

In the previous discussion we have focused on a 3-level mixed-effects probit regression model, however, many researchers are more familiar with the logistic regression model. Fortunately, modification of the response function and associated likelihood equations is trivial as Gibbons et al. [1994] and Hedeker and Gibbons [1994] have shown for 2-level logistic regression models.

Following Gibbons and Hedeker [1994] we replace the normal response function (*i.e.*, the cdf, or cumulative distribution function) $\Phi(\mathrm{z}_{ijk})$ with

$$\Psi(\mathrm{z}_{ijk}) = \frac{1}{1 + \exp[-\mathrm{z}_{ijk}]}, \tag{13.14}$$

and the normal density function $\phi(\mathrm{z}_{ijk})$ (*i.e.*, the pdf, or probability density function) with the product

$$\Psi(\mathrm{z}_{ijk})(1 - \Psi(\mathrm{z}_{ijk})). \tag{13.15}$$

As in the normal case, we let $\gamma = 0$, however, the residual variance corresponding to the standard logistic distribution is $\pi^2/3$. Application of the logistic response function is attractive in many cases in which the response probability is small because the logistic distribution has greater tail probability than the normal distribution.

13.2.3 Illustration

The Television School and Family Smoking Prevention and Cessation Project (TVSFP) study [Flay et al., 1988] was designed to test independent and combined effects of a school-based social-resistance curriculum and a television-based program in terms of tobacco use prevention and cessation. The study involved seventh-grade students from 135 classrooms from 28 schools, where the schools were randomized to one of four study conditions: (a) a social-resistance classroom curriculum, (b) a media (television) intervention, (c) a social-resistance classroom curriculum combined with a mass-media intervention, and (d) a no-treatment control group. These conditions form a 2×2 design of social-resistance classroom curriculum (CC = yes or no) by mass-media intervention (TV = yes or no). A tobacco and health knowledge scale (THKS) was used in classifying subjects as knowledgeable or not. Data from 1600 students with pre and post-intervention data were available. The resulting dataset was unbalanced with one to 13 classrooms per school, and two to 28 students per classroom. Student frequencies for positive and negative THKS results, broken down by condition subgroups, are given in Table 13.6.

Table 13.6. Tobacco and Health Knowledge Scale. Post-Intervention Results Subgroup Frequencies (and Percentages)

Subgroup		THKS Score		
CC	TV	Pass	Fail	Total
No	No	175 (41.6)	246 (58.4)	421
No	Yes	201 (48.3)	215 (51.7)	416
Yes	No	240 (63.2)	140 (36.8)	380
Yes	Yes	231 (60.3)	152 (39.7)	383
	Total	847 (52.9)	753 (47.1)	1600

Two 3-level probit regression models were fit to these data. In the first analysis pre- and post-intervention THKS responses were treated as a within-subject effect (*i.e.*, level 2) and classroom was treated as a cluster (level 3). In the second analysis, post-intervention THKS scores were considered and subjects were clustered within classrooms (level 2) and schools (level 3). In both cases, THKS knowledge was modeled in terms of CC, TV, and CC by TV interaction. In the second model, pre-intervention knowledge was also used as a covariate.

Results from the first analysis are presented in Table 13.7. The first column of Table 13.7 lists results for a probit regression analysis of student-level data ignoring clustering of students and treating each student and measurement occasion as an independent observation. This analysis indicates the positive effect of the social-resistance classroom curriculum as well as the television part of the intervention, but no interaction. As compared to the mixed-effects models, the standard errors are clearly underestimated suggesting that we have greater precision than is actually the case given the dependence of measurements within students and students within classrooms. Results are somewhat similar for the two 2-level models, where standard errors for the model with a subject random effect are somewhat smaller than those for the model with classroom random effect.

Table 13.7. THKS Pre vs. Post Intervention (Binary) Scores Comparison of Two- and Three-Level Mixed-Effects Probit Model (MEPMs) Estimates (Standard Errors)

	Fixed-Effects Model		Two-Level MEPM Class		Two-Level MEPM Student		Three-Level MEPM	
Constant	−0.3116	***	−0.3377	***	−0.3778	***	−0.3916	***
	(0.062)		(0.104)		(0.076)		(0.116)	
Pre vs. Post	0.0986		0.1007		0.1198		0.1181	
	(0.088)		(0.103)		(0.098)		(0.100)	
CC (pre)	−0.1095		−0.1427		−0.1333		−0.1608	
	(0.091)		(0.135)		(0.110)		(0.151)	
TV (pre)	−0.0776		−0.1154		−0.0918		−0.1251	
	(0.089)		(0.129)		(0.107)		(0.144)	
CC × TV (pre)	−0.0026		0.0523		0.0030		0.0534	
	(0.129)		(0.184)		(0.157)		(0.208)	
CC (post−pre)	0.6585	***	0.6922	***	0.7995	***	0.8065	***
	(0.128)		(0.143)		(0.144)		(0.143)	
TV (post−pre)	0.2484	**	0.2632	**	0.2997	**	0.3045	***
	(0.124)		(0.123)		(0.137)		(0.121)	
CC × TV (post−pre)	−0.2428		−0.2691		−0.3066		−0.3204	*
	(0.181)		(0.184)		(0.197)		(0.182)	
Class sd			0.2872	***			0.3103	***
			(0.039)				(0.049)	
Student sd					0.6798	***	0.6048	***
					(0.059)		(0.066)	
Log L	−2108.63		−2082.12		−2077.89		−2061.92	

$^{***}p < 0.01$ $^{**}p < 0.05$ $^{*}p < 0.10$

For the 3-level model (*i.e.*, measurements within subjects and subjects within classrooms) the standard errors are typically the largest, as expected. However, the 3-level model indicates an even stronger effect for the TV intervention and a result that approaches significance for the interaction. The 3-level analysis suggests that while TV intervention is effective in increasing THKS scores for those not receiving the CC component, it has a slight negative effect on those exposed to both components (see Table 13.6). Similarly, while the CC intervention is effective in increasing THKS scores, the effect is more pronounced for those not receiving the TV component than for those receiving it. Both 2-level models provide significant improvement in fit relative to the fixed-effects model ($X_1^2 = 53.02$, $p < .0001$ and $X_1^2 = 61.48$, $p < .0001$ for classroom and student level mixed-effects models, respectively). Similarly, the 3-level model provided significant improvement in fit relative to the two 2-level models ($X_1^2 = 40.40$, $p < .0001$ and $X_1^2 = 31.946$, $p < .0001$ for classroom and student level mixed-effects models, respectively). A 3-level model with two random effects at the subject level (*i.e.*, baseline and post−pre change) did not significantly improve the fit relative to the 3-level model with a single subject level random effect ($X_2^2 = 0.06$, $p = ns$). The intraclass (classroom) correlation equals .066 (*i.e.*, $.3103^2/[.3103^2 + .6048^2 + 1]$) and the intra-student correlation equals

.250 (*i.e.*, $.6048^2/[.3103^2 + .6048^2 + 1]$). Thus approximately 6.6% of the variance is attributable to classrooms and 25.0% is attributable to students.

To aid in interpreting the estimated model parameters, estimated response proportions are computed from the latent response vectors $\boldsymbol{y}_i$ which are normally distributed with mean $\boldsymbol{X}_i\boldsymbol{\beta}$ and variance–covariance matrix $\boldsymbol{Z}_i\Sigma_{\upsilon^*}\boldsymbol{Z}_i' + \sigma_\varepsilon \boldsymbol{I}_{n_i}$. For example, for students in the no-treatment control condition, the estimated response proportion equals $\Phi(-.3916/\sqrt{1 + .3103^2 + .6048^2}) = .373$ at pre-intervention and $\Phi((-.3916 + .1181)/\sqrt{1 + .3103^2 + .6048^2}) = .397$ at post-intervention. Similar pre- and post-intervention estimates are .324 and .621 for CC, .335 and .469 for TV, and .303 and .593 for CC plus TV. These estimates corroborate what the estimated model parameters indicate, namely, that there is a considerable difference in change between controls and CC students, and less of a difference comparing controls to either TV or CC plus TV students. Also note the close agreement with the observed post-intervention response proportions listed in Table 13.6. Results from the second analysis (*i.e.*, random class and school effects on post-intervention THKS knowledge) are presented in Table 13.8.

Table 13.8. THKS Post-Intervention (Binary) Scores Comparison of Two- and Three-Level Mixed-Effects Probit Model (MEPM) Estimates (Standard Errors)

	Fixed-Effects Model		Two-Level MEPM Class		Two-Level MEPM School		Three-Level MEPM	
Constant	−0.4209	***	−0.4412	***	−0.4394	***	−0.4410	***
	(0.069)		(0.100)		(0.153)		(0.154)	
Pre	0.5335	***	0.5310	***	0.5106	***	0.5211	***
	(0.068)		(0.071)		(0.075)		(0.079)	
CC	0.5913	***	0.5916	***	0.6698	***	0.6277	***
	(0.092)		(0.128)		(0.226)		(0.228)	
TV	0.1931	**	0.1718		0.2369		0.2078	
	(0.089)		(0.124)		(0.172)		(0.171)	
CC × TV	−0.2596	**	−0.2224		−0.3591		−0.2921	
	(0.128)		(0.178)		(0.259)		(0.260)	
Class sd			0.3005	***			0.2532	***
			(0.059)				(0.086)	
School sd					0.2143	***	0.1690	**
					(0.071)		(0.086)	
Log L	−1050.26		−1040.86		−1043.96		−1039.95	

$^{***}p < 0.01$ $^{**}p < 0.05$ $^{*}p < 0.10$

The picture is somewhat different when analysis focuses on post-intervention THKS knowledge. Here, the fixed-effects model identified a significant main effect of TV and a CC by TV interaction, however, neither of these two effects were significant for 2-level or 3-level mixed-effects models. All three mixed-effects models (*i.e.*, the 2-level classroom, 2-level school and 3-level school by classroom models) provided significant improvement in fit relative to the fixed-effects model ($X_1^2 = 18.80$, $p < .0001$, $X_1^2 = 12.60$, $p < .0004$

and $X_2^2 = 20.62$, $p < .0001$, respectively); however, the 3-level model did not provide significant improvement in fit relative to the 2-level (classroom) model ($X_1^2 = 1.82$, $p = ns$). Estimates of the intra-unit correlations based on the 3-level model are .059 for classrooms and .026 for schools.

Tables 13.9 and 13.10 present results for 3-level models with alternate response function (*i.e.*, logistic) and random-effects distribution (normal versus rectangular). Table 13.9 displays results for the first example (*i.e.*, clustered and longitudinal—class and pre- versus post-intervention), whereas Table 13.10 displays results for the second example (*i.e.*, two levels of clustering—school and classroom).

Table 13.9. THKS Pre- vs. Post-Intervention (Binary) Scores Three-Level Mixed-Effects Logistic Regression Models Estimates (Standard Errors)

	Normal Prior		Rectangular Prior	
Constant	−0.6533	***	−0.6681	***
	(.196)		(0.201)	
Pre vs. Post	0.1981		0.1982	
	(0.168)		(0.167)	
CC (pre)	−0.2716		−0.2622	
	(0.254)		(0.252)	
TV (pre)	−0.2129		−0.1680	
	(0.243)		(0.243)	
CC × TV (pre)	0.0863		0.0416	
	(0.350)		(0.349)	
CC (post−pre)	1.3450	***	1.3408	***
	(0.241)		(0.240)	
TV (post−pre)	0.5095	***	0.5093	***
	(0.203)		(0.203)	
CC × TV (post−pre)	−0.5242	**	−0.5148	**
	(0.306)		(0.305)	
Class sd	0.5228	***	0.1727	***
	(0.082)		(0.024)	
Student sd	1.0135	***	0.3230	***
	(0.115)		(0.024)	
Log L	−2061.68		−2061.48	

***$p < 0.01$ **$p < 0.05$ *$p < 0.10$

Table 13.10. THKS Post-Intervention (Binary) Scores Three-Level Mixed-Effects Logistic Regression Models Estimates (Standard Errors)

	Normal Prior	Rectangular Prior
Constant	−0.7172 ***	−0.7365 ***
	(0.252)	(0.266)
Pre	0.8502 ***	0.8498 ***
	(0.131)	(0.130)
CC	1.0179 ***	1.0585 ***
	(0.375)	(0.391)
TV	0.3345	0.3458
	(0.280)	(0.290)
CC × TV	−0.4625	−0.5046
	(0.426)	(0.442)
Class sd	0.4154 ***	0.1381 ***
	(0.141)	(0.040)
School sd	0.2742 **	0.0914 ***
	(0.140)	(0.039)
Log L	−1039.97	−1039.75

***$p < 0.01$ **$p < 0.05$ *$p < 0.10$

Comparison of Tables 13.7 and 13.8 with Tables 13.9 and 13.10 reveal that probit and logistic regression models lead to virtually identical conclusions regarding significance of fixed and random effects in the 3-level models. The MLEs and standard errors for the fixed effects are approximately 40% larger for the logistic regression model; however, estimates of the random-effect standard deviations were quite similar when both models used a normal random effect distribution. Fit of the probit and logistic models were virtually identical.

When a rectangular prior was used, estimates of the fixed effects and their standard errors exhibited little, if any, effect; however, MLEs and standard errors of the random effects decreased by approximately two-thirds. In both examples, fit of the models with alternative random effect distributions was virtually identical.

Taken as a whole, these results are quite encouraging because they demonstrate that while a normal random effect distribution is a good choice, fit of the model, and parameter estimates and standard errors of at least the structural parameters are reasonably robust to model misspecification. As in the case of fixed-effects models, selection of probit versus logistic response functions appears to have more to do with custom or practice within a particular discipline than differences in statistical properties.

13.2.4 More General Outcomes

Generalizations of the three-level model for dichotomous outcomes presented above for ordinal outcomes can be found in Raman and Hedeker [2005] and Liu and Hedeker [2006]. Here, we also present generalizations for nominal and count outcome variables. For these extensions, let us describe the estimation method given above in a slightly more general way. Essentially, one must obtain the derivatives of the model parameters with respect to the log-likelihood function:

$$\log L = \sum_{i}^{N} \log h(\mathbf{y}_i). \tag{13.16}$$

Notice that

$$h(\mathbf{y}_i) = \int_{\theta_{(3)}} \ell_i(\theta_{(3)})\, g(\theta_{(3)})\, d(\theta_{(3)}), \tag{13.17}$$

where

$$\ell_i(\theta_{(3)}) = \prod_{j=1}^{n_i} h(\mathbf{y}_{ij}) \tag{13.18}$$

and

$$h(\mathbf{y}_{ij}) = \int_{\boldsymbol{\theta}_{(2)}} \ell_{ij}(\boldsymbol{\theta})\, g(\boldsymbol{\theta}_{(2)})\, d(\boldsymbol{\theta}_{(2)}). \tag{13.19}$$

Here, $\ell_{ij}(\boldsymbol{\theta}) = \ell_{ij}(\boldsymbol{\theta}_{(2)}, \theta_{(3)})$ represents the conditional likelihood of the response vector of level-2 subject j within level-3 cluster i, namely,

$$\ell_{ij}(\boldsymbol{\theta}) = \prod_{k=1}^{n_{ij}} \left[\Phi(\mathrm{z}_{ijk})\right]^{1-\mathrm{y}_{ijk}} \left[1 - \Phi(\mathrm{z}_{ijk})\right]^{\mathrm{y}_{ijk}}, \tag{13.20}$$

where y_{ijk} is the dichotomous outcome coded 0 or 1, and the response model is given as

$$\mathrm{z}_{ijk} = \sigma_{(3)}\theta_{0i} + \boldsymbol{z}'_{ijk}\boldsymbol{T}\boldsymbol{\theta}_{ij} + \boldsymbol{x}'_{ijk}\boldsymbol{\beta}. \tag{13.21}$$

It is this likelihood function $\ell_{ij}(\boldsymbol{\theta})$ that changes as we move from dichotomous to ordinal and other types of categorical responses. Thus, it is the specification of this likelihood function, along with the derivatives with respect to (the log of) it, that needs to be modified for these other types of outcomes. This can be seen be noting that

$$\begin{aligned} \frac{\partial \log L}{\partial \boldsymbol{\eta}} &= \sum_{i=1}^{N} h^{-1}(\mathbf{y}_i) \frac{\partial h(\mathbf{y}_i)}{\partial \boldsymbol{\eta}} \\ &= \sum_{i=1}^{N} h^{-1}(\mathbf{y}_i) \int_{\theta_{(3)}} \ell_i(\theta_{(3)}) \frac{\partial \log \ell_i(\theta_{(3)})}{\partial \boldsymbol{\eta}}\, g(\theta_{(3)})\, d(\theta_{(3)}), \end{aligned} \tag{13.22}$$

where

$$\frac{\partial \log \ell_i(\theta_{(3)})}{\partial \boldsymbol{\eta}} = \sum_{j=1}^{n_i} h^{-1}(\mathbf{y}_{ij}) \int_{\boldsymbol{\theta}_{(2)}} \ell_{ij}(\boldsymbol{\theta}) \frac{\partial \log \ell_{ij}(\boldsymbol{\theta})}{\partial \boldsymbol{\eta}}\, g(\boldsymbol{\theta}_{(2)})\, d(\boldsymbol{\theta}_{(2)}). \tag{13.23}$$

Note that in Fisher's scoring solution, estimates for $\boldsymbol{\eta}$ on iteration ι are improved by

$$\boldsymbol{\eta}_{\iota+1} = \boldsymbol{\eta}_{\iota} - \mathcal{E}\left[\frac{\partial^2 \log L}{\partial \boldsymbol{\eta}_{\iota}\, \partial \boldsymbol{\eta}'_{\iota}}\right]^{-1} \frac{\partial \log L}{\partial \boldsymbol{\eta}_{\iota}}, \tag{13.24}$$

where the empirical information matrix is given by

$$\mathcal{E}\left[\frac{\partial^2 \log L}{\partial \boldsymbol{\eta}_{\iota}\, \partial \boldsymbol{\eta}'_{\iota}}\right] = -\sum_{i=1}^{N} h^{-2}(\mathbf{y}_i)\, \frac{\partial h(\mathbf{y}_i)}{\partial \boldsymbol{\eta}_{\iota}} \left(\frac{\partial h(\mathbf{y}_i)}{\partial \boldsymbol{\eta}_{\iota}}\right)'. \tag{13.25}$$

Thus, it is only the first derivatives that are needed for the algorithm.

13.2.4.1 Ordinal Outcomes The results presented here follow those from Hedeker and Gibbons [1994]. Let the C ordered response categories be coded as $c = 1, 2, \ldots, C$. Also, assume there are $C-1$ strictly increasing model thresholds γ_c (*i.e.*, $\gamma_1 < \gamma_2 \ldots < \gamma_{C-1}$). Again, for identification $\gamma_1 = 0$ and $\gamma_0 = -\infty$ and $\gamma_C = \infty$. For the ordinal probit model,

$$\ell_{ij}(\boldsymbol{\theta}) = \prod_{k=1}^{n_{ij}} \prod_{c=1}^{C} \left[\Phi(\gamma_c - \mathrm{z}_{ijk}) - \Phi(\gamma_{c-1} - \mathrm{z}_{ijk})\right]^{\mathrm{d}_{ijkc}}, \tag{13.26}$$

where $\mathrm{d}_{ijkc} = 1$ if $\mathrm{y}_{ijk} = c$ and 0 otherwise (*i.e.*, for each ijkth observation, $\mathrm{d}_{ijkc} = 1$ for only one of the C categories. Here, $\Phi(\gamma_c - \mathrm{z}_{ijk})$ represents a cumulative probability, namely, $\Pr(\mathrm{y}_{ijk} \leq c)$, and so the subtraction is needed to obtain the probability of a response in a single category c.

Differentiating first for the parameters that do not vary across the categories ($\sigma_{(3)}$, $\mathrm{v}(\boldsymbol{T})$ and $\boldsymbol{\beta}$), we get

$$\frac{\partial \log \ell_{ij}(\boldsymbol{\theta})}{\partial \boldsymbol{\eta}} = \sum_{k=1}^{n_i} \sum_{c-1}^{C} \mathrm{d}_{ijkc} \frac{\phi(\gamma_{c-1} - \mathrm{z}_{ijk}) - \phi(\gamma_c - \mathrm{z}_{ijk})}{\Phi(\gamma_c - \mathrm{z}_{ijk}) - \Phi(\gamma_{c-1} - \mathrm{z}_{ijk})} \frac{\partial \mathrm{z}_{ijk}}{\partial \boldsymbol{\eta}} \tag{13.27}$$

with, as in the dichotomous case,

$$\frac{\partial \mathrm{z}_{ijk}}{\partial \boldsymbol{\beta}} = \boldsymbol{x}_{ijk}, \qquad \frac{\partial \mathrm{z}_{ijk}}{\partial \mathrm{v}(\boldsymbol{T})} = \mathsf{J}_r(\boldsymbol{\theta}_{(2)} \otimes \boldsymbol{z}_{ijk}), \qquad \frac{\partial \mathrm{z}_{ijk}}{\partial \sigma_{(3)}} = \theta_{(3)},$$

where, as before, J_r is a transformation matrix eliminating elements above the main diagonal (see [Magnus, 1988]), and $\mathrm{v}(\boldsymbol{T})$ is the vector containing the unique elements of the Cholesky factor $\boldsymbol{T}$.

Differentiating with respect to the threshold parameters, we get for a particular $\gamma_{c'}$,

$$\frac{\partial \log \ell_{ij}(\boldsymbol{\theta})}{\partial \gamma_{c'}} = \sum_{k=1}^{n_i} \sum_{c=1}^{C} \mathrm{d}_{ijkc} \frac{\phi(\gamma_c - \mathrm{z}_{ijk})\delta_{c,c'} - \phi(\gamma_{c-1} - \mathrm{z}_{ijk})\delta_{c-1,c'}}{\Phi(\gamma_c - \mathrm{z}_{ijk}) - \Phi(\gamma_{c-1} - \mathrm{z}_{ijk})} \tag{13.28}$$

$$\delta_{c,c'} = \begin{cases} 1 & \text{if } c = c' \\ 0 & \text{if } c \neq c'. \end{cases}$$

For an ordinal logistic regression model, again, we replace the normal cdf $\Phi(\gamma_c - \mathrm{z}_{ijk})$ with its logistic counterpart, namely,

$$\Psi(\gamma_c - \mathrm{z}_{ijk}) = \frac{1}{1 + \exp[-(\gamma_c - \mathrm{z}_{ijk})]}, \tag{13.29}$$

and the normal pdf $\phi(\gamma_c - \mathrm{z}_{ijk})$ with the product $\Psi(\gamma_c - \mathrm{z}_{ijk})(1 - \Psi(\gamma_c - \mathrm{z}_{ijk}))$.

13.2.4.2 Nominal Outcomes As described in Hedeker [2003], the nominal mixed model is typically written in terms of the logistic representation. Also, it's common to use the first category as a reference cell and to estimate the covariate effects, which vary across categories, relative to this reference cell. We will also follow these conventions. For the nominal model, the conditional probability for a level-2 subject j within a level-3 cluster i is given as

$$\ell_{ij}(\boldsymbol{\theta}) = \prod_{k=1}^{n_{ij}} \prod_{c=1}^{C} (p_{ijc})^{\mathrm{d}_{ijkc}} \tag{13.30}$$

where, just as in the ordinal case, $\mathrm{d}_{ijkc} = 1$ if $\mathrm{y}_{ijk} = c$ and 0 otherwise. Here, the probability that $\mathrm{y}_{ijk} = c$ (a response occurs in category c), conditional on the random effects, is denoted as p_{ijc} and given by

$$p_{ijc} = \frac{\exp(\mathrm{z}_{ijkc})}{1 + \sum_{h=2}^{C} \exp(\mathrm{z}_{ijkh})} \quad \text{for } c = 2, 3, \ldots, C, \tag{13.31}$$

$$p_{ij1} = \frac{1}{1 + \sum_{h=2}^{C} \exp(\mathrm{z}_{ijkh})}, \tag{13.32}$$

where the multinomial logit z_{ijkc} is expressed as

$$\mathrm{z}_{ijkc} = \sigma_{c\,(3)}\theta_{0i} + \boldsymbol{z}'_{ijk}\boldsymbol{T}_c\boldsymbol{\theta}_{ij} + \boldsymbol{x}'_{ijk}\boldsymbol{\beta}_c. \tag{13.33}$$

Comparing this to the model for ordered responses, we see that all of the covariate effects $\boldsymbol{\beta}_c$ vary across categories ($c = 2, 3, \ldots, C$), as do the random-effect variance terms $\sigma_{c\,(3)}$ and $\boldsymbol{T}_c$. Using $\boldsymbol{\eta}_c$ to represent an arbitrary parameter vector, differentiating yields

$$\frac{\partial \log \ell_{ij}(\boldsymbol{\theta})}{\partial \boldsymbol{\eta}_c} = \sum_{k=1}^{n_{ij}} (d_{ijkc} - p_{ijkc}) \frac{\partial \mathrm{z}_{ijkc}}{\partial \boldsymbol{\eta}_c}, \tag{13.34}$$

where

$$\frac{\partial \mathrm{z}_{ijkc}}{\partial \boldsymbol{\beta}_c} = \boldsymbol{x}_{ijk}, \qquad \frac{\partial \mathrm{z}_{ijkc}}{\partial \mathrm{v}(\boldsymbol{T}_c)} = \mathrm{J}_r(\boldsymbol{\theta} \otimes \boldsymbol{z}_{ijk}), \qquad \frac{\partial \mathrm{z}_{ijkc}}{\partial \sigma_{c\,(3)}} = \theta_{(3)}. \tag{13.35}$$

13.2.4.3 Count Outcomes Siddiqui [1996] describes the statistical development of the 2-level mixed-effects Poisson regression model. Here, we will present the key computational features for the extension to three levels. Let t_{ijk} represent the follow-up time associated with units i, j, and k. Let y_{ijk} be the value of the count variable (where y_{ijk} can equal $0, 1, \ldots$) associated with level-3 unit i, level-2 unit j and level-1 unit k. If this count is assumed to be drawn from a Poisson distribution, then the mixed-effects Poisson regression model indicates the expected number of counts in t_{ijk} as

$$\lambda_{ijk} = t_{ijk} \exp(\mathrm{z}_{ijk}) \tag{13.36}$$

with

$$\mathrm{z}_{ijk} = \sigma_{(3)}\theta_{0i} + \boldsymbol{z}'_{ijk}\boldsymbol{T}\boldsymbol{\theta}_{ij} + \boldsymbol{x}'_{ijk}\boldsymbol{\beta}. \tag{13.37}$$

Assuming the Poisson process for the count y_{ijk}, the probability that $\mathrm{Y}_{ijk} = \mathrm{y}_{ijk}$ (Y is a random variable whose realization in data is y), conditional on the random effects is given as

$$P(\mathrm{Y}_{ijk} = \mathrm{y}_{ijk} \mid \boldsymbol{\theta}) = \exp(-\lambda_{ijk}) \frac{(\lambda_{ijk})^{\mathrm{y}_{ijk}}}{\mathrm{y}_{ijk}!} \tag{13.38}$$

The conditional probability associated with the level-2 unit j and level-3 unit i is then

$$\begin{aligned} \ell_{ij}(\boldsymbol{\theta}) &= \prod_{k=1}^{n_{ij}} \left[\exp(-\lambda_{ijk}) \frac{(\lambda_{ijk})^{\mathrm{y}_{ijk}}}{\mathrm{y}_{ijk}!} \right] \\ &= \exp\left[-\sum_{k=1}^{n_{ij}} \lambda_{ijk} + \sum_{k=1}^{n_{ij}} \mathrm{y}_{ijk} \log t_{ijk} + \sum_{k=1}^{n_{ij}} \mathrm{y}_{ijk}\mathrm{z}_{ijk} - \sum_{k=1}^{n_{ij}} \log(\mathrm{y}_{ijk}!) \right]. \end{aligned} \tag{13.39}$$

The derivatives are derived as

$$\frac{\partial \log \ell_{ij}(\boldsymbol{\theta})}{\partial \boldsymbol{\eta}} = \sum_{k=1}^{n_{ij}} (y_{ijk} - \lambda_{ijk}) \frac{\partial \mathrm{z}_{ijk}}{\partial \boldsymbol{\eta}} \tag{13.40}$$

with, as in the dichotomous case,

$$\frac{\partial \mathrm{z}_{ijk}}{\partial \boldsymbol{\beta}} = \boldsymbol{x}_{ijk}, \qquad \frac{\partial \mathrm{z}_{ijk}}{\partial \mathrm{v}(\boldsymbol{T})} = \mathsf{J}_r(\boldsymbol{\theta}_{(2)} \otimes \boldsymbol{z}_{ijk}), \qquad \frac{\partial \mathrm{z}_{ijk}}{\partial \sigma_{(3)}} = \theta_{(3)}.$$

13.3 SUMMARY

Mixed models can be extremely useful in analysis of categorical clustered multivariate data. This chapter describes these models for analysis of three-level data (*i.e.*, two levels of clustering). The approach taken here advances previous work of Longford [1988, 1994] and Goldstein [1991] which provide approximate solutions based on first and second-order Taylor series expansions of the logarithm of the conditional likelihood required to linearize these models so that closed form solutions of the likelihood equations exist. In practice, most three-level clustered problems will require only two random effects (*i.e.*, one for each level of clustering). For longitudinal data, typically one to three random effects are required to model the time trends (*e.g.*, random intercept and linear trend), leaving one additional random effect due to clustering. In all of these cases, the method described here is appropriate and numerical evaluation of the likelihood equations is computationally tractable. An additional advantage of the numerical solution is that it can accommodate alternate random effect distributions, including empirical estimation of the random effect distribution [Bock and Aitkin, 1981].

Alternatively, the models presented here and even more complicated models could be evaluated using Gibbs sampling [Geman and Geman, 1984; Gelfand and Smith, 1990; Gelfand et al., 1990; Tanner, 1996]. Although potentially more computationally intensive, the Gibbs sampler should have no practical limitation on number of random effects, whereas the numerical solution is computationally intractable for models with more than five or six random effects.

CHAPTER 14

MISSING DATA IN LONGITUDINAL STUDIES

14.1 INTRODUCTION

Even in well-controlled situations, missing data invariably occur in longitudinal studies. Subjects can be missed at a particular measurement wave, with the result that these subjects provide data at some, but not all, study timepoints. Alternatively, subjects who are assessed at a given study timepoint, might only provide responses to a subset of the study variables, again resulting in incomplete data. Finally, subjects might dropout of the study, or be lost to follow-up, thus providing no data beyond a specific point in time.

An increasing number of articles reviewing methods for handling missing data in longitudinal studies have been published [Demirtas, 2004b; Gornbein et al., 1992; Hogan and Laird, 1997b; Molenberghs et al., 2004]. In particular, the seminal article by Little [1995] provides an important statistical overview and framework for these methods, and the recent tutorial article by Hogan et al. [2004] illustrates several applications of these methods. Also, several texts include a wealth of material on this topic [Diggle et al., 2002; Little and Rubin, 2002; Verbeke and Molenberghs, 2000]. This area continues to develop at a rapid pace, and the availability of software to deal with missing data has also increased.

An attractive and important feature of many of the models considered in this book is their flexibility in handling missing data. For example, for both MRMs and GEE models, subjects are not assumed to be measured at the same number of timepoints, and in fact, can be measured at different timepoints. Since there are no restrictions on the number of observations per individual, subjects who are missing at a given interview wave are not excluded from the analysis. However, these models make different assumptions about the missing data, and so it is important to consider these missing data assumptions when performing a particular analysis.

In this chapter, we present most of the missing data methods in terms of longitudinal modeling using mixed-effects regression models (MRMs) for continuous outcomes. The application of these methods, however, extends more generally to other methods allowing incomplete data across time (*e.g.*, GEE and covariance pattern models) and also to non-normal outcomes (*e.g.*, categorical outcomes). Additionally, we will focus on ways in which the basic MRM can be augmented to more generally deal with missing data, detailing how selection and pattern-mixture MRMs can be constructed using standard software. In terms of omission, we will ignore at least two important areas of methods for missing data: multiple imputation and weighting approaches. For the former, the interested reader should consult Rubin [1987, 1996] and Schafer [1997]; for the latter, Robins et al. [1995], Rotnitzky and Robins [1999], and Demirtas [2004a] describe weighted-estimating equations models for incomplete data.

14.2 MISSING DATA MECHANISMS

The missing data mechanism is what characterizes the reasons for the missing data. In other words, the mechanism addresses the basic question of "why are the data missing?" As we will see, the performance of longitudinal data analysis models can depend critically on the missing data mechanism, so it is vital to consider the mechanism in choosing an appropriate analysis. These mechanisms were introduced in a typology for missing data by Rubin [1976]. As noted by Schafer and Graham [2002], who provide an accessible review of missing data methods, while this typology is widely cited, it is less widely understood. Perhaps part of the reason for this is that there are some relatively subtle, but important, aspects which distinguish the various mechanisms. Here, in addition to defining these mechanisms, we will also present some simulation results to aid in their understanding.

To describe these mechanisms, we need to introduce some notation for the missing data. Let R_{ij} be an indicator variable which takes on value 1 if subject i is observed at time j, and 0 if the subject was not observed at this timepoint. Here, we are considering whether the dependent variable y is observed or not. If a study calls for measurement at n timepoints, then the $n \times 1$ complete dependent variable vector is

$$\boldsymbol{y}_i' = (y_{i1}, y_{i2}, \ldots, y_{in}).$$

The $n \times 1$ missing data indicator vector for a subject is then

$$\boldsymbol{R}_i' = (R_{i1}, R_{i2}, \ldots, R_{in}),$$

where, the specific R_{ij} values equal 1 or 0 depending on whether y_{ij} is observed or not (*i.e.*, $R_{ij} = 1$ if subject i is observed at time j, or $R_{ij} = 0$ if subject i is missing at time j). Based on $\boldsymbol{R}_i$, we can partition the complete dependent variable vector $\boldsymbol{y}_i$ into its observed $\boldsymbol{y}_i^O$ and unobserved $\boldsymbol{y}_i^M$ components for a given subject i. Here, we are thinking of $\boldsymbol{y}_i$ as the planned or potential dependent variable vector for subject i, which differs notation-wise from our usual treatment of this vector, and $\boldsymbol{y}_i^O$ as the actually observed dependent variable vector for subject i. Analogously, $\boldsymbol{y}_i^M$ is the component of the dependent variable vector that we planned on measuring and observing, but did not, for subject i.

In longitudinal studies, it is not uncommon to have subjects drop out of a study and never return. If the missing data are only due to such dropout, then the missing data indicators can be summarized simply by the time of dropout variable D_i, where $D_i = j'$ if subject i drops out between the $(j' - 1)$th and j'th timepoint; namely, $y_{i1}, \ldots, y_{i,j'-1}$ are observed and $y_{i,j'}, \ldots, y_{in}$ are missing. Thus, if a subject was missing at the last timepoint only, $D_i = n$ for that subject. For subjects with complete data across time (*i.e.*, subject who do not drop out), D_i can be defined equal to 0, as in Little [1995]. Alternatively, for some purposes, D_i can be designated as $n + 1$; the notion being that the time of dropout exceeds the last measurement timepoint. In what follows, the particular analysis will dictate which of these two treatments of D_i for completers we use.

14.2.1 Missing Completely at Random (MCAR)

The most basic assumption about the missing data is to assume that they are missing completely at random. That is, a subject is missing at a particular timepoint for completely random reasons. This mechanism is termed "missing completely at random" or MCAR. This implies that the missing data indicators $\boldsymbol{R}_i$ are independent of both $\boldsymbol{y}_i^O$ and $\boldsymbol{y}_i^M$; they do not depend on the dependent variable values that were observed or those that were not observed.

A less stringent case of MCAR is what Little [1995] refers to as covariate-dependent missingness. Here, the missing data indicators can be related to fully observed covariates $\boldsymbol{X}_i$. This is a very important special case, for example, because it allows missingness to increase across time (assuming that the variable time is included in $\boldsymbol{X}_i$). Almost all longitudinal studies have more missing data as time goes on, so allowing for this within MCAR is extremely useful. Under covariate-dependent missingness, MCAR can be thought of as an assumption of conditional independence. Namely, given the covariates $\boldsymbol{X}_i$, missingness $\boldsymbol{R}_i$ is independent of the observed $\boldsymbol{y}_i^O$ and unobserved dependent variable $\boldsymbol{y}_i^M$ vectors. As Fitzmaurice et al. [2004] point out, this raises a subtle but important point. For a given analysis to be consistent with MCAR, it is vital to include, as covariates in $\boldsymbol{X}_i$, variables that are thought to be related to missingness and that can mediate any possible relationship between missingness $\boldsymbol{R}_i$ and $\boldsymbol{y}_i^O$ or $\boldsymbol{y}_i^M$. As an example, suppose that both missingness and the dependent variable increase with time, but that conditional on time, missingness and the dependent variable are unrelated (*i.e.*, at a particular timepoint, there is no relationship between who is missing and either past values of y or the value of y that would be obtained if there were no missing values). In this case, if one did not include time as a covariate in the analysis, then the data analysis would no longer be consistent with MCAR.

Because it is important to include predictors of missingness in a given analysis, Schafer and Graham [2002] and Demirtas and Schafer [2003] recommend including a question like "How likely is it that you will remain in this study through the next measurement period?" in longitudinal questionnaires. To the extent that this question is related to subsequent missingness, including this variable as a covariate in analyses could convert a non-MCAR situation to one that is essentially MCAR.

14.2.2 Missing at Random (MAR)

Missing at random (MAR) goes one important step further by allowing the missingness to depend on both fully observed model covariates $\boldsymbol{X}_i$ *and* the observed dependent variable vector $\boldsymbol{y}_i^O$. MAR assumes that conditional on these two, the missing data are not related to the unobserved dependent variable vector $\boldsymbol{y}_i^M$. Again, one can think of this as a conditional independence assumption. Namely, conditional on $\boldsymbol{X}_i$ and $\boldsymbol{y}_i^O$, $\boldsymbol{R}_i$ is independent of $\boldsymbol{y}_i^M$. An example of MAR is when subjects drop out of the study because their value of the dependent variable falls below (or exceeds) some critical value. For instance, if subjects in a depression study who have Hamilton depression scores below 15 drop out of the study (*i.e.*, they are measured at a particular timepoint with a score below 15, but then are not measured at any future timepoints). Allowing missingness to additionally depend on the observed dependent variable vector $\boldsymbol{y}_i^O$ is the major difference between MCAR and MAR, and results in far less restrictive assumptions about the missing data. For this reason, many experts on missing data advocate use of MAR analysis as the default approach, unless there are strong reasons to support the MCAR assumption (see Fitzmaurice et al. [2004]). As we will see, one can test whether MCAR is reasonable or not, relative to MAR, because the distinction involves the observed data vector $\boldsymbol{y}_i^O$.

MAR posits that the missing data are related to the observed data (both $\boldsymbol{X}_i$ and $\boldsymbol{y}_i^O$), but that the missingness is not additionally related to the unobserved data $\boldsymbol{y}_i^M$. As a result, for longitudinal models that assume MAR (*e.g.*, MRMs and CPMs), it is vital that both the appropriate covariates are included in $\boldsymbol{X}_i$, and that the variance–covariance structure of $\boldsymbol{y}_i$ is correctly specified. If either condition is not satisfied, then the given analysis is not necessarily consistent with the underlying MAR mechanism. Our simulation results in Section 15.3.2 will illustrate how misspecification of the variance–covariance structure of $\boldsymbol{y}_i$ can yield biased results under MAR missingness.

As detailed in Schafer [1997], Demirtas [2004b], and elsewhere, closely related to MAR is the concept of ignorability. A missing data mechanism is ignorable if (1) the missing data are MAR and (2) the parameters of the data model (*i.e.*, the longitudinal model for the dependent variable) and the parameters of the missingness mechanism are distinct [Little and Rubin, 2002]. This latter condition is termed the "distinct parameters" condition and is described in detail by Shih [1992]. Essentially, this condition means that the parameters of these two processes are independent. The term ignorable suggests that the missing data mechanism can be ignored in the longitudinal modeling of $\boldsymbol{y}$. This is the case for full likelihood-based models (*e.g.*, MRMs and CPMs) if the distinct parameters condition holds and the missing data are MAR, while for GEE1 models, the stronger assumption of MCAR is required.

14.2.3 Missing Not at Random (MNAR)

Missing not at random (MNAR) is the situation where the missingness is related to the unobserved dependent variable vector $\boldsymbol{y}_i^M$ after taking observed variables (*i.e.*, $\boldsymbol{X}_i$ and $\boldsymbol{y}_i^O$) into account. The notion here is that there is a relationship between what would have been observed (*i.e.*, the values of $\boldsymbol{y}_i^M$) and the missingness R_{ij}. MNAR can occur if subjects are not measured at a given timepoint because their value of the dependent variable falls below (or exceeds) some critical value. For instance, to contrast MNAR with MAR, MNAR occurs if subjects who have Hamilton depression scores below 15 are not measured *at that timepoint*.

An example of where researchers sometimes assume MNAR can be found in smoking cessation research. In these studies, the dependent variable is whether a person is smoking or not at a given timepoint, and researchers often assume that if a subject is missing at a particular timepoint it is because they are smoking. Of course, this is a rather strong form of MNAR because it implies that the association between missingness R and unobserved y is a perfect one.

Because the distinction between MAR and MNAR involves the unobserved data $\boldsymbol{y}_i^M$, there is no way to confirm or reject MAR versus MNAR. As we will see, one can confirm or reject a particular MAR model relative to a particular MNAR model, but this does not settle the more global issue because, again, the crucial piece of information ($\boldsymbol{y}_i^M$) does not exist. Use of MNAR models is typically done in situations where there is strong suspicion that the data violate MAR, and in this case, it is often useful to do a sensitivity analysis by varying the MNAR missingness assumptions. Here, we will present some relatively simple MNAR approaches that can be used for sensitivity analyses, but the reader should realize that MNAR missingness is a topic of intense and emerging statistical research.

14.3 MODELS AND MISSING DATA MECHANISMS

In terms of the primary methods for analysis of incomplete longitudinal data considered in this text, as Laird [1988] points out, MRMs using maximum likelihood estimation, and correctly modeling both the mean and variance–covariance structure of the dependent variable, provide valid inferences in the presence of ignorable nonresponse. This also holds for the covariance pattern models (CPMs), described in Chapter 6, when estimated under full likelihood procedures. Ignorable nonresponse can include both MCAR and MAR mechanisms, depending on model of analysis, but does not include MNAR. Alternatively, GEE1 models only allow for covariate-dependent MCAR. As noted, the essential distinction between MAR and covariate-dependent MCAR is that in addition to allowing dependency between the missing data and the model covariates, MAR allows the missing data to be related to observed values of the dependent variable. Thus, ordinary GEE1 models make more restrictive assumptions about the missing data than do MRMs and CPMs. As a result, within the GEE class of models, several authors have proposed weighted-estimating equations (WEE) that allow for non-MCAR missingness [Robins et al., 1995; Rotnitzky and Robins, 1999]; Hogan et al. [2004] show how such models can be estimated using standard software.

14.3.1 MCAR Simulations

To give a sense of the treatment of missing data by longitudinal data analysis models, we describe here a small simulation study. A much more complete simulation study can be found in Touloumi et al. [2001]. Here, we simulated data according to the following model:

$$y_{ij} = \beta_0 + \beta_1 Time_j + \beta_2 Grp_i + \beta_3 (Grp_i \times Time_j) + \upsilon_{0i} + \upsilon_{1i} Time_j + \varepsilon_{ij}, \quad (14.1)$$

where $Time_j$ was coded 0, 1, 2, 3, 4 for five timepoints, and Grp_i was a dummy-coded (*i.e.*, 0 or 1) grouping variable with half of the total subjects in each group. The regression coefficients were defined to be: $\beta_0 = 25$, $\beta_1 = -1$, $\beta_2 = 0$, and $\beta_3 = -1$. Thus, the population means for the two groups across the five timepoints were

$$25, 24, 23, 22, 21 \text{ for } Grp = 0$$

and

$$25, 23, 21, 19, 17 \text{ for } Grp = 1.$$

The random subject effects υ_{0i} and υ_{1i} were assumed normal with zero means, variances $\sigma^2_{\upsilon_0} = 4$ and $\sigma^2_{\upsilon_1} = .25$, and covariance $\sigma_{\upsilon_{01}} = -.1$ (it also equals $-.1$ expressed as a correlation). The errors ε_i were assumed to be normal with mean 0 and variance $\sigma^2 = 4$. The population variance–covariance matrix, $V(\boldsymbol{y}) = \boldsymbol{Z}\boldsymbol{\Sigma}_\upsilon\boldsymbol{Z}' + \sigma^2\boldsymbol{I}$, was thus

$$V(\boldsymbol{y}) = \begin{bmatrix} 8.00 & 3.90 & 3.80 & 3.70 & 3.60 \\ 3.90 & 8.05 & 4.20 & 4.35 & 4.50 \\ 3.80 & 4.20 & 8.60 & 5.00 & 5.40 \\ 3.70 & 4.35 & 5.00 & 9.65 & 6.30 \\ 3.60 & 4.50 & 5.40 & 6.30 & 11.20 \end{bmatrix},$$

or expressed as a correlation matrix:

$$\begin{bmatrix} 1.00 & 0.49 & 0.46 & 0.42 & 0.38 \\ 0.49 & 1.00 & 0.50 & 0.49 & 0.47 \\ 0.46 & 0.50 & 1.00 & 0.55 & 0.55 \\ 0.42 & 0.49 & 0.55 & 1.00 & 0.61 \\ 0.38 & 0.47 & 0.55 & 0.61 & 1.00 \end{bmatrix}.$$

These defined mean and variance–covariance parameters are based on and similar in structure to their estimated counterparts from several of the datasets we analyzed in previous chapters. Essentially, we are comparing two groups which both improve across time, though one group ($Grp = 1$) has a faster rate of improvement. The variance of the dependent variable increases across time, and the data are moderately correlated within subjects, though the correlation diminishes as the time-lag increases. Within a time-lag (*i.e.*, a particular band of the matrix), the correlation increases a bit over time.

To begin, data from 5,000 subjects were simulated, each with 5 timepoints of data, according to the defined mean and variance–covariance structure. Some of the data were then classified as missing, according to specifications described below, and the statistical models were estimated on the resulting incomplete dataset. Here, we contrast two models: a MRM allowing for random intercepts and time trends, and a GEE1 model with an unstructured working correlation structure.

To begin, these models were estimated on the full dataset, without any missing data. Table 14.1 presents the MRM estimates for this first situation and three other missing data situations. The first missing data situation allowed 50% missing data at every timepoint, however these missing data were completely random and unrelated to any variable. This is the basic MCAR situation. The next missing data situation, time-related dropout, simulated data with dropout rates of 0%, 25%, 50%, 75%, and 87.5% for the five respective timepoints. In this situation, if subjects were missing at a given timepoint, then they were missing at all later timepoints as well, and these rates indicate the percentage of the original sample that were missing at each of these timepoints. Notice, that this is a case of covariate-dependent MCAR with time as the covariate. Also, admittedly, these rates of missing data are rather high; such high rates were chosen to provide a more telling picture of what happens under various missing data situations. The final missing data situation in Table 14.1 allowed the missingness to vary by time and group by time, such that dropout rates of 0%, 23%, 46%, 70%, and 83% were posited for the first group (*i.e.*, group = 0), and droprout rates of 0%,

27%, 55%, 81%, and 91% were specified for the second group (*i.e.*, group = 1). Again, this is a covariate-dependent MCAR situation with covariates of time and group by time.

Table 14.1. MCAR Simulation Results—MRM Estimates (Standard Errors)

	β_0 (i)	β_1 (t)	β_2 (g)	β_3 (g × t)	$\sigma^2_{\upsilon 0}$ (i)	$\sigma_{\upsilon 01}$ (i, t)	$\sigma^2_{\upsilon 1}$ (t)	σ^2 (e)
Simulated value	25	−1	0	−1	4	−.1	.25	4
Complete data	24.969 (.050)	−.994 (.016)	−.001 (.071)	−.986 (.023)	3.918 (.129)	−.057 (.032)	.239 (.014)	3.991 (.046)
50% random missing	24.991 (.063)	−1.024 (.023)	−.087 (.089)	−.933 (.032)	3.811 (.193)	−.020 (.056)	.199 (.025)	4.070 (.083)
Time-related dropout	24.989 (.053)	−.968 (.028)	.019 (.075)	−1.021 (.040)	3.853 (.150)	−.062 (.060)	.229 (.032)	4.000 (.074)
Group by time dropout	24.991 (.053)	−.977 (.026)	.041 (.075)	−1.014 (.041)	3.872 (.150)	−.048 (.059)	.234 (.031)	3.994 (.073)

Note. i = intercept, t = time, g = group, e = error.

As can be seen from Table 14.1, the parameter values are all reasonably recovered under these MCAR situations. While this is expected given the theory, because the covariates upon which dropout depends are included in the model, it is nice to confirm this with simulation results under these different types of MCAR situations. In particular, as the last situation makes clear, it is useful to see that differential dropout for the two treatment groups, which investigators often fear, is not, in and of itself, a problem. Though not presented, GEE1 gave very similar estimates for the regression coefficients in all of these MCAR situations.

While the estimates are essentially the same throughout, the standard errors do indicate the effect of missing data. Namely, these are larger when missing data are present (*i.e.*, contrasting the latter three situations to the first one). Also, note that the standard errors for the intercept (β_0) and the group difference at time 0 (β_2) are not greatly changed in the time-related dropout situations relative to the complete data situation, whereas they are in the 50% random missing situation. This makes sense because it is only the latter situation that does not have complete data at the first timepoint (*i.e.*, when time = 0).

14.3.2 MAR and MNAR Simulations

Continuing with this example, Table 14.2 presents simulation results under this basic model, but with MAR and MNAR missingness. For MAR, two situations were specified. In the first, denoted MAR(a) in the table, if the value of the dependent variable was lower than 23, then the subject dropped out at the next timepoint (*i.e.*, they were missing at the next and all subsequent timepoints). For the second MAR situation, denoted MAR(b) in the table, we simulated a case where the MAR specification was different for the two groups. Specifically, for one group ($Grp = 1$), we retained the criterion that if the dependent variable was lower than 23, then the subject dropped out at the next timepoint, however for the other group ($Grp = 0$) we specified that if the dependent variable was greater than 25.5, then the subject dropped out at the next timepoint. Thus, for one group a decrease in the dependent variable signaled a subsequent dropout, whereas in the other an increase signaled dropout. For MNAR, after the first timepoint, if the value of the dependent variable was lower than 21.5, then the subject was missing *at that timepoint* and all subsequent timepoints. All of these cutoff values were selected to yield approximate dropout rates of of 0%, 25%, 50%, 75%, and 87.5% for the five timepoints, so the amount of missing data is very similar to the earlier MCAR simulations.

Table 14.2. MAR and MNAR Simulation Results—Estimates (Standard Errors)

	β_0 (i)	β_1 (t)	β_2 (g)	β_3 (g × t)	$\sigma^2_{\upsilon_0}$ (i)	$\sigma_{\upsilon_{01}}$ (i, t)	$\sigma^2_{\upsilon_1}$ (t)	σ^2 (e)
Simulated value	25	−1	0	−1	4	−.1	.25	4
MAR(a)								
MRM	24.996	−1.039	−.010	−.969	3.981	−.064	.233	3.873
	(.053)	(.025)	(.075)	(.041)	(.158)	(.065)	(.032)	(.078)
GEE1	25.281	−1.164	.019	−1.001				
	(.058)	(.037)	(.097)	(.085)				
MAR(b)								
MRM	24.999	−1.003	−.016	−1.004	4.050	−.082	.229	3.812
	(.053)	(.022)	(.075)	(.039)	(.154)	(.064)	(.027)	(.073)
GEE1	24.635	−.714	.634	−1.532				
	(.055)	(.030)	(.097)	(.090)				
MNAR								
MRM	24.956	−.233	.027	−.552	3.856	−.943	.319	3.020
	(.049)	(.020)	(.070)	(.035)	(.131)	(.051)	(.025)	(.053)
GEE1	25.051	−.386	.016	−.583				
	(.049)	(.020)	(.071)	(.034)				

Note. i = intercept, t = time, g = group, e = error.

As can be seen, when the data are MAR, the parameters are well-covered by MRM, but not as well by GEE1. In this regard, Touloumi et al. [2001] observed, in their more extensive simulation study, that the degree of bias in GEE1 estimates increased with the severity of non-randomness and with the amount of missing data.

Figure 14.1 displays results from the MAR(b) simulation. In this figure, the lines represent the population trends for the two groups, and the symbols represent estimated means for the two groups across time. Thus, Figure 14.1a presents only the population trends, whereas Figure 14.1b presents the population trends and the calculated observed averages across time for the two groups. Similarly, Figures 14.1c and 14.1d present the population means and the estimated means from MRM and GEE1.

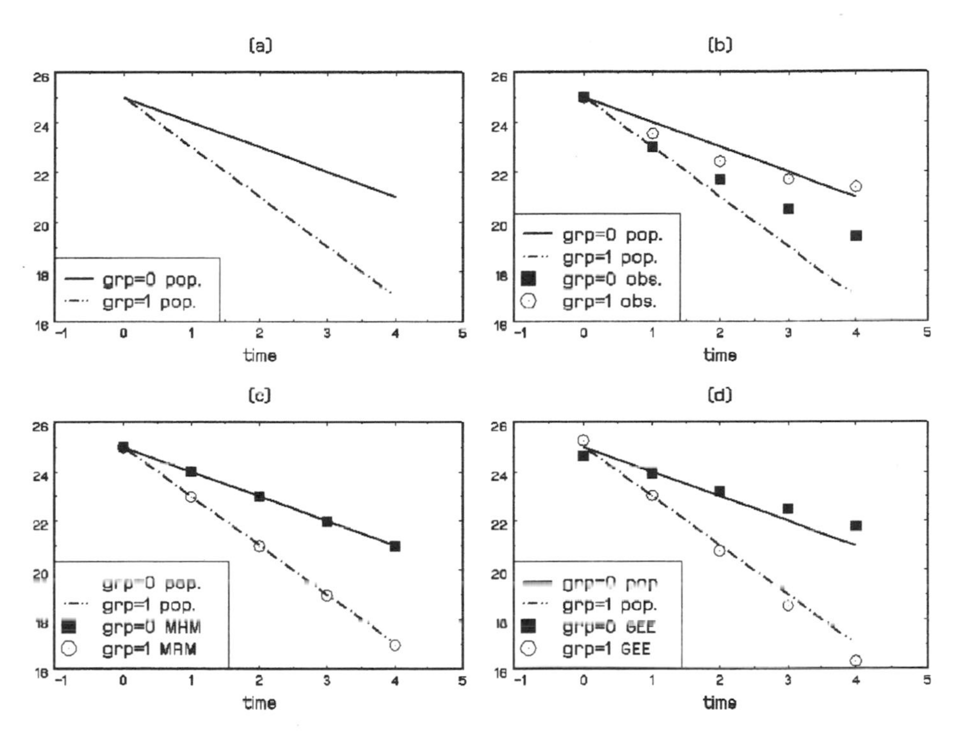

Figure 14.1. MAR(b) simulation: (a) Population means across time by group; (b) observed versus population means across time by group; (c) MRM estimates versus population means across time by group; (d) GEE estimates versus population means across time by group.

As can be seen, under MAR, MRM (Figure 14.1c) fits the simulated trends very well, while GEE1 (Figure 14.1d) yields estimated trends above and below the population trends of groups 0 and 1, respectively. Figure 14.1b presents the observed group means across time. These observed means are interesting because they give a very false impression of the actual population trends. From the observed means, and relying solely on eyeball statistics, one might wrongly conclude that group 0 has the greater decrease across time, relative to group 1. Why do these observed means show this pattern? It is because of the differential reasons for dropout in these two groups for this missing data situation. Under the MAR(b) situation, group 0 subjects with higher values (above 25.5) subsequently dropped out, leaving in subjects with low values and thus producing observed means that are artificially low. Alternatively, under MAR(b), group 1 subjects with lower values (below 23) subsequently dropped out, and so the observed means are artificially high. This simulation example clearly shows the danger of relying on observed means when missing data are present. Of course, this point is somewhat exaggerated because of the great deal of missing data here.

Turning to the results for MNAR in Table 14.2, we see major differences for both the MRM and GEE1 analyses. In fact, in this particular case, the results from these two are very similar, though there is no compelling reason for why that should occur. For both, the negative trend parameters are appreciably underestimated. Figure 14.2 displays results from the MNAR simulation. Here, as can be seen, the trends of both groups are significantly less pronounced than the underlying population trends. In fact, based on these analyses, one would conclude that the trend of group 1 was approximately equal to the true trend for group 0, and that the trend of group 0 was only somewhat negative. Clearly, analysis of MNAR data can be problematic with MAR (Figure 14.2c) and MCAR (Figure 14.2d) data analysis models. Likewise, the observed means in Figure 14.2b are very misleading and do not give a correct impression of the true population trends.

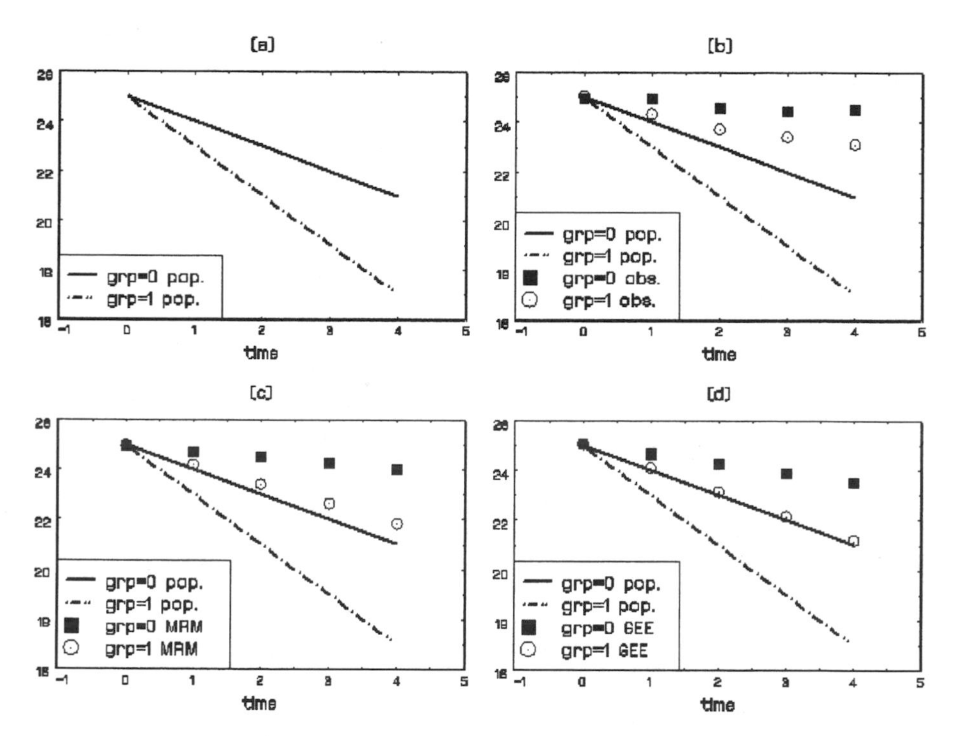

Figure 14.2. MNAR simulation: (a) Population means across time by group; (b) observed versus population means across time by group; (c) MRM estimates versus population means across time by group; (d) GEE estimates versus population means across time by group.

This simulation shows that if the data are MAR, then the MAR analysis does yield the correct results. An important and subtle proviso, pointed out by Fitzmaurice et al. [2004], is that both the mean and the variance–covariance structure of the longitudinal data must be correctly specified. Notice that our MRM analysis model did correctly specify both structures. For instance, in terms of the mean structure, the simulated data included effects of time, group, and group by time. These effects were also included in the MRM analysis model, and so the mean structure was correctly specified. Similarly, the simulated data included both a random subject intercept and time-trend, as did our analysis models. Thus, the variance–covariance structure of the dependent variable was correctly specified. To give a sense of violation of the latter, we re-analyzed the simulated MAR data, but using only a random-intercepts model:

$$y_{ij} = \beta_0 + \beta_1 Time_j + \beta_2 Grp_i + \beta_3 (Grp_i \times Time_j) + \upsilon_{0i} + \varepsilon_{ij}. \tag{14.2}$$

In other words, we have a misspecified model for the variance–covariance structure because the random trend term was omitted from the analysis. Table 14.3 presents the results for the MAR(a) and MAR(b) simulated datasets. As can be seen, these random-intercepts analyses yield biased results, in particular for the time-related parameters β_1 and β_3. Thus, performing any MAR analysis, even with MAR data, does not guarantee that the correct results will be obtained. The selection of an appropriate variance–covariance structure for a given dataset is critical, and so one must pay particular attention to this when missing data are present.

Table 14.3. Misspecified MAR Simulation Results—MRM Estimates (Standard Errors)

	β_0 (i)	β_1 (t)	β_2 (g)	β_3 (g × t)	$\sigma^2_{\upsilon 0}$ (i)	$\sigma_{\upsilon 01}$ (i, t)	$\sigma^2_{\upsilon 1}$ (t)	σ^2 (e)
Simulated value	25	-1	0	-1	4	$-.1$	.25	4
MRM with MAR(a)	24.938	$-.891$	$-.009$	$-.949$	3.722			4.329
	(.053)	(.021)	(.075)	(.036)	(.136)			(.072)
MRM with MAR(b)	24.998	-1.048	$-.069$	$-.805$	3.880			4.288
	(.053)	(.018)	(.076)	(.035)	(.138)			(.070)

Note. i = intercept, t = time, g = group, e = error.

14.4 TESTING MCAR

Consistent with the theory, as the above simulation has shown, if the missing data are MCAR then analysis using either GEE1 or MRM is fine, provided that the covariate matrix $\boldsymbol{X}_i$ includes any predictors of missingness. However, if the missing data are MAR, then the GEE1 analysis does not perform well, whereas MRM analysis is acceptable as long as the mean and variance–covariance structures are correctly modeled. Therefore, in choosing a method for a given analysis, it is helpful to determine whether MCAR is acceptable or not. Again, the essential distinction between MCAR and MAR is that missingness cannot depend on observed values of the dependent variable, $\boldsymbol{y}_i^O$, in the former, but can in the latter. Tests of whether MCAR is reasonable or not can therefore be based on analyses involving $\boldsymbol{y}_i^O$. Such tests are described by Little [1988] and Diggle [1989], and here we illustrate their application.

As an example, suppose that all subjects have data at time 1, but some are missing at time 2. Define the variable $D_i = 0$ for subjects with data at both timepoints and $D_i = 1$

for those that only have data at the first timepoint. Then if we compare the y_1 data between these two groups (*i.e.*, $D_i = 0$ versus $D_i = 1$), then MCAR would posit that the y_1 data should not differ on average. Thus, a simple t-test comparing the y_1 means between the two groups could be used to address this distinction. More generally, because MCAR does allow missingness to depend on covariates, we can perform the following regression

$$y_{i1} = \beta_0 + \beta_1 D_i + \boldsymbol{\beta}_2 \boldsymbol{x}_i + \varepsilon_i, \tag{14.3}$$

where $\boldsymbol{\beta}_2$ is the vector of regression coefficients for the set of covariates included in $\boldsymbol{x}_i$. We can also interact the dropout variable with covariates to yield

$$y_{i1} = \beta_0 + \beta_1 D_i + \boldsymbol{\beta}_2 \boldsymbol{x}_i + \boldsymbol{\beta}_3 (D_i \times \boldsymbol{x}_i) + \varepsilon_i, \tag{14.4}$$

where $\boldsymbol{\beta}_3$ is the vector of regression coefficients for the interactions of dropout with covariates. In this model, MCAR would specify that $\beta_1 = \boldsymbol{\beta}_3 = 0$.

As noted by Ridout [1991], it is beneficial to turn this question around and to specifically model dropout in terms of a logistic regression, namely,

$$\log\left[\frac{\mathrm{P}(D_i = 1)}{1 - \mathrm{P}(D_i = 1)}\right] = \alpha_0 + \alpha_1 y_{i1} + \boldsymbol{\alpha}_2 \boldsymbol{x}_i + \boldsymbol{\alpha}_3 (y_{i1} \times \boldsymbol{x}_i). \tag{14.5}$$

Again, here, $\boldsymbol{\alpha}_2$ and $\boldsymbol{\alpha}_3$ represent vectors of regression coefficients for the set of covariates $\boldsymbol{x}_i$ and their interactions with y_{i1}. MCAR would dictate that $\alpha_1 = \boldsymbol{\alpha}_3 = 0$.

If there are more than two timepoints in a study, then this model can be generalized both in terms of the left and right side of the equation. Let's first consider generalization on the right side of the equation. In terms of the predictor, one can modify the above models by using, for example, a subject's final observed measurement of y_{ij} in place of y_{i1}. Alternatively, in this case, one could also use a function of the observed dependent variable values, say $h(\boldsymbol{y}_i^O)$. An obvious choice for this is simply an average, namely,

$$\bar{y}_i = \frac{1}{n_i} \sum_{j=1}^{n_i} y_{ij}. \tag{14.6}$$

Or, more generally, one could consider h to be a weighted average of the observed values,

$$h(\boldsymbol{y}_i^O) = \sum_{j=1}^{n_i} w_j y_{ij}. \tag{14.7}$$

Depending on the the specifications for w_j, this allows a linear trend across the observed timepoints, a difference between the first and last observed timepoints, a difference between the second to last and last timepoint, etc. Choice of which function(s) of the observed dependent variable values to use depends on the given study, but one might consider several of these functions in performing this testing.

Turning to the left-hand side of the equation, with more than two timepoints, modeling of dropout can be reframed as modeling of time to dropout. For this, the variable $D_i = j'$ if subject i drops out between the $(j' - 1)$th and j'th timepoint; namely, $y_{i,j'-1}$ is observed, but $y_{i,j'}, \ldots, y_{in}$ are all missing. Note, that we are ignoring intermittent missingness here and concentrating on whether time to dropout is MCAR or not. This simplification of missingness is reasonable to the extent that intermittent missingness is MCAR, but dropout is potentially not. Because it is often reasonable to assume that dropout is less likely to be MCAR than intermittent missingness, this simplification is a natural way of proceeding first. Further testing could be done using the indicators R_{ij}, but we will not consider that here. In terms of the model for time until dropout, we can use either a discrete-time or grouped-time survival analysis model as described in several sources [Allison, 1982; D'Agostino et al., 1990; Singer and Willett, 1993]. Using a logistic link function, this model is written as

$$\log\left[\frac{\mathrm{P}(D_i = j \mid D_i \geq j)}{1 - \mathrm{P}(D_i = j \mid D_i \geq j)}\right] = \alpha_{0j} + \alpha_1 h(\boldsymbol{y}_i^O) + \boldsymbol{\alpha}_2 \boldsymbol{x}_i + \boldsymbol{\alpha}_3 (h(\boldsymbol{y}_i^O) \times \boldsymbol{x}_i), \quad (14.8)$$

where the timepoints are indexed by $j = 1, \ldots, n$, and we further define completers as $D_i = n + 1$ (where n is the last timepoint). Here, the logit is for the probability of dropout at a given timepoint, given that it has not already occurred. As before, MCAR is rejected if $\alpha_1 = \boldsymbol{\alpha}_3 = 0$ is rejected.

As in survival analysis models, we can also consider time-varying covariates in this model. Thus, more generally, we have

$$\log\left[\frac{\mathrm{P}(D_i = j \mid D_i \geq j)}{1 - \mathrm{P}(D_i = j \mid D_i \geq j)}\right] = \alpha_{0j} + \alpha_1 h(\boldsymbol{y}_{ij}^O) + \boldsymbol{\alpha}_2 \boldsymbol{x}_{ij} + \boldsymbol{\alpha}_3 (h(\boldsymbol{y}_{ij}^O) \times \boldsymbol{x}_{ij}). \quad (14.9)$$

For example, the function $h(\boldsymbol{y}_{ij}^O)$ might represent the changing average of the dependent variable across time: $h(\boldsymbol{y}_{i1}^O) = y_{i1}$, $h(\boldsymbol{y}_{i2}^O) = 1/2(y_{i1}+y_{i2})$, $h(\boldsymbol{y}_{i3}^O) = 1/3(y_{i1}+y_{i2}+y_{i3})$, etc. Similarly, x_{ij} might represent a level of stress or mood across time, or for that matter, the variable suggested by Demirtas and Schafer [2003]: "How likely are you to remain in the study?"

To estimate this model, standard software for logistic regression can be used, but the dataset needs to be created as a "person–period dataset" [Singer and Willett, 2003]. This is described in Section 10.2.3, here we will go into more specific details. Essentially, in this type of dataset, each person contains as many records as the number of timepoints (or periods) that they are at risk of dropping out.

As an example, suppose that a study has four timepoints and that all subjects are measured at the first timepoint (*i.e.*, baseline), but some subjects drop out at each of the next three follow-ups. Then, we can define $D_i = 1, 2,$ or 3 for dropouts at the three follow-up timepoints, and $D_i = 4$ for subjects who complete the study. Table 14.4 lists data for four subjects, each representing a different value of D_i in this example with a baseline and three follow-up timepoints. Notice that the variable y_{ij} equals 1 if subject i drops out at timepoint j, and in this case, the subject has no further records in the dataset. Similarly, y_{ij} equals 0 if subject i was measured at timepoint j, or if the subject has not yet dropped out (*i.e.*, the case of intermittent missingness). Remember, here we are only considering time until dropout, the point at which a subject provides no further data on the dependent variable.

Table 14.4. Example of a Person–Period Dataset

ID	D_i	Period	y_{ij}
101	1	1	1
—	—	—	—
102	2	1	0
102	2	2	1
—	—	—	—
103	3	1	0
103	3	2	0
103	3	3	1
—	—	—	—
104	3	1	0
104	3	2	0
104	3	3	0

For a given analysis, the above data records could also contain information on covariates, either time-varying or time-invariant, and also functions of the observed values of the dependent variable. As an example, suppose that the covariates of sex (time-invariant) and stress level (time-varying) are available, and also the time-varying average of the observed values of the dependent variable. Also, let $j = 0$ denote the baseline timepoint. Then, the dataset would be augmented as indicated in Table 14.5.

Table 14.5. Example of a Person–Period Dataset with Covariates

ID	D_i	Period	y_{ij}	Sex	Stress	y Average
101	1	1	1	Sex $_{101}$	$\text{Stress}_{101,0}$	$y_{101,0}$
—	—	—	—	—	—	—
102	2	1	0	Sex $_{102}$	$\text{Stress}_{102,0}$	$y_{102,0}$
102	2	2	1	Sex $_{102}$	$\text{Stress}_{102,1}$	$(y_{102,0} + y_{102,1})/2$
—	—	—	—	—	—	—
103	3	1	0	Sex $_{103}$	$\text{Stress}_{103,0}$	$y_{103,0}$
103	3	2	0	Sex $_{103}$	$\text{Stress}_{103,1}$	$(y_{103,0} + y_{103,1})/2$
103	3	3	1	Sex $_{103}$	$\text{Stress}_{103,2}$	$(y_{103,0} + y_{103,1} + y_{103,2})/3$
—	—	—	—			
104	3	1	0	Sex $_{104}$	$\text{Stress}_{104,0}$	$y_{104,0}$
104	3	2	0	Sex $_{104}$	$\text{Stress}_{104,1}$	$(y_{104,0} + y_{104,1})/2$
104	3	3	0	Sex $_{104}$	$\text{Stress}_{104,2}$	$(y_{104,0} + y_{104,1} + y_{104,2})/3$

Once the data are organized in this manner, a logistic regression analysis can be performed regressing the dropout indicators y_{ij} on period, sex, stress, y average, and interactions. Although it is common to use logistic regression, and therefore the logit link, the complementary log–log (clog–log) link function can also be used here. As described in

Section 10.2.3, use of the clog–log link is advantageous because it yields a grouped-time proportional hazards model [Prentice and Gloeckler, 1978; Hedeker et al., 2000]. This model is written as

$$\log(-\log(1 - \mathrm{P}(D_i = j \mid D_i \geq j))) = \alpha_{0j} + \alpha_1 h(\boldsymbol{y}_{ij}^O) + \boldsymbol{\alpha}_2 \boldsymbol{x}_{ij} + \boldsymbol{\alpha}_3 (h(\boldsymbol{y}_{ij}^O) \times \boldsymbol{x}_{ij}).$$

The only change to the model is that a different link function is specified, namely clog–log instead of logit. In practice, it often doesn't matter greatly if the logit or clog–clog link is selected for this purpose, see Singer and Willett [2003] for more discussion on this issue.

14.4.1 Example

Consider the schizophrenia study described previously in Chapter 9. In that study, subjects were measured at a baseline timepoint and weekly for up to 6 weeks. The study protocol specified the primary measurement weeks as 0 (baseline), 1, 3, and 6; however, some subjects were also measured at weeks 2, 4, and 5. There was some intermittent missingness in this study, however, dropout was a much more common pattern of missingness. In all, 102 of 437 subjects did not complete the trial. Table 14.6 lists a cross-tabulation of treatment group (denoted `Drug`) by last measurement wave (denoted `Maxweek`). As can be seen, dropout is more common among the placebo group; a Pearson χ^2 test yields $p < .025$, while a Mantel–Haenszel χ^2 test for trend yields $p < .0013$. Thus, dropout is at least covariate-dependent.

Table 14.6. Crosstabulation of Drug by Maxweek—Cell Frequencies (Row Percentages)

	Maxweek						
Drug	1	2	3	4	5	6	Total
Placebo	13	5	16	2	2	70	108
	(.12)	(.05)	(.15)	(.02)	(.02)	(.65)	
Drug	24	5	26	3	6	265	329
	(.07)	(.02)	(.08)	(.01)	(.02)	(.81)	

Using the variable `Maxweek` as D_i, we created a person–period dataset in the manner described. The possible dropout times were weeks 1 to 5, and so a given subject could have up to five records in this dataset. For instance, if a subject completed the trial (*i.e.*, `Maxweek` = 6; the subject was measured at week 6), then they contributed five records to this dataset with $y_{ij} = 0$ for all five. Alternatively if, for instance, a subject was measured at week 5 but not at week 6 (*i.e.*, `Maxweek` = 5), the subject also contributed five records to the dataset with $y_{i1} = y_{i2} = y_{i3} = y_{i4} = 0$ and $y_{i5} = 1$. Finally, if a subject was only measured at week 1, then they only contributed a single record to the dataset with $y_{i1} = 1$.

In terms of covariates, we included `Week`, `Drug`, `MeanY` (a subject's cumulative mean of the observed values of the dependent variable at a given timepoint), and their interactions, We fit several clog–log regression models in a sequential manner; Table 14.7 lists the deviances (*i.e.*, $-2 \log L$) for the various models considered. Based on the deviances, it is clear that the `Drug` × `MeanY` interaction is significant ($X_1^2 = 728.13 - 706.77 = 21.36, p < .0001$), however none of the other interactions are significant.

Table 14.7. Time to Dropout Models

Covariates	p	Deviance ($-2 \log L$)
Week, Drug, MeanY	7	729.44
+ Week × Drug	11	728.13
+ Drug × MeanY	12	706.77
+ Week × MeanY	16	700.50
+ Week × Drug × MeanY	20	697.71

p = number of regression coefficients (including one for the intercept).

We refit the model including the three main effects and only the Drug × MeanY interaction. The results from this model, with a deviance of 708.91, are listed in Table 14.8. In this model, MeanY represents the effect of observed data on dropout for the placebo group. It is positive and highly signifiant, indicating that placebo subjects with higher IMPS79 scores were more likely to drop out. Exponentiating the estimate .635 yields 1.89, indicating that a unit increase in MeanY yields an approximate doubling in the hazard of dropping out for the placebo group. Conversely, the Drug × MeanY interaction is negative and significant, indicating that this pattern is reversed for drug patients. For the drug group, the effect of observed IMPS79 data is obtained as $.635 - 1.108 = -.473$, which has a standard error of .131 and is significant at $p = .0003$ (these results can be obtained simply by reversing the coding of the drug variable and re-estimating the model). Thus, for the drug group, subjects with lower observed IMPS79 data are more likely to drop out. Exponentiating $-.473$ yields .623, which inverted equals 1.60, indicating that a unit *decrease* in MeanY is associated with a 1.6 increase in the dropout hazard for the drug group. In summary, the effect of observed y, as represented by MeanY, on dropout is highly significant when interacted with drug group, and so MCAR is rejected.

Table 14.8. Final Time to Dropout Model

Term	Estimate	SE	$p <$
Intercept	−6.573	1.208	.0001
Week 1	1.327	.393	.0007
Week 2	0.096	.476	.84
Week 3	1.549	.386	.0001
Week 4	−0.494	.570	.39
Drug	4.765	1.297	.0002
MeanY	0.635	.214	.003
Drug × MeanY	−1.108	.249	.0001

What is also interesting here is that if one excludes Drug × MeanY from the model (*i.e.*, the main effects model with deviance 729.44), the effect of MeanY is estimated to be $-.147$ and is not significant ($p = .18$). Thus, based on this main effects analysis, one might conclude that MCAR is reasonable! Of course, in this model, the effect of MeanY is the averaged effect across both groups, and the effect clearly varies across groups. This illustrates that the examination of interactions is important in the testing of MCAR.

14.5 MODELS FOR NONIGNORABLE MISSINGNESS

As the simulation indicated, when the data are nonignorable (*i.e.*, MNAR), standard statistical models can yield badly biased results. Of course, in the simulation we know that the data are nonignorable, because they were created that way, however considering a real dataset, one never knows whether the data are ignorable or not. As many authors have pointed out, the observed data provide no information to either confirm or refute ignorability. With that said, assuming a particular model for nonignorability and ignorability, one can test for ignorability, but this test is completely dependent on the proposed models for nonignorability and ignorability (*e.g.*, see Kenward [1998]). Again, the data cannot address this important point independent of assumed models.

Because one sometimes suspects that the missing data are nonignorable, development of methods for dealing with such missing data has been a very active area of statistical work. Little [1995] described much of these approaches in terms of two broad model classes: selection and pattern-mixture models. Here, we will describe and illustrate use of these models primarily in terms of longitudinal MRMs, though their usage extends to other longitudinal data analysis models as well. Besides the article by Little [1995], further discussion on differences between pattern-mixture and selection models can be found in Glynn et al. [1986], Hogan and Laird [1997b], Michiels et al. [2002], Little [1993, 1994], and in the discussion of the Diggle and Kenward [1994] article.

Use of nonignorable models can be helpful in conducting a sensitivity analysis; to see how the conclusions might vary as a function of what is assumed about the missing data. However, several authors warn against use of a particular nonignorable model as "the" model, because these models make assumptions about the missing data that are essentially impossible to verify with the observed data. In what follows, we will describe particular nonignorable models that can be fit using standard software. Our aim is to provide practical methods for exploring the possibility of nonignorability. The reader should realize, however, that the class of nonignorable models is broad and the methods described here are a relatively small part of this class.

14.5.1 Selection Models

The use of selection models for dealing with missing data in longitudinal studies has a relatively long history, being first proposed by Heckman [1976] in the econometric literature. More recently, Leigh et al. [1993] present a useful tutorial article on implementation of this approach. In its original formulation, selection modeling involves two stages which are either performed separately or iteratively. The first stage is to develop a predictive model for whether or not a subject drops out, using variables usually obtained prior to the dropout, often the variables measured at baseline or time-varying covariates. This model of dropout provides a predicted dropout probability or propensity for each subject; these dropout propensity scores are then used in the (second stage) longitudinal data model as a covariate to adjust for the potential influence of dropout. By modeling dropout, selection models provide valuable information regarding the predictors of study dropout. Diggle and Kenward [1994] extend this approach by augmenting the dropout model with past values of the dependent variable and the unobserved dependent variable at the time of dropout.

Selection models have often been criticized because results can depend greatly on distributional assumptions of the missing data that are impossible to verify [Little, 1995; Little and Rubin, 2002]. To address this, Kenward [1998] describes how the distributional assumptions can be varied, allowing one to assess, to some degree, the sensitivity of the results to the distributional assumptions.

14.5.1.1 Mixed-Effects Selection Models The models described in this section, mixed-effects selection models, augment the usual mixed-effects regression model (MRM) for longitudinal data with a model of dropout (or missingness), in which dropout depends on the random effects of the longitudinal MRM. They are is a bit different than Heckman's original selection model, because the dropout propensity, or a function of this propensity, is not included as a covariate in the longitudinal model. However, they share the property of having two models, one for the longitudinal process and one for dropout, that are linked together. These models have also been called random-coefficient selection models [Little, 1995], random-effects-dependent models [Hogan and Laird, 1997b], and shared parameter models [De Gruttola and Tu, 1994; Wu and Carroll, 1988; Wu and Bailey, 1989; Schluchter, 1992; Ten Have et al., 1998] in the literature. As in ordinary selection models, one specifies both a model for the longitudinal outcome and a model for the dropout. However, as mentioned, both models depend on random subject effects, most or all of which are shared by both models; hence, the term "shared parameter" models. Appealing aspects of this class of models is that they can be used for nonignorable missingness and can be fit using some standard software.

To describe the mixed-effects selection model, following Ten Have et al. [1998], let $f_y(\boldsymbol{y}_i \mid \boldsymbol{v})$ represent the conditional model for the longitudinal outcome $\boldsymbol{y}_i$ given the random subject effects $\boldsymbol{v}$. Notice that this is simply shorthand notation for a mixed-effects regression model of the longitudinal outcome vector $\boldsymbol{y}_i$, that is,

$$\boldsymbol{y}_i = \boldsymbol{X}_i\boldsymbol{\beta} + \boldsymbol{Z}_i\boldsymbol{v}_i + \boldsymbol{\varepsilon}_i. \tag{14.10}$$

Similarly, let $f_D(D_i \mid \boldsymbol{v})$ represent the conditional model for time to dropout given the same subject random effects. Here, for simplicity, we will describe the model in terms of the time to dropout variable D_i, rather than the missing data indicators R_{ij}, though the logic is the same. For modeling dropout there are many possibilities for how the random effects can be included in the model, several of these are described in Little [1995]. Here, similar to Wu and Carroll [1988], we will include them as covariates. The notion is that dropout may be related to a individual's underlying starting point and time-trend of the longitudinal outcome. Using a complementary log–log regression of the time to dropout variable, we have

$$\log(-\log(1 - \mathrm{P}(D_i = j \mid D_i \geq j))) = \boldsymbol{W}_i\boldsymbol{\alpha} + \boldsymbol{v}_i\boldsymbol{\alpha}^*, \tag{14.11}$$

where, $\boldsymbol{W}_i$ would include predictors of time to dropout, some or all of which may be also in $\boldsymbol{X}_i$. We additionally posit that dropout depends on the random subject effects, $\boldsymbol{v}_i$, which characterize both the unobserved and observed components of the dependent variable vector $\boldsymbol{y}_i$. As such, to the extent that the regression coefficients $\boldsymbol{\alpha}^*$ are nonzero, this is a nonignorable model because missingness, here characterized simply as dropout, is dependent on both $\boldsymbol{y}_i^O$ *and* $\boldsymbol{y}_i^M$.

This mixed-effects selection, or shared parameter, model posits that conditional on the random effects, the outcomes $\boldsymbol{y}_i$ and dropout times D_i are independent. That is, the random effects account for all of the association in these observations within a subject (*i.e.*, both $\boldsymbol{y}_i$ and D_i). Thus, the marginal likelihood for a subject for both the longitudinal and dropout components is given by

$$f(\boldsymbol{y}_i, D_i) = \int_{\boldsymbol{v}} f_y(\boldsymbol{y}_i \mid \boldsymbol{v})\, f_D(D_i \mid \boldsymbol{v})\, f(\boldsymbol{v})\, d\boldsymbol{v}, \tag{14.12}$$

where $f(\boldsymbol{v})$ is the distribution of the random effects. Typically, these are multivariate normal with mean 0 and variance–covariance $\boldsymbol{\Sigma}_v$. The marginal likelihood for the sample of N subjects is then given by the sum

$$\log L = \sum_i^N f(\boldsymbol{y}_i, D_i), \tag{14.13}$$

which can be maximized to obtain the maximum likelihood solution. As in the models for non-normal outcomes (*e.g.*, Chapters 9 and 10), this maximization requires evaluation of the integration over the random effects. As in the non-normal models, this can be done using numerical integration, and in particular Gauss–Hermite quadrature because the random effects are assumed to be normally distributed. In doing this, it is convenient to standardize the random effects, $\boldsymbol{v}_i = \boldsymbol{S}\boldsymbol{\theta}_i$ (where $\boldsymbol{\theta}_i$ is distributed as a multivariate standard normal), using the Cholesky factorization, namely $\boldsymbol{\Sigma}_v = \boldsymbol{S}\boldsymbol{S}'$, where $\boldsymbol{S}$ is a lower triangular matrix. The likelihood for a given individual is now of the form

$$f(\boldsymbol{y}_i, D_i) = \int_{\boldsymbol{\theta}} f_y(\boldsymbol{y}_i \mid \boldsymbol{\theta})\, f_D(D_i \mid \boldsymbol{\theta})\, f(\boldsymbol{\theta})\, d\boldsymbol{\theta}. \tag{14.14}$$

14.5.1.2 Example We will illustrate application of mixed-effects selection models using the Schizophrenia dataset. Specifically, in terms of the longitudinal model, we will consider the following model for IMPS79 over time:

$$\begin{aligned} \texttt{IMPS79}_{ij} &= \beta_0 + \beta_1 \texttt{Drug}_i + \beta_2 \texttt{SWeek}_j + \beta_3 (\texttt{Drug}_i \times \texttt{SWeek}_j) \\ &\quad + v_{0i} + v_{1i}\texttt{SWeek}_j + \varepsilon_{ij}, \end{aligned} \tag{14.15}$$

where SWeek is the square root of week, which is used here to linearize the relationship between IMPS79 and time. The random effects, v_{0i} and v_{1i} are assumed to be normally distributed in the population, and so the above model is an ordinary random intercept and trend model. The variance–covariance matrix of the random effects is given by $\boldsymbol{\Sigma}_v$. As mentioned, we will utilize the Cholesky factorization of this matrix, $\boldsymbol{\Sigma}_v = \boldsymbol{S}\boldsymbol{S}'$, where in the two-dimensional case,

$$\boldsymbol{S} = \begin{bmatrix} s_0 & 0 \\ s_{01} & s_1 \end{bmatrix} = \begin{bmatrix} \sigma_{v_0} & 0 \\ \sigma_{v_{01}}/\sigma_{v_0} & \sqrt{\sigma^2_{v_1} - \sigma^2_{v_{01}}/\sigma^2_{v_0}} \end{bmatrix}. \tag{14.16}$$

Using the Cholesky, we have $\boldsymbol{v}_i = \boldsymbol{S}\boldsymbol{\theta}_i$, where the $r \times 1$ vector $\boldsymbol{\theta}_i$ contains standardized random effects which are uncorrelated and have unit variance. Incorporating the design matrix of the random effects, $\boldsymbol{Z}_i$, further yields $\boldsymbol{Z}_i\boldsymbol{v}_i = \boldsymbol{Z}_i\boldsymbol{S}\boldsymbol{\theta}_i$, and so we can write the random effects part of the above model as

$$\begin{aligned} v_{0i} + v_{1i}\texttt{SWeek}_j &= (s_0 + s_{01}\texttt{SWeek}_j)\theta_{0i} + (s_1\texttt{SWeek}_j)\theta_{1i} \\ &= (\sigma_{v_0} + (\sigma_{v_{01}}/\sigma_{v_0})\texttt{SWeek}_j)\theta_{0i} + \left(\sqrt{\sigma^2_{v_1} - \sigma^2_{v_{01}}/\sigma^2_{v_0}}\,\texttt{SWeek}_j\right)\theta_{1i}. \end{aligned}$$

The longitudinal model is thus

$$\begin{aligned}\texttt{IMPS79}_{ij} &= \beta_0 + \beta_1\texttt{Drug}_i + \beta_2\texttt{SWeek}_j + \beta_3(\texttt{Drug}_i \times \texttt{SWeek}_j) \\ &\quad + (\sigma_{\upsilon_0} + (\sigma_{\upsilon_{01}}/\sigma_{\upsilon_0})\,\texttt{SWeek}_j)\theta_{0i} \\ &\quad + \left(\sqrt{\sigma^2_{\upsilon_1} - \sigma^2_{\upsilon_{01}}/\sigma^2_{\upsilon_0}}\,\texttt{SWeek}_j\right)\theta_{1i}. \qquad (14.17)\end{aligned}$$

For time to dropout, consider the following proportional hazards survival model:

$$\begin{aligned}\log(-\log(1 - \mathrm{P}(D_i = j \mid D_i \geq j))) &= \alpha_{0j} + \alpha_1\texttt{Drug}_i + \alpha_2\theta_{0i} + \alpha_3\theta_{1i} \\ &\quad + \alpha_4(\texttt{Drug}_i \times \theta_{0i}) \\ &\quad + \alpha_5(\texttt{Drug}_i \times \theta_{1i}). \qquad (14.18)\end{aligned}$$

This model specifies that the time of dropout is influenced by a person's drug group, their intercept and trend in IMPS79, and in the interaction between these random subject effects and their drug group. The interaction between these random effects and drug group would seem to be important because our previous analyses indicated that drug group interacted with the mean of the observed data in predicting subsequent dropout. Here, the random effects are summaries of a person's observed *and unobserved* data, and so this shared parameter model is a nonignorable model if $\alpha_2 = \alpha_3 = \alpha_4 = \alpha_5 = 0$ is rejected. Notice in this context, we are testing whether *a particular model of ignorability is reasonable versus a particular model of nonignorability*, so it is important that this test is not construed as a general test of ignorability, which it is not.

Because the above model for dropout does not include any time-varying covariates, besides the intercept terms representing the baseline hazard, we can take advantage of the equivalence of certain models under the clog–log link [Engel, 1993; L¨aärä and Matthews, 1985]. Namely, the above dichotomous regression model utilizing person–period indicators of dropout is equivalent to the following ordinal regression model:

$$\begin{aligned}\log(-\log(1 - \mathrm{P}(D_i \leq j))) &= \alpha_{0j} + \alpha_1\texttt{Drug}_i + \alpha_2\theta_{0i} + \alpha_3\theta_{1i} \\ &\quad + \alpha_4(\texttt{Drug}_i \times \theta_{0i}) \\ &\quad + \alpha_5(\texttt{Drug}_i \times \theta_{1i}). \qquad (14.19)\end{aligned}$$

This equivalency holds for all regression coefficients of time-invariant covariates. Thus, in either represenation, the coefficients α_1 to α_5 are identical, and only the baseline hazard parameters α_{0j} vary between these two representations. In the ordinal version, we simply treat D_i as an ordinal outcome, taking on values 1 to 6 (*i.e.*, values 1 to 5 represent dropout week and 6 indicates study completion), and regress it on the covariates using the clog–log link function. This is simpler, from a data analytic perspective, because we do not have to create a person–period dataset. Instead, we have one outcome per person (D_i) and several person-level covariates ($\texttt{Drug}_i, \theta_{0i}, \theta_{1i}$). This equation can also be written in terms of the cumulative probability of response as

$$\begin{aligned}\mathrm{P}(D_i \leq j) &= 1 - \exp(-\exp(\alpha_{0j} + \alpha_1\texttt{Drug}_i + \alpha_2\theta_{0i} + \alpha_3\theta_{1i} \\ &\quad + \alpha_4(\texttt{Drug}_i \times \theta_{0i}) + \alpha_5(\texttt{Drug}_i \times \theta_{1i}))). \qquad (14.20)\end{aligned}$$

This model can be estimated using SAS PROC NLMIXED, which is a general program for estimation of many kinds of mixed-effects model. For this, the first step is to create a dataset in which a single vector contains, for each subject, the dependent variable vector $\boldsymbol{y}_i$ and the time to dropout variable D_i as one vector, say $\boldsymbol{y}_i^*$. Code for accomplishing this is presented, using the schizophrenia dataset, in Table 14.9. In this listing, uppercase letters are used to represent specific SAS syntax, and lowercase letters to represent user-defined information. Notice that the SAS dataset `all` includes the $(n_i + 1) \times 1$ outcome vector $\boldsymbol{y}_i^*$, named `outcome`, which contains $\boldsymbol{y}_i$ as its first n_i elements and D_i as its final element. An indicator variable, named `ind` with values of 0 or 1, is also defined; this variable will be used subsequently to distinguish between the $\boldsymbol{y}_i$ and D_i elements.

Table 14.9. SAS Code: Setting Up Data for Shared Parameter Model

```
TITLE1 analysis of schizophrenic data with SAS ;
DATA one; INFILE 'c:\schizrep.dat'; INPUT id imps79 week drug sex ;

/* The coding for the variables is as follows:
id = subject id number
imps79 = overall severity (1=normal, ..., 7=most extremely ill)
week = 0,1,2,3,4,5,6 (most of the obs. are at weeks 0,1,3, and 6)
drug 0=placebo 1=drug (chlorpromazine, fluphenazine, or thioridazine)
sex 0=female 1=male
*/

/* compute the square root of week to linearize relationship */
sweek = SQRT(week);

/* calculate the maximum value of WEEK for each subject
and get drug in this aggregated dataset too */
PROC MEANS NOPRINT; CLASS id; VAR week drug;
OUTPUT OUT=two MAX(week drug)=maxweek drug;
RUN;

/* setting up IMPS79 across time and MAXWEEK as one outcome vector */
DATA daty; SET one; outcome = imps79; ind = 0;
DATA datr; SET two; outcome = maxweek; ind = 1; IF id NE .;
DATA all; SET daty datr; BY id;
```

Table 14.10 lists the SAS PROC NLMIXED code that can be used to run the shared parameter model described herein. Because NLMIXED is a fairly general procedure, the user must specify many of the particulars of a given model. In the code below, the `PARMS` statement lists the parameters to be estimated, and their starting values for the iterative maximum likelihood estimation procedure. These starting values can be obtained based on simpler analyses (*i.e.*, separate analyses of the IMPS79 outcome and the time to dropout). In terms of the longitudinal model of $\boldsymbol{y}_i$ (*i.e.*, the IMPS scores), the `b` terms are for the regression coefficients (*i.e.*, the β's), `sde` is the error standard deviation, and the `v` terms are for the random-effects variance–covariance parameters. Notice that, as in (14.17), the

Cholesky factorization of the variance–covariance matrix of the random effects is used. Specifically, v0 = σ_{υ_0}, v1 = σ_{υ_1}, and v01 = $\sigma_{\upsilon_{01}}$. This yields random random subject intercepts u1 and time-trends u2 that are independent standard normals. In terms of the time to dropout model, the a terms represent the regression coefficients (*i.e.*, the α's), and the i terms correspond to the parameters of the baseline cumulative hazard function (*i.e.*, the α_{0j}'s). The ind variable is used to distinguish the code for the longitudinal model of $\boldsymbol{y}_i$ (ind = 0) and the time to dropout model of D_i (ind = 1). For the former, z represents the error associated with an observation, and p is its probability based on the normal distribution.

Table 14.10. SAS PROC NLMIXED Code: Shared Parameter Model

```
PROC NLMIXED DATA=all;
PARMS b0=6 b1=0 b2=-1 b3=-1 sde=1 v0=1 v01=0 v1=.5
a1=0 a2=.5 a3=.2 a4=.1 a5=.1 i1=-1 i2=-.7 i3=-.5 i4=0 i5=.2;
IF (ind = 0) THEN
DO;
    z = (outcome - (b0 + b1*drug + b2*sweek + b3*drug*sweek
       + (v0 + v01*sweek/v0)*u1
       + SQRT(v1*v1 - (v01*v01)/(v0*v0))*sweek*u2));
    p = (1 / SQRT(2*3.14159*sde*sde)) * EXP(-.5 * (z*z) / (sde*sde));
END;
IF (ind = 1) THEN
DO;
    z = a1*drug + a2*u1 + a3*u2 + a4*u1*drug + a5*u2*drug;
    IF (outcome=1) THEN
       p = 1 - EXP(0 - EXP(i1+z));
    ELSE IF (outcome=2) THEN
       p = (1 - EXP(0 - EXP(i2+z))) - (1 - EXP(0 - EXP(i1+z)));
    ELSE IF (outcome=3) THEN
       p = (1 - EXP(0 - EXP(i3+z))) - (1 - EXP(0 - EXP(i2+z)));
    ELSE IF (outcome=4) THEN
       p = (1 - EXP(0 - EXP(i4+z))) - (1 - EXP(0 - EXP(i3+z)));
    ELSE IF (outcome=5) THEN
       p = (1 - EXP(0 - EXP(i5+z))) - (1 - EXP(0 - EXP(i4+z)));
    ELSE IF (outcome=6) THEN
       p = 1 - (1 - EXP(0 - EXP(i5+z)));
END;
IF (p > 1e-8) THEN ll = LOG(p);
else ll = -1e100;
MODEL outcome ~ GENERAL(ll);
RANDOM u1 u2 ~ NORMAL([0,0], [1,0,1]) SUBJECT=id;
RUN;
```

The code for the time to dropout is in terms of the complementary log–log link function as in (14.20). Here, z represents the response model for time to dropout, and p is the probability for a given observation based on the complementary log–log model. As the outcome is ordinal, with values of 1 to 6 indicating the final week of measurement for the individual, (14.20) yields cumulative probabilities representing the probability of response in a given category and below. Individual category probabilities, the `p`'s, are therefore obtained by subtraction. As noted, the `i1` to `i5` parameters represent the cumulative baseline hazard, and the parameters `a1` to `a5` are the effects on time to dropout. In particular, the latter four, `a2` to `a5`, indicate the effect of the subject intercepts and time-trends on time to dropout.

Table 14.11 lists the regression coefficient estimates from separate and shared parameter modeling of these data. The separate parameter model sets $\alpha_2 = \alpha_3 = \alpha_4 = \alpha_5 = 0$; in terms of the SAS PROC NLMIXED code in Table 14.10, the response model for time to dropout is therefore indicated simply as `z = a1*drug`, and the parameters `a2` to `a5`, and their starting value designations, are omitted on the `PARMS` statement.

Table 14.11. Separate and Shared Parameter Models

	Separate			Shared		
Parameter	Estimate	SE	$p <$	Estimate	SE	$p <$
Outcome						
Intercept β_0	5.348	.088	.0001	5.320	.088	.0001
`Drug` β_1	.046	.101	.65	.088	.102	.87
`SWeek` β_2	$-.336$	.068	.0001	$-.272$	.073	.0002
`Drug` $\times$ `Sweek` β_3	$-.641$	.078	.0001	$-.737$	.083	.0001
Dropout						
`Drug` α_1	$-.693$	.205	.0008	$-.703$	.301	.02
Random intercept α_2				.447	.333	.18
Random slope α_3				.891	.467	.06
`Drug` $\times$ intercept α_4				$-.592$	.398	.14
`Drug` $\times$ slope α_5				-1.638	.536	.003
Deviance	5380.2			5350.1		

As expected, the one separate parameter model yields identical parameter estimates and standard errors as running these two models, one for $\boldsymbol{y}_i$ and one for D_i, separately (not shown). Thus, the results for the longitudinal component represent the usual MAR mixed-effects model with random subject intercepts and time-trends. As can be seen, both the `SWeek` and `Drug` $\times$ `Sweek` terms are significant. This indicates, respectively, that the placebo group is improving across time (with an estimated slope of $-.336$), and that the drug group is improving at an even faster rate across time (with an estimated slope of $-.336 + -.641 = -.977$). Additionally, because the `Drug` term is not significant, these groups are not different when week=0 (*i.e.*, at baseline). The dropout component indicates that the drug group has a significantly diminished hazard. Thus, consistent with the previous results, placebo subjects are more likely to dropout than drug patients. Exponentiating the estimate of $-.693$ yields approximately .5, indicating that the hazard for dropout is double in the placebo group, relative to the drug group.

The shared paramter model fits these data significantly better, as evidenced by the likelihood ratio test, $X_4^2 = 30.1, p < .0001$. As mentioned, this is not necessarily a rejection of MAR, but it is a rejection of this particular MAR model in favor of this particular MNAR shared parameter model. In terms of the longitudinal component, we see that the conclusions are the same as in the MAR model. If anything, the results are slightly stronger for the drug group. Based on the estimates, the placebo group is improving across time with an estimated slope of $-.272$, while the drug group's estimated slope is $-.272 + -.737 = -1.009$. In terms of the dropout component, the significant terms are `Drug` and the `Drug` × slope interaction. Additionally, there is a marginally significant effect of the random slope. The significant `Drug` effect indicates that when the time trend is 0 (which is the population average), the drug group has a significantly diminished hazard of dropping out. As in the separate model, there is an approximate two-fold increase in the hazard for the placebo group, relative to the drug group. The marginally significant slope effect indicates that for the placebo group there is a tendency to dropout of the study as the slope increases. Namely, among placebo subjects, those who are not improving, or improving at a slower rate, are more likely to drop out. The significant negative `Drug` × slope interaction indicates that the slope effect is opposite for the drug group, where drug patients with more negative slopes (*i.e.*, greater improvement) are more likely to drop out.

14.5.2 Pattern-Mixture Models

As an alternative to selection models, Little [1993, 1994, 1995] formulated a general class of models under the rubric "pattern-mixture models" for the analysis of missing, or incomplete, data. There were earlier developments on this topic [Glynn et al., 1986; Marini et al., 1980], as well as similar procedures for structural equation models [Allison, 1987; McArdle and Hamagami, 1992; Muthén et al., 1987]. The articles by Little, however, provide a statistically rigorous and thorough treatment of these models in a general way. Since then, there have been many articles describing and further developing pattern-mixture models for longitudinal data [Daniels and Hogan, 2000; Demirtas and Schafer, 2003; Demirtas, 2005a,b; Fitzmaurice et al., 2001; Guo et al., 2004; Hedeker and Gibbons, 1997; Lin et al., 2004; Molenberghs et al., 1998; Roy, 2003; Thijs et al., 2002].

In these models, subjects are divided into groups depending on their missing-data pattern. These groups then can be used, for example, to examine the effect of the missing-data pattern on the outcome(s) of interest. Using the pattern-mixture approach, a model can be specified that does not require the missing-data mechanism to be ignorable. Also, this approach provides assessment of degree to which important model terms (*i.e.*, group and group by time interaction) depend on a subject's missing-data pattern. Overall estimates can also be obtained by averaging over the missing-data patterns.

The first step in applying the pattern-mixture approach to handling missing data is to divide the subjects into groups depending on their missing-data pattern. For example, suppose that subjects are measured at three timepoints, then there are eight (2^3) possible missing-data patterns:

Pattern Group	Time 1	Time 2	Time 3
1	O	O	O
2	M	O	O
3	O	M	O
4	M	M	O
5	O	O	M
6	M	O	M
7	O	M	M
8	M	M	M

where, O denotes being observed and M being missing. By grouping the subjects this way, we have created a between-subjects variable, the missing-data pattern, which can then be used in subsequent longitudinal data analysis, just as one might include a subject's gender as a variable in the data analysis.

Next, to utilize the missing-data pattern as a grouping variable in analysis of longitudinal data there is one important criterion: the method of analysis must allow subjects to have complete or incomplete data across time. Thus, this approach will not work with a method which requires complete data across time, since in this case, only subjects with the complete data pattern (*i.e.*, OOO) would be included in the analysis. For example, software for multivariate repeated measures analysis of variance usually only includes subjects with complete data across time (*i.e.*, pattern OOO) and so using the missing-data pattern as a grouping variable is not possible. Also, since the last pattern (MMM) provides no data, for practical purposes, this pattern is often ignored in data analysis, though this is not a requirement of the general pattern-mixture method (see Little [1993]). For simplicity, however, in what follows we will exclude this last pattern.

These seven patterns (*i.e.*, excluding MMM from the analysis) can be represented by six dummy-coded variables, for example the (general) codings D1 to D6 given in Table 14.12. As given, these six dummy-coded variables represent deviations from the non-missing pattern (OOO). Other coding schemes can be used to provide alternative comparisons among the seven pattern groups, for example, "effect" or "sequential" coding [Darlington, 1990]. The dummy, or "effect" or "sequential," coded variables are then entered, into the model as a main effect and as interactions with other model variables. In this way, one can examine: (a) the degree to which the groups defined by the missing-data patterns differ in terms of the outcome variable (*i.e.*, a main effect of the missing-data pattern dummy-coded variables), and (b) the degree to which the missing-data pattern moderates the influence of other model terms (*i.e.*, interactions with missing-data pattern). Also, from the model with the main effect and interactions of missing-data pattern, submodels can be obtained for each of the missing-data pattern groups, and overall averaged estimates (*i.e.*, averaging over the missing-data patterns) derived for the model parameters.

Table 14.12. Dummy Codes for Missing-Data Patterns: A Three Timepoint Study

Pattern	D1	D2	D3	D4	D5	D6
OOO	0	0	0	0	0	0
MOO	1	0	0	0	0	0
OMO	0	1	0	0	0	0
MMO	0	0	1	0	0	0
OOM	0	0	0	1	0	0
MOM	0	0	0	0	1	0
OMM	0	0	0	0	0	1

Modeling differences between all potential missing-data patterns may not always be possible. For example, some of the patterns, either by design or by chance, may not be realized in the sample. In some longitudinal studies, once a subject is missing at a given wave they are missing at all later waves. In this case, the number of patterns with available data equals the number of measurement waves. For our example with three waves, the three missing-data patterns (OMM, OOM, and OOO) would then represent a monotone pattern of dropout. In this case, the two dummy-coded variables M1 and M2 given in Table 14.13 could be used to represent differences between each of the two dropout groups and the group of subjects observed at all timepoints. Again, other coding schemes are possible.

Even when there are data at intermittent waves, one might want to combine some of the patterns to increase interpretability. For example, one might combine the patterns into groups based on the last available measurement wave. For this, with three timepoints, Table 14.13 lists dummy-codes L1 and L2: L1 is a dummy-coded variable that contrasts those individuals who were not measured after the first timepoint with those who were measured at the last timepoint, and L2 contrasts those subjects not measured after the second timepoint with those who were measured at the last timepoint. Other recodings that may be reasonable include a simple grouping of complete data vs. incomplete data, as given by contrast I1 in Table 14.13, or missing at the final timepoint vs. available at the final timepoint, as given by contrast F1 in Table 14.13.

Table 14.13. Other Coding Schemes for Missing-Data Patterns: A Three Timepoint Study

	Monotone		Last Wave		Incomplete	Not at Final
Pattern	M1	M2	L1	L2	I1	F1
OOO	0	0	0	0	0	0
MOO			0	0	1	0
OMO			0	0	1	0
MMO			0	0	1	0
OOM	0	1	0	1	1	1
MOM			0	1	1	1
OMM	1	0	1	0	1	1

In deciding on an appropriate grouping of the missing-data patterns, a few things need to be considered. One is the sparseness of the patterns. If a pattern has very few observations, it may not make sense to treat it as a separate group in the analysis. In this case, recoding the patterns to obtain fewer groupings is reasonable. Another consideration is the potential influence of the missing-data pattern on the response variable. In longitudinal studies, it is often reasonable to assume that the intermittent missing observations are randomly missing. In this case, recoding the patterns into groups based on the last available measurement wave is a sensible option. If a large percentage of subjects complete the study, it may be reasonable to simply contrast completers versus dropouts. Another consideration is whether one is interested only in estimating the main effects of the missing-data patterns, or also in interactions with the missing-data patterns. For example, if one is interested in examining whether the trends across time differ by the missing-data pattern (a missing-data pattern by time interaction), it is important to realize that the patterns with only one available observation (OMM, MOM, and MMO) provide no information for the assessment of this interaction.

14.5.2.1 Example Using the schizophrenia dataset, we will augment the basic MRM of IMPS79 over time:

$$\begin{aligned} \texttt{IMPS79}_{ij} &= \beta_0 + \beta_1 \texttt{Drug}_i + \beta_2 \texttt{SWeek}_j + \beta_3 (\texttt{Drug}_i \times \texttt{SWeek}_j) \\ &\quad + v_{0i} + v_{1i} \texttt{SWeek}_j + \varepsilon_{ij}, \end{aligned} \tag{14.21}$$

with variables based on the missing data patterns. In Hedeker and Gibbons [1997], we simply contrasted those that completed the trial ($N = 335$) versus those that did not ($N = 102$). Defining the variable $\texttt{Drop} = 0$ or 1 for those that did not or did dropout from the trial (*i.e.*, was not measured at the final study timepoint), respectively, yields the pattern-mixture model:

$$\begin{aligned}\texttt{IMPS79}_{ij} \quad = \quad & \beta_0 + \beta_1\texttt{Drug}_i + \beta_2\texttt{SWeek}_j + \beta_3(\texttt{Drug}_i \times \texttt{SWeek}_j) \\ & + \beta_0^D\texttt{Drop}_i + \beta_1^D(\texttt{Drop}_i \times \texttt{Drug}_i) + \beta_2^D(\texttt{Drop}_i \times \texttt{Sweek}_j) \\ & + \beta_3^D(\texttt{Drop}_i \times \texttt{Drug}_i \times \texttt{Sweek}_j) \\ & + \upsilon_{0i} + \upsilon_{1i}\texttt{SWeek}_j + \varepsilon_{ij}. \end{aligned} \tag{14.22}$$

In this model, one set of regression parameters (β_0, β_1, β_2, and β_3) are for the completer subsample, and the other (β_0^D, β_1^D, β_2^D, and β_3^D) indicate how the dropouts differ from the completers. A main focus of the study is whether the changes across time are the same for the two treatment groups, placebo and drug, and so the $\texttt{Drug}_i \times \texttt{SWeek}$ interaction is of primary interest. In this pattern-mixture model, the $\texttt{Drug}_i \times \texttt{SWeek}$ interaction is for the completers only, whereas the three-way interaction of $\texttt{Drop}_i \times \texttt{Drug}_i \times \texttt{Sweek}_j$ (with regression coefficient β_3^D) indicates whether any differential change across time for drug, relative to placebo, varies between dropouts and completers. Thus, in this pattern-mixture model it is the three-way interaction that is particularly of interest.

In addition to the above simple pattern-mixture model, we will also consider a less restrictive version that uses the week of dropout variable $\texttt{D}_i$ in forming the missing data patterns. Specifically, in this broader model, there are six missing data patterns formed by the five dropout weeks and the completers. Let $\texttt{D}_m = \texttt{D}_1, \ldots, \texttt{D}_5$ denote dummy-variables which contrast each dropout pattern to the completers. Thus, completer is the reference cell, coded as zero on all of these dummy-codes, and each dropout group is coded as 1 for the week of their dropout and zeros for all others. The model is now:

$$\begin{aligned}\texttt{IMPS79}_{ij} \quad = \quad & \beta_0 + \beta_1\texttt{Drug}_i + \beta_2\texttt{SWeek}_j + \beta_3(\texttt{Drug}_i \times \texttt{SWeek}_j) \\ & + \sum_{m=1}^{5} [\beta_0^m\texttt{D}_m + \beta_1^m(\texttt{D}_m \times \texttt{Drug}_i) + \beta_2^m(\texttt{D}_m \times \texttt{Sweek}_j) \\ & + \beta_3^m(\texttt{D}_m \times \texttt{Drug}_i \times \texttt{Sweek}_j)] \\ & + \upsilon_{0i} + \upsilon_{1i}\texttt{SWeek}_j + \varepsilon_{ij}. \end{aligned} \tag{14.23}$$

Here, as in the previous model, we are assuming that the different missing data patterns have the same random-effect variance–covariance structure and the same error variance. Using the dummy codes described above, the parameters β_0, β_1, β_2, and β_3 represent the mean structure for the completer group, and the parameters β_0^m, β_1^m, β_2^m, and β_3^m indicate how the mean structure varies across the missing data patterns. In particular, because the focus of the analysis is on the $\texttt{Drug}_i \times \texttt{SWeek}$ interaction, the β_3^m parameters are of great interest. These indicate the degree to which the differential improvement rate of drug, relative to placebo, varies across missing data patterns.

Table 14.14 lists the estimates of the fixed effects (*i.e.*, the regression coefficients) from the ordinary MRM, the simple pattern-mixture model using only dropout/completion status, and the more general pattern-mixture model allowing for all dropout weeks. Note, that only the ordinary MRM assumes MAR for the missing data, the pattern-mixture models are MNAR models.

Table 14.14. Ordinary and Pattern-Mixture MRMs—Estimates (est), Standard Errors (se), and p-Values

	MRM			Pattern-Mixture MRMs					
Parameter	est	se	$p <$	est	se	$p <$	est	se	$p <$
Intercept β_0	5.348	.088	.0001	5.221	.108	.0001	5.221	.107	.0001
Drug β_1	.046	.101	.65	.202	.121	.096	.202	.120	.094
SWeek β_2	−.336	.068	.0001	−.393	.076	.0001	−.393	.075	.0001
Drug × Sweek β_3	−.641	.078	.0001	−.539	.086	.0001	−.539	.085	.0001
Dropout				Drop = 1 ($N = 102$)			D = 1 ($N = 37$)		
Intercept β_0^1				.320	.186	.086	.471	.288	.102
Drug β_1^1				−.399	.227	.079	−.456	.353	.20
SWeek β_2^1				.252	.159	.115	.240	.334	.47
Drug × Sweek β_3^1				−.635	.196	.002	−.412	.412	.32
							D = 2 ($N = 10$)		
Intercept β_0^2							.524	.437	.23
Drug β_1^2							−.703	.613	.25
SWeek β_2^2							.338	.398	.40
Drug × Sweek β_3^2							−.735	.562	.19
							D = 3 ($N = 42$)		
Intercept β_0^3							.047	.256	.85
Drug β_1^3							−.198	.318	.53
SWeek β_2^3							.377	.208	.07
Drug × Sweek β_3^3							−.835	.261	.002
							D = 4 ($N = 5$)		
Intercept β_0^4							.801	.653	.22
Drug β_1^4							−.237	.841	.78
SWeek β_2^4							−.101	.485	.84
Drug × Sweek β_3^4							−1.210	.625	.054
							D = 5 ($N = 8$)		
Intercept β_0^5							.337	.645	.60
Drug β_1^5							−.842	.746	.26
SWeek β_2^5							−.157	.466	.74
Drug × Sweek β_3^5							.231	.538	.67
$-2 \log L$	4649.0			4623.3			4607.8		

Comparing the ordinary MRM with the full pattern-mixture model, yields a likelihood-ratio statistic of $4649.0 - 4607.8 = 41.2$, 20 df, $p < .004$, indicating that the model terms do vary by missing data pattern. This is not a test of the MAR assumption, and the fact that the pattern-mixture model fits the data significantly better than the ordinary MRM model does not indicate that the assumption of MAR is necessarily violated. As noted by Demirtas and Schafer [2003], "relationships between R and pre-dropout responses cannot disprove the hypothesis of ignorability, which states that there is no residual relationship between R and the *post-dropout responses* given the pre-dropout values." In their notation, R represents the missing data pattern indicators. These authors also indicate that ignorable dropout mechanisms can yield data patterns like those in the subsequent Figure 14.3, which displays the very different treatment group trend-lines for study completers and dropouts.

Comparing the two pattern-mixture models yields a likelihood-ratio statistic of $4623.3 - 4607.8 = 15.5$, df= 16, $p < .49$. Thus, as in Hedeker and Gibbons [1997], it is reasonable for these data to collapse the dropout groups, and go with the simpler pattern-mixture model that only classifies subjects as completers or dropouts. Notice also that this simple pattern-mixture model does fit better than the ordinary MRM: $4649.0 - 4623.3 = 25.7$, df = 4, $p < .0001$.

In terms of the parameter estimates of the simple pattern-mixture model, we see that among the completers the time effect is significant for placebo patients ($Z = -.393/.076 = -5.15$, $p < .0001$), and that the drug by time effect is also significant ($Z = -.539/.086 = -6.28$, $p < .0001$), indicating that the drug patients improved at a faster rate. The nonsignificant dropout by time interaction indicates that for placebo subjects, improvement across time was not significantly different for dropouts as compared to completers. However, the significant three-way interaction indicates that the drug by time interaction (which indicates a more dramatic improvement over time for drug relative to placebo subjects) is significantly more pronounced for dropouts, relative to completers ($Z = -.635/.196 = -3.24$, $p < .002$).

Based on the simple pattern-mixture model, Figure 14.3 portrays the estimated treatment group trend-lines for study completers and dropouts, respectively. The observed sample means are also plotted along with the estimated trend-lines. These estimated trend-lines fit the observed means very well, and illustrate that although all groups start the study, on average, between markedly and severely ill (*i.e.*, between IMPS79 values of 5 and 6), the estimated improvement rate over time depends on both the treatment and completion status. In particular, the estimated improvement rate is most pronounced for dropouts in the drug treatment, and least pronounced for dropouts in the placebo treatment.

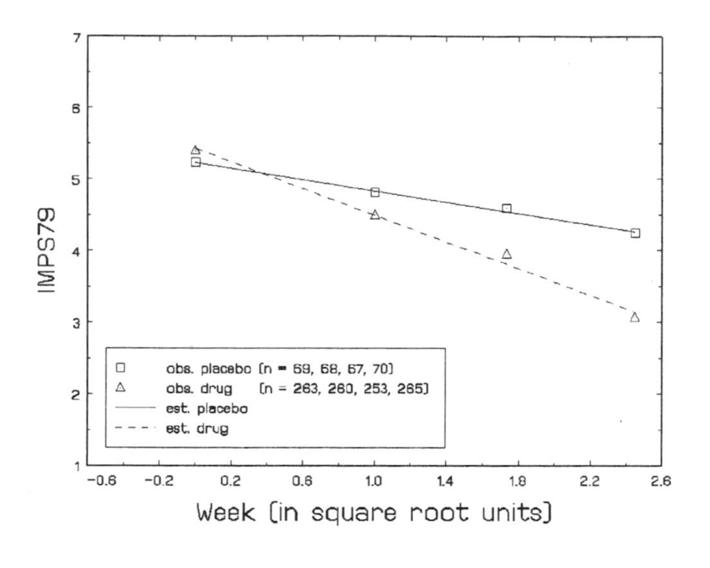

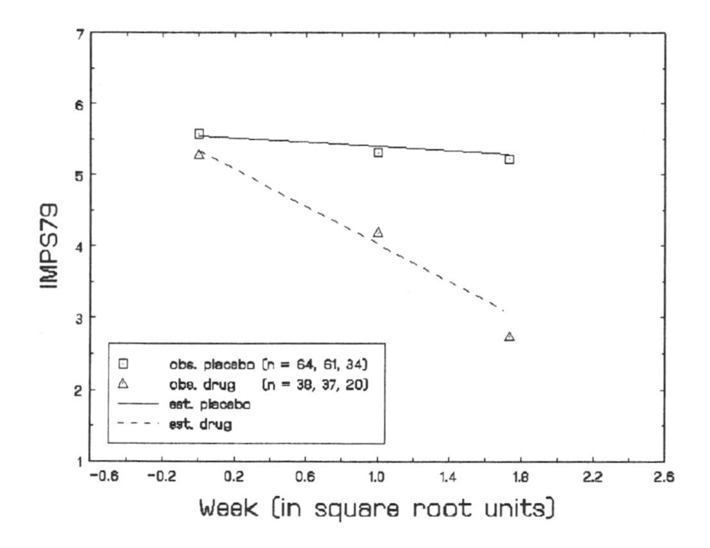

Figure 14.3. Estimated and observed IMPS79 means across time for completers (top figure) and dropouts (bottom figure), based on the simple pattern-mixture model.

Thus far, the analysis has indicated that the treatment effect across time varies by completion status. An additional step in the pattern-mixture approach is necessary to yield overall population estimates averaging over the missing-data patterns. Little [1993, 1995] and Hogan and Laird [1997a] discuss this step of averaging over pattern to yield estimates for the whole population. In the present case, based on the simple pattern-mixture model, we can obtain estimates for the four fixed effects (`Intercept`, `Drug`, `Sweek`, and `Drug` × `Sweek`) separately for completers

$$\hat{\boldsymbol{\beta}}^{(c)} = \begin{bmatrix} 5.221 & .202 & -.393 & -.539 \end{bmatrix}' \tag{14.24}$$

and dropouts

$$\begin{aligned} \hat{\boldsymbol{\beta}}^{(d)} &= \hat{\boldsymbol{\beta}}^{(c)} + \begin{bmatrix} .320 & -.399 & .252 & -.635 \end{bmatrix}' \\ &= \begin{bmatrix} 5.541 & -.197 & -.141 & -1.174 \end{bmatrix}'. \end{aligned} \tag{14.25}$$

Averaged estimates for these four parameters (denoted $\hat{\hat{\boldsymbol{\beta}}}$) are then equal to

$$\hat{\hat{\boldsymbol{\beta}}} = \pi^{(c)} \hat{\boldsymbol{\beta}}^{(c)} + \pi^{(d)} \hat{\boldsymbol{\beta}}^{(d)}, \tag{14.26}$$

where $\pi^{(c)}$ and $\pi^{(d)}$ represent the population weights for completers and dropouts, respectively. Although these weights are not usually known, they can be estimated by the sample proportions (335/437 and 102/437 for completers and dropouts, respectively). This yields

$$\hat{\hat{\boldsymbol{\beta}}} - \begin{bmatrix} 5.296 & .109 & .335 & .687 \end{bmatrix}' \tag{14.27}$$

as the averaged overall estimates.

To obtain corresponding estimates of the standard errors for these overall estimates, the delta method as described in Hogan and Laird [1997a] can be used:

$$\hat{V}(\hat{\hat{\beta}}_h) = (\hat{\pi}^{(c)})^2 \, \hat{V}(\hat{\beta}_h^{(c)}) + (\hat{\pi}^{(d)})^2 \, \hat{V}(\hat{\beta}_h^{(d)}) + \frac{\hat{\pi}^{(c)} \hat{\pi}^{(d)}}{N} (\hat{\beta}_h^{(c)} - \hat{\beta}_h^{(d)})^2, \tag{14.28}$$

where $h = 1, 2, 3, 4$ denotes the four fixed effects, $N = 437$ is the total number of subjects, and $\hat{V}(\hat{\beta}_h)$ denotes the estimate of the variance of $\hat{\beta}_h$ (*i.e.*, the square of its estimated standard error). The last term in the sum is the contribution to the variance that is added because the proportion of completers (and dropouts) is estimated in the sample. The estimated standard errors for these four overall terms are then .090, .103, .067, and .079.

The averaged results can also be obtained a bit more directly. For this, the dropout dummy code is replaced using a weighted effect coding approach (see Darlington [1990], pages 238–239). Specifically, define the variable `DropW` $= -1 \times (\hat{\pi}^{(d)} / \hat{\pi}^{(c)})$ for completers and 1 for dropouts. Then, running the model

$$\begin{aligned} \texttt{IMPS79}_{ij} = \; & \beta_0 + \beta_1 \texttt{Drug}_i + \beta_2 \texttt{SWeek}_j + \beta_3 (\texttt{Drug}_i \times \texttt{SWeek}_j) \\ & + \beta_0^{DW} \texttt{DropW}_i + \beta_1^{DW} (\texttt{DropW}_i \times \texttt{Drug}_i) + \beta_2^{DW} (\texttt{DropW}_i \times \texttt{Sweek}_j) \\ & + \beta_3^{DW} (\texttt{DropW}_i \times \texttt{Drug}_i \times \texttt{Sweek}_j) \\ & + \upsilon_{0i} + \upsilon_{1i} \texttt{SWeek}_j + \varepsilon_{ij}, \end{aligned} \tag{14.29}$$

yields the pattern-mixture averaged estimates as $\hat{\beta}_0, \hat{\beta}_1, \hat{\beta}_2$, and $\hat{\beta}_3$. The standard errors associated with these estimates will be too small because they are obtained assuming that the sample proportions for completers and dropouts are known, and not estimated. However, a simple modification of these standard errors, based on (14.28), yields the correct results. Specifically, denote the square of the standard error obtained from the above model using the weighted effect coding as $V_F(\hat{\beta}_h)$, where $h = 1, 2, 3, 4$ denotes the four fixed effects of interest in the averaging (*i.e.*, $\hat{\beta}_0$ to $\hat{\beta}_3$), Then,

$$\hat{V}(\hat{\bar{\beta}}_h) = V_F(\hat{\beta}_h) + \frac{\hat{\pi}^{(c)}\hat{\pi}^{(d)}}{N}(\hat{\beta}_h^{(c)} - \hat{\beta}_h^{(d)})^2. \tag{14.30}$$

The standard errors are obtained as the square root of these values.

This approach can be generalized to situations where there are more than two missing data pattterns. For example, there were six patterns considered in the earlier pattern-mixture analysis of these data. With six patterns, we would have the following contrast matrix for the weighted effect coding:

Missing Data Pattern	Contrast Variables				
	DW1	DW2	DW3	DW4	DW5
$D_i = 1$	1	0	0	0	0
$D_i = 2$	0	1	0	0	0
$D_i = 3$	0	0	1	0	0
$D_i = 4$	0	0	0	1	0
$D_i = 5$	0	0	0	0	1
$D_i = 6$	$-1 \times \frac{\hat{\pi}_1}{\hat{\pi}_6}$	$-1 \times \frac{\hat{\pi}_2}{\hat{\pi}_6}$	$-1 \times \frac{\hat{\pi}_3}{\hat{\pi}_6}$	$-1 \times \frac{\hat{\pi}_4}{\hat{\pi}_6}$	$-1 \times \frac{\hat{\pi}_5}{\hat{\pi}_6}$

Here, the completer subgroup (*i.e.*, $D_i = 6$) is treated as the reference cell. Estimating the model

$$\begin{aligned} \texttt{IMPS79}_{ij} &= \beta_0 + \beta_1 \texttt{Drug}_i + \beta_2 \texttt{SWeek}_j + \beta_3(\texttt{Drug}_i \times \texttt{SWeek}_j) \\ &\quad + \sum_{m=1}^{5} [\beta_0^m \texttt{DW}_m + \beta_1^m(\texttt{DW}_m \times \texttt{Drug}_i) + \beta_2^m(\texttt{DW}_m \times \texttt{Sweek}_j) \\ &\quad + \beta_3^m(\texttt{DW}_m \times \texttt{Drug}_i \times \texttt{Sweek}_j)] \\ &\quad + \upsilon_{0i} + \upsilon_{1i}\texttt{SWeek}_j + \varepsilon_{ij}, \end{aligned} \tag{14.31}$$

yields the pattern-mixture averaged estimates as $\hat{\beta}_0, \hat{\beta}_1, \hat{\beta}_2$, and $\hat{\beta}_3$. Again, the standard errors associated with these estimates will be too small, but can be corrected. Denote the square of the standard errors of the P averaged fixed effects from the above model as $V_F(\hat{\boldsymbol{\beta}})$. Note, here $P = 4$ and $V_F(\hat{\boldsymbol{\beta}})$ contains the square of the standard errors associated with the estimates of $\hat{\beta}_0, \hat{\beta}_1, \hat{\beta}_2$, and $\hat{\beta}_3$ from the above model. Also, denote the number of missing data patterns, besides the reference cell, as M (*e.g.*, here $M = 5$). Define $\hat{\boldsymbol{B}}$ as the $P \times M$ matrix containing the pattern-specific regression estimates, expressed as deviations relative to the reference cell. For example, in our case with six missing data patterns, based on the results in Table 14.14:

$$\hat{\boldsymbol{B}} = \begin{bmatrix} .471 & .524 & .047 & .801 & .337 \\ -.456 & -.703 & -.198 & -.237 & -.842 \\ .240 & .338 & .377 & --.101 & -.157 \\ -.412 & -.735 & -.835 & -1.210 & .231 \end{bmatrix}.$$

Next, form the variance–covariance matrix of the M sample proportions. From Agresti [2002], this is given by

$$V(\hat{\boldsymbol{\pi}}) = \frac{1}{N} \begin{bmatrix} \hat{\pi}_1(1-\hat{\pi}_1) & -\hat{\pi}_1\hat{\pi}_2 & -\hat{\pi}_1\hat{\pi}_3 & -\hat{\pi}_1\hat{\pi}_4 & -\hat{\pi}_1\hat{\pi}_5 \\ -\hat{\pi}_2\hat{\pi}_1 & \hat{\pi}_2(1-\hat{\pi}_2) & -\hat{\pi}_2\hat{\pi}_3 & -\hat{\pi}_2\hat{\pi}_4 & -\hat{\pi}_2\hat{\pi}_5 \\ -\hat{\pi}_3\hat{\pi}_1 & -\hat{\pi}_3\hat{\pi}_2 & \hat{\pi}_3(1-\hat{\pi}_3) & -\hat{\pi}_3\hat{\pi}_4 & -\hat{\pi}_3\hat{\pi}_5 \\ -\hat{\pi}_4\hat{\pi}_1 & -\hat{\pi}_4\hat{\pi}_2 & -\hat{\pi}_4\hat{\pi}_3 & \hat{\pi}_4(1-\hat{\pi}_4) & -\hat{\pi}_4\hat{\pi}_5 \\ -\hat{\pi}_5\hat{\pi}_1 & -\hat{\pi}_5\hat{\pi}_2 & -\hat{\pi}_5\hat{\pi}_3 & -\hat{\pi}_5\hat{\pi}_4 & \hat{\pi}_5(1-\hat{\pi}_5) \end{bmatrix}$$

for the present case of $M = 5$. For the schizophrenia data, as indicated in Table 14.14, we have $\hat{\pi}_1 = 37/437$, $\hat{\pi}_2 = 10/437$, $\hat{\pi}_3 = 42/437$, $\hat{\pi}_4 = 5/437$, and $\hat{\pi}_5 = 8/437$. Then,

$$\hat{V}(\hat{\bar{\boldsymbol{\beta}}}) = V_F(\hat{\boldsymbol{\beta}}) + \text{vecdiag}\left[\hat{\boldsymbol{B}}\, V(\hat{\boldsymbol{\pi}})\, \hat{\boldsymbol{B}}'\right], \tag{14.32}$$

where vecdiag $[\boldsymbol{A}]$ represents the vector of diagonal elements of the matrix $\boldsymbol{A}$. The standard errors are obtained as the square root of the values in $\hat{V}(\hat{\bar{\boldsymbol{\beta}}})$. Thus, the standard errors are augmented by the term to the right of the addition sign in the above equation to take into account the fact that the pattern probabilities are estimated and not known.

Finally, Table 14.15 lists the regression coefficient estimates from the ordinary MRM, as well as the pattern-mixture averaged results based on the two pattern mixture models considered in this section. As can be seen, the estimates are very similar across these three models. Again, it is the `Sweek` and `Drug` × `Sweek` terms that are significant, and the `Drug` term is not. Thus, the previously described conclusions, based on the ordinary MRM, are the same for the pattern-mixture models. These same conclusions were also observed in the selection modeling of these data. Thus, we have a greater degree of confidence in these results, having allowed for several types of nonignorable models. In the process of doing this extended modeling, we have also demonstrated that dropout varied by treatment group: placebo patients who were not improving were more likely to dropout, whereas drug patients with rapid improvement were more likely to dropout. Thus, though the basic conclusions have not changed, we have a more complete understanding of the responses across time and the role of missing data.

Table 14.15. Ordinary and Averaged Pattern-Mixture MRMs—Estimates (est), Standard Errors (se), and p-Values

				Pattern-Mixture MRMs					
	MRM			Simple			Complete		
Parameter	est	se	$p <$	est	se	$p <$	est	se	$p <$
`Intercept` β_0	5.348	.088	.0001	5.296	.090	.0001	5.293	.090	.0001
`Drug` β_1	.046	.101	.65	.109	.103	.30	.110	.103	.29
`SWeek` β_2	−.336	.068	.0001	−.335	.067	.0001	−.333	.068	.0001
`Drug` × `Sweek` β_3	−.641	.078	.0001	−.687	.079	.0001	−.680	.080	.0001

"Simple" refers to the model only including the two patterns of completer or dropout;
"Complete" refers to the model including all six dropout patterns.

14.6 SUMMARY

This chapter has described practial aspects of longitudinal modeling with missing data. In particular, we have illustrated how one can test for MCAR, and do sensitivity analysis by fitting both selection and pattern-mixture MRMs. Whereas the ordinary MRM assumes MAR, these extended models do not assume MAR and so researchers can assess the degree to which their conclusions vary as a function of the assumed underlying missing data mechanism. Several researchers warn against reliance of a single MNAR model, because the assumptions regarding the missing data are impossible to assess with the observed data. For this reason, one should use MNAR models with caution and perhaps examine several types of such models for a given dataset.

Finally, we have focused on longitudinal modeling using mixed-effects regression models (MRMs) for continuous outcomes in this chapter. However, the methods considered extend more generally to other models. In particular, the same kinds of approaches can be applied to categorical MRMs, covariance pattern models, and GEE models.

Bibliography

R. J. Adams, M. Wilson, and M. Wu. Multilevel item response models: an approach to errors in variable regression. *Journal of Educational and Behavioral Statistics*, 22: 47–76, 1997.

A. Agresti. *Categorical Data Analysis, 2nd edition*. Wiley, New York, 2002.

A. Agresti and J. B. Lang. A proportional odds model with subject-specific effects for repeated ordered categorical responses. *Biometrika*, 80:527–534, 1993.

H. Akaike. Information theory and an extension of the maximum likelihood principle. In B. N. Petrov and F. Csaki, editors, *Second International Symposium on Information Theory*, pages 267–281. Academiai Kiado, Budapest, 1973.

J. Albert. A Bayesian analysis of a Poisson random effects model for home run hitters. *The American Statistician*, 46:246–253, 1992.

P. D. Allison. Discrete-time methods for the analysis of event histories. In S. Leinhardt, editor, *Sociological Methodology 1982*, pages 61–98. Jossey-Bass, San Francisco, 1982.

P. D. Allison. Estimation of linear models with incomplete data. *Sociology Methodology, 1987*, pages 71–103, 1987.

P. D. Allison. Change scores as dependent variables in regression analysis. In C Clogg, editor, *Sociological Methodology*, pages 93–114. Jossey-Bass, San Francisco, 1990.

P. D. Allison. *Survival Analysis Using the SAS System: A Practical Guide*. SAS Institute Inc., Cary, NC, 1995.

T. Amemiya. Qualitative response models: a survey. *Journal of Econometric Literature*, 19:483–536, 1981.

D. A. Anderson and M. Aitkin. Variance component models with binary response: interviewer variability. *Journal of the Royal Statistical Society*, 47:203–210, 1985.

J. D. Angrist, G. W. Imbens, and D. B. Rubin. Identification of causal effects using instrumental variables. *Journal of the American Statistical Association*, 91:444–455, 1996.

G. A. Ballinger. Using generalized estimating equations for longitudinal data analysis. *Organizational Research Methods*, 7:127–150, 2004.

D. J. Bartholomew and M. Knott. *Latent Variable Models and Factor Analysis, 2nd edition*. Oxford University Press, New York, 1999.

J. Berkhof and T. A. B. Snijders. Variance component testing in multilevel models. *Journal of Educational and Behavioral Statistics*, 26:133–152, 2001.

E. Berndt, B. Hall, R. Hall, and J. Hausman. Estimation and inference in nonlinear structural models. *Annals of Economic and Social Measurement*, 3:653–665, 1974.

J. C. Biesanz, N. Deeb-Sossa, A. A. Papadakis, K. A. Bollen, and P. J. Curran. The role of coding time in estimating and interpreting growth curve models. *Psychological Methods*, 9:30–52, 2004.

R. C. Blair, J. J. Higgins, M. H. Topping, and A. L. Mortimer. An investigation of the robustness of the t-test to unit-of-analysis violations. *Educational and Psychological Measurement*, 43:69–80, 1983.

R. D. Bock. Estimating item parameters and latent ability when responses are scored in two or more nominal categories. *Psychometrika*, 37:29–51, 1972.

R. D. Bock. *Multivariate Statistical Methods in Behavioral Research*. McGraw-Hill, New York, 1975.

R. D. Bock. The discrete Bayesian. In H. Wainer and S. Messick, editors, *Modern Advances in Psychometric Research*, pages 103–115. Erlbaum, Hillsdale, NJ, 1983a.

R. D. Bock. Within-subject experimentation in psychiatric research. In R. D. Gibbons and M. W. Dysken, editors, *Statistical and Methodological Advances in Psychiatric Research*, pages 59–90. Spectrum, New York, 1983b.

R. D. Bock. Measurement of human variation: a two stage model. In R. D. Bock, editor, *Multilevel Analysis of Educational Data*. Academic Press, New York, 1989.

R. D. Bock and M. Aitkin. Marginal maximum likelihood estimation of item parameters: an application of the EM algorithm. *Psychometrika*, 46:443–459, 1981.

R. D. Bock and H. C. du Toit. Parameter estimation in the context of non-linear longitudinal growth models. In R. C. Hauspie, N. Cameron, and L. Molinari, editors, *Methods in Human Growth Research*. Cambridge University Press, New York, 2004.

R. D. Bock, R. D. Gibbons, and E. Muraki. Full-information item factor analysis. *Applied Psychological Measurement*, 12:261–280, 1988.

R. D. Bock and S. Shilling. High-dimensional full-information item factor analysis. In M. Berkane, editor, *Latent Variable Modeling and Applications to Causality*, pages 163–176. Springer, New York, 1997.

N. E. Breslow. Extra-Poisson variation in log-linear models. *Applied Statistics*, 33:38–44, 1984.

N. E. Breslow and X. Lin. Bias correction in generalised linear mixed models with a single component of dispersion. *Biometrika*, 82:81–91, 1995.

A. S. Bryk and S. W. Raudenbush. *Hierarchical Linear Models: Applications and Data Analysis Methods*. Sage Publications, Inc., Newbury Park, CA, 1992.

L. Burstein. The analysis of multilevel data in educational research and evaluation. In D. Berliner, editor, *Review of Research in Education, volume 8*, pages 158–233. American Educational Research Association, Washington, DC, 1980.

P. Burton, L. Gurrin, and P. Sly. Tutorial in biostatistics: extending the simple linear regression model to account for correlated responses: an introduction to generalized estimating equations and multi-level modelling. *Statistics in Medicine*, 17:1261–1291, 1998.

A. C. Cameron and P. K. Trivedi. Econometric models based on count data: comparisons and applications of some estimators and tests. *Journal of Applied Econometrics*, 1: 29–54, 1986.

A. C. Cameron and P. K. Trivedi. *Regression Analysis of Count Data*. Cambridge University Press, New York, 1998.

G. Camilli and L. A. Shepard. *Methods for Identifying Biased Test Items*. Sage Publications, Thousand Oaks, CA, 1994.

S. K. Campbell and D. Hedeker. Validity of the test of infant motor performance for discriminating among infants with varying risks for poor motor outcome. *The Journal of Pediatrics*, 139:546–551, 2001.

K. M. Carroll, B. J. Rounsaville, C. Nich, L.T. Gordon, P.W. Wirtz, and F. Gawin. One-year follow-up of psychotherapy and pharmacotherapy for cocaine dependence. *Archives of General Psychiatry*, 51:989–997, 1994.

E. M. Chi and G. C. Reinsel. Models for longitudinal data with random effects and AR(1) errors. *Journal of the American Statistical Society*, 84:452–459, 1989.

P. D. Cleary and R. Angel. The analysis of relationship involving dichotomous dependent variables. *Journal of Health and Social Behavior*, 25(3):334–48, 1984.

W. G. Cochran. Errors of measurement in statistics. *Technometrics*, 10:55–83, 1968.

A. Cohen. Estimation of the Poisson parameter from truncated samples and from censored samples. *Journal of the American Statistical Association*, 49:158–168, 1954.

M. R. Conaway. Analysis of repeated categorical measurements with conditional likelihood methods. *Journal of the American Statistical Association*, 84:53–61, 1989.

R. J. Cook and V. T. Farewell. Multiplicity considerations in the design and analysis of clinical trials. *Journal of the Royal Statistical Society, Series A*, 159:93–110, 1996.

C. Corcoran, B. Coull, and A. Patel. *EGRET for Windows User Manual*. CYTEL Software Corporation, Cambridge, MA, 1999.

C. Cox. Location-scale cumulative odds models for ordinal data: a generalized non-linear model approach. *Statistics in Medicine*, 14:1191–1203, 1995.

M. J. Crowder and D. J. Hand. *Analysis of Repeated Measures*. Chapman and Hall, New York, 1990.

L. A. Cupples, R. B. D'Agostino, K. Anderson, and W. B. Kannel. Comparison of baseline and repeated measure covariate techniques in the framingham heart study. *Statistics in Medicine*, 7:205–218, 1985.

P. J. Curran, E. Stice, and L. Chassin. The relation between adolescent and peer alcohol use: a longitudinal random coefficients model. *Journal of Consulting and Clinical Psychology*, 65:130–140, 1997.

A. R. Cushny and A. R. Peebles. The action of optical isomers. II. Hyoscines. *Journal of Physiology*, 32:501–510, 1905.

R. B. D'Agostino, M.-L. Lee, A. J. Belanger, L. A. Cupples, K. Anderson, and W. B. Kannel. Relation of pooled logistic regression to time dependent Cox regression analysis: the Framingham Heart Study. *Statistics in Medicine*, 9:1501–1515, 1990.

M. J. Daniels and C. Gatsonis. Hierarchical polytomous regression models with applications to health services research. *Statistics in Medicine*, 16:2311–2325, 1997.

M. J. Daniels and J. W. Hogan. Reparameterizing the pattern mixture model for sensitivity analyses under informative dropout. *Biometrics*, 56:1241–1248, 2000.

R. B. Darlington. *Regression and Linear Models*. McGraw-Hill, New York, 1990.

M. Davidian and D. M. Giltinan. *Nonlinear Models for Repeated Measurement Data*. Chapman and Hall, New York, 1995.

C. S. Davis. A computer program for regression analysis of repeated measures using generalized estimating equations. *Computer Methods and Programs in Biomedicine*, 40: 15–31, 1993.

F. Davis. *Cooperative Achievement Tests*. Educational Testing Service, Princeton, New Jersey, 1950.

V. De Gruttola and X. M. Tu. Modelling progression of CD4-lymphocyte count and its relationship to survival time. *Biometrics*, 50:1003–1014, 1994.

J. de Leeuw and I. Kreft. Random coefficient models for multilevel analysis. *Journal of Educational Statistics*, 11:57–85, 1986.

J. de Leeuw and I. Kreft. Software for multilevel analysis. In A. H. Leyland and H. Goldstein, editors, *Multilevel Modelling of Health Statistics*, pages 187–204. Wiley, New York, 2001.

C. Dean and F. Lawless. Testing for overdispersion in Poisson regression model. *Journal of the American Statistical Association*, 84:98–106, 1989.

H. Demirtas. Assessment of relative improvement due to weights within generalized estimating equations framework for incomplete clinical trials data. *Journal of Biopharmaceutical Statistics*, 14:1085–1098, 2004a.

H. Demirtas. Modeling incomplete longitudinal data. *Journal of Modern Applied Statistical Methods*, 3:305–321, 2004b.

H. Demirtas. Bayesian analysis of hierarchical pattern-mixture models for clinical trials data with attrition and comparisons to commonly used ad-hoc and model-based approaches. *Journal of Biopharmaceutical Statistics*, 15:383–402, 2005a.

H. Demirtas. Multiple imputation under Bayesian smoothed pattern-mixture models for non-ignorable drop-out. *Statistics in Medicine*, 24:2345–2363, 2005b.

H. Demirtas and J. L. Schafer. On the performance of random-coefficient pattern-mixture models for nonignorable dropout. *Statistics in Medicine*, 22:2553–2575, 2003.

A. P. Dempster, D. B. Rubin, and R. K. Tsutakawa. Estimation in covariance component models. *Journal of the American Statistical Society*, 76:341–353, 1981.

P. Dempster, N. M. Laird, and D. B. Rubin. Maximum likelihood from incomplete data via the EM algorithm. *Journal of Royal Statistical Society*, 39:1–38, 1977.

DHHS. *Federal Register*, 63:16296–16338, 1998.

P. Diggle and M. G. Kenward. Informative drop-out in longitudinal data analysis (with discussion). *Applied Statistics*, 43:49–93, 1994.

P. J. Diggle. Testing for random dropouts in repeated measurement data. *Biometrics*, 45: 1255–1258, 1989.

P. J. Diggle, P. Heagerty, K.-Y. Liang, and S. L. Zeger. *Analysis of Longitudinal Data, 2nd edition*. Oxford University Press, New York, 2002.

A. Dobson. *An Introduction to Generalized Linear Models*. Chapman Hall, London, 1990.

K. A. Doksum and M. Gasko. On a correspondence between models in binary regression analysis and in survival analysis. *International Statistical Review*, 58:243–252, 1990.

A. Donner. An empirical-study of cluster randomization. *International Journal of Epidemiology*, 11:283–286, 1982.

A. Donner and N. Klar. *Design and Analysis of Cluster Randomization Trials in Health Research*. Oxford University Press, New York, 2000.

D. M. Dos Santos and D. M. Berridge. A continuation ratio random effects model for repeated ordinal responses. *Statistics in Medicine*, 19:3377–3388, 2000.

N. R. Draper and H. Smith. *Applied Regression Analysis, 2nd edition*. Wiley, New York, 1981.

D. D. Dunlop. Regression for longitudinal data: a bridge from least squares regression. *The American Statistician*, 48:299–303, 1994.

B. Efron. Logistic regression, survival analysis, and the kaplan-meier curve. *Journal of the American Statistical Association*, 83:414–425, 1988.

I. Elkin, R. D. Gibbons, M. T. Shea, S. M. Sotsky, J. T. Watkins, P. A. Pilkonis, and D. Hedeker. Initial severity and differential treatment outcome in the nimh treatment of depression collaborative research program. *Journal of Consulting and Clinical Psychology*, 63:841–847, 1995.

J. Engel. On the analysis of grouped extreme-value data with GLIM. *Applied Statistics*, 42:633–640, 1993.

F. Ezzet and J. Whitehead. A random effects model for ordinal responses from a crossover trial. *Statistics in Medicine*, 10:901–907, 1991.

A. Fielding. Why use arbitrary points scores?: ordered categories in models of educational progress. *Journal of the Royal Statistical Society, Series A*, 162:303–330, 1999.

J. D. Finn. *A General Model for Multivariate Analysis*. Holt, Rinehart and Winston, Inc., New York, 1974.

D. J. Finney. *Probit Analysis, 2nd edition*. Cambridge University Press, Cambridge, 1971.

G. M. Fitzmaurice. A caveat concerning independence estimating equations with multivariate binary data. *Biometrics*, 51:309–317, 1995.

G. M. Fitzmaurice, N. M. Laird, and A. G. Rotnitzky. Regression models for discrete longitudinal responses. *Statistical Science*, 8:284–309, 1993.

G. M. Fitzmaurice, N. M. Laird, and L. Shneyer. An alternative parameterization of the general linear mixture model for longitudinal data with nonignorable dropouts. *Statistics in Medicine*, 20:1009–1021, 2001.

G. M. Fitzmaurice, N. M. Laird, and J. H. Ware. *Applied Longitudinal Analysis*. Wiley, New York, 2004.

B. R. Flay, B. R. Brannon, C. A. Johnson, W. B. Hansen, A. Ulene, D. A. Whitney-Saltiel, L. R. Gleason, S. Sussman, D. M. Gavin, K. M. Glowacz, D. F. Sobol, and D. C. Spiegel. The television, school and family smoking prevention/cessation project: I. theoretical basis and program development. *Preventive Medicine*, 17(5):585–607, 1988.

J. L. Fleiss. *Design and Analysis of Clinical Experiments*. Wiley, New York, 1986.

S. A. Freels, R. B. Warnecke, T. P. Johnson, and B. R. Flay. Evaluation of the effects of a smoking cessation intervention using the multilevel thresholds of change model. *Evaluation Review*, 26:40–58, 2002.

T. J. Gallagher, L. B. Cottler, W. M. Compton, and E. Spitznagel. Changes in HIV/AIDS risk behaviors in drug users in St. Louis: applications of random regression models. *Journal of Drug Issues*, 27:399–416, 1997.

A. E. Gelfand, S. E. Hills, A. Racine-Poon, and A. F. M. Smith. Illustration of Bayesian inference in normal data models using Gibbs sampling. *Journal of the American Statistical Association*, 85:972–985, 1990.

A. E. Gelfand and A. F. M. Smith. Sampling-based approaches to calculation marginal densities. *Journal of the American Statistical Association*, 85:398–409, 1990.

S. Geman and D. Geman. Stochastic relaxation, Gibbs distributions and the Bayesian restoration of images. *IEEE Transactions on Pattern Analysis and Machine Intelligence*, 6:721–741, 1984.

R. D. Gibbons. *Trend in Correlated Proportions*. Ph.D. thesis, University of Chicago, Department of Psychology, 1981.

R. D. Gibbons and R. D. Bock. Trend in correlated proportions. *Psychometrika*, 52: 113–124, 1987.

R. D. Gibbons, N. Duan, D. Meltzer, A. Pope, E. D. Penhoet, N. N. Dubler, C. K. Francis, B. Gill, E. Guinan, M. Henderson, S. T. Ildstad, P. A. King, M. Martinez-Maldonado, G. E. Mclain, J. E. Murray, D. Nelkin, M. W. Spellman, and S. Pitluck. Waiting for organ transplantation: results of an analysis by an Institute of Medicine Committee. *Biostatistics*, 42:207–222, 2003.

R. D. Gibbons and D. Hedeker. Application of random-effects probit regression models. *Journal of Consulting and Clinical Psychology*, 62:285–296, 1994.

R. D. Gibbons and D. Hedeker. Random effects probit and logistic regression models for three-level data. *Biometrics*, 53:1527–1537, 1997.

R. D. Gibbons, D. Hedeker, S. Charles, and Frisch P. A random-effects probit model for predicting medical malpractice claims. *Journal of the American Statistical Association*, 89:760–767, 1994.

R. D. Gibbons, D. Hedeker, I. Elkin, C. M. Waternaux, H. C. Kraemer, J. B. Greenhouse, M. T. Shea, S. D. Imber, S. M. Sotsky, and J. T. Watkins. Some conceptual and statistical issues in analysis of longitudinal psychiatric data. *Archives of General Psychiatry*, 50: 739–750, 1993.

R. D. Gibbons, D. Hedeker, C. M. Waternaux, and J. M. Davis. Random regression models: a comprehensive approach to the analysis of longitudinal psychiatric data. *Psychopharmacology Bulletin*, 24:438–443, 1988.

R. D. Gibbons, K. Hur, D. Bhaumik, and J. J. Mann. The relationship between antidepressant medication use and rate of suicide. *Archives of General Psychaitry*, 62:165–172, 2005.

R. D. Gibbons, D. Meltzer, N. Duan, E. D. Penhoet, N. N. Dubler, C. K. Francis, B. Gill, E. Guinan, M. Henderson, S. T. Ildstad, P. A. King, M. Martinez-Maldonado, G. E. Mclain, J. E. Murray, D. Nelkin, M. W. Spellman, A. Pope, and S. Pitluck. Waiting for organ transplantation. *Science*, 287:237–238, 2000.

R. J. Glynn, N. M. Laird, and D. B. Rubin. Selection modeling versus mixture modeling with nonignorable nonresponse. In H. Wainer, editor, *Drawing Inferences from Self-Selected Samples*, pages 115–142. Springer-Verlag, New York, 1986.

S. K. Goldsmith, T. C. Pellmar, A. M. Kleinman, and W. E. Bunney. *Reducing Suicide. A National Imperative.* The National Academies Press, Washington, DC, 2002.

H. Goldstein. Nonlinear multilevel models, with an application to discrete response data. *Biometrika*, 78:45–51, 1991.

H. Goldstein. *Multilevel Statistical Models, 2nd edition*. Halstead Press, New York, 1995.

H. Goldstein and J. Rasbash. Improved approximations for multilevel models with binary responses. *Journal of the Royal Statistical Society, Series B*, 159:505–513, 1996.

H. Goldstein, J. Rasbash, I. Plewis, D. Draper, W. Browne, M. Yang, G. Woodhouse, and M. Healy. *A User's Guide to MLwiN*. Institute of Education, University of London, London, 1998.

A. Goraski. Distribution z-Poisson. *Publications de l'Institut de Statistique de l'Universit é de Paris*, 12:45–53, 1977.

J. A. Gornbein, C. G. Lazaro, and R. J. A. Little. Incomplete data in repeated measures analysis. *Statistical Methods in Medical Research*, 1:275–295, 1992.

J. M. Gottman. *Time-Series Analysis: A Comprehensive Introduction for Social Scientists*. Cambridge University Press, New York, 1981.

J. J. Grady and R. W. Helms. Model selection techniques for the covariance matrix for incomplete longitudinal data. *Statistics in Medicine*, 14:1397–1416, 1995.

W. H. Greene. *Econometric Analysis, 2nd edition*. Prentice Hall, Englewood Cliffs, NJ, 1993.

W. H. Greene. Accounting for excess zeros and sample selection in Poisson and negative binomial regression models. *Working Paper No. EC-94-10, Stern School of Business, New York University*, 1994.

W. H. Greene. *LIMDEP Version 7.0 User's Manual, revised edition*. Econometric Software, Inc., Plainview, NY, 1998.

S. W. Greenhouse and S. Geisser. On methods in the analysis of profile data. *Psychometrika*, 24:95111, 1959.

J. Grogger. The deterrent effect of capital punishment: an analysis of daily homicide counts. *Journal of the American Statistical Association*, 85:295–303, 1990.

C. L. Gruder, R. J. Mermelstein, S. Kirkendol, D. Hedeker, S. C. Wong, J. Schreckengost, R. B. Warnecke, R. Burzette, and T. Q. Miller. Effects of social support and relapse prevention training as adjuncts to a televised smoking cessation intervention. *Journal of Consulting and Clinical Psychology*, 61:113–120, 1993.

J. P. Guilford. *Psychometric Methods, 2nd edition*. McGraw-Hill, New York, 1954.

W. Guo, S. J. Ratcliffe, and T. R. Ten Have. A random pattern-mixture model for longitudinal data with dropouts. *Journal of the American Statistical Association*, 99:929–937, 2004.

J. A. Halikas, R. D. Crosby, V. L. Pearson, and N .M. Graves. A randomized double-blind study of carbamazepine in the treatment of cocaine abuse. *Clinical Pharmacology and Therapeutics*, 62:89–105, 1997.

D. Hall. Zero-inflated Poisson and binomial regression with random effects: a case study. *Biometrics*, 56:1030–1039, 2000.

M. Hamilton. A rating scale for depression. *Journal of Neurology and Neurosurgical Psychiatry*, 23:56–62, 1960.

A. Han and J. A. Hausman. Flexible parametric estimation of duration and competing risk models. *Journal of Applied Econometrics*, 5:1–28, 1990.

J. W. Hardin and J. M. Hilbe. *Generalized Estimating Equations*. Chapman and Hall, New York, 2003.

J. Hartzel, A. Agresti, and B. Caffo. Multinomial logit random effects models. *Statistical Modelling*, 1, 2001.

D. A. Harville and R. W. Mee. A mixed-model procedure for analyzing ordered categorical data. *Biometrics*, 40:393–408, 1984.

J. Hausman, H. Bronwyn, and Z. Griliches. Econometric models for count data with an application to the patents–R&D relationship. *Econometrica*, 52:909–938, 1984.

P. J. Heagerty and S. L. Zeger. Marginalized multilevel models and likelihood inference. *Statisical Science*, 15:1–19, 2000.

J. Heckman. The common structure of statistical models of truncation, sample selection, and limited dependent variables and a simple estimator for such models. *Annals of Economic and Social Measurement*, 5:475–492, 1976.

J. Heckman. New evidence on the dynamics of female labor supply. In C. B. Lloyd, E. Andrews, and C. Gilroy, editors, *Handbook of Econometrics*. Elseiver, Amsterdam, 1979.

D. Hedeker. *Random Regression Models with Autocorrelated Errors*. Ph.D. thesis, University of Chicago, Department of Psychology, 1989.

D. Hedeker. MIXNO: a computer program for mixed-effects nominal logistic regression. *Journal of Statistical Software*, 4(5):1–92, 1999.

D. Hedeker. A mixed-effects multinomial logistic regression model. *Statistics in Medicine*, 21:1433–1446, 2003.

D. Hedeker, M. Berbaum, and R. J. Mermelstein. Location-scale models for multilevel ordinal data: between- and within-subjects variance modeling. *Journal of Probability and Statistical Science*, 4, 2006.

D. Hedeker, B. R. Flay, and J. Petraitis. Estimating individual differences of behavioral intentions: an application of random-effects modeling to the theory of reasoned action. *Journal of Consulting and Clinical Psychology*, 64:109–120, 1996.

D. Hedeker and R. D. Gibbons. A random effects ordinal regression model for multilevel analysis. *Biometrics*, 50:933–944, 1994.

D. Hedeker and R. D. Gibbons. MIXOR: a computer program for mixed-effects ordinal probit and logistic regression analysis. *Computer Methods and Programs in Biomedicine*, 49:157–176, 1996a.

D. Hedeker and R. D. Gibbons. MIXREG: a computer program for mixed-effects regression analysis with autocorrelated errors. *Computer Methods and Programs in Biomedicine*, 49:229–252, 1996b.

D. Hedeker and R. D. Gibbons. Application of random-effects pattern-mixture models for missing data in longitudinal studies. *Psychological Methods*, 2:64–78, 1997.

D. Hedeker, R. D. Gibbons, and J. M. Davis. Random regression models for multi-center clinical trials data. *Psychopharmacology Bulletin*, 27:73–77, 1991.

D. Hedeker, R. D. Gibbons, and B. R. Flay. Random-effects regression models for clustered data: with an example from smoking prevention research. *Journal of Consulting and Clinical Psychology*, 62:757–765, 1994.

D. Hedeker and R. J. Mermelstein. A multilevel thresholds of change model for analysis of stages of change data. *Multivariate Behavioral Research*, 33:427–455, 1998.

D. Hedeker and R. J. Mermelstein. Analysis of longitudinal substance use outcomes using random-effects regression models. *Addiction (Supplement 3)*, 95:S381–S394, 2000.

D. Hedeker, O. Siddiqui, and F. B. Hu. Random-effects regression analysis of correlated grouped-time survival data. *Statistical Methods in Medical Research*, 9:161–179, 2000.

D. Heilbron. Generalized linear models for altered zero probabilities and overdispersion in count data. *Technical Report, Department of Epidemiology and Biostatistics, University of California, San Francisco*, 1989.

D. Heilbron. Zero-altered and other regression models for count data with added zeros. *Biometrical Journal*, 36:531–547, 1994.

J. W. Hogan and N. M. Laird. Mixture models for the joint distribution of repeated measures and event times. *Statistics in Medicine*, 16:239–258, 1997a.

J. W. Hogan and N. M. Laird. Model-based approaches to analysing incomplete longitudinal and failure time data. *Statistics in Medicine*, 16:259–272, 1997b.

J. W. Hogan, J. Roy, and C. Korkontzelou. Handling drop-out in longitudinal studies. *Statistics in Medicine*, 23:1455–1497, 2004.

N. J. Horton and S. R. Lipsitz. Review of software to fit generalized estimating equation regression models. *The American Statistician*, 53:160–169, 1999.

R. I. Horwitz, C. M. Viscoli, J. D. Clemens, and R. T. Sadock. Developing improved observational methods for evaluating therapeutic effectiveness. *American Journal of Medicine*, 89:630–638, 1990.

D. W. Hosmer and S. Lemeshow. *Applied logistic regression, 2nd edition*. Wiley, New York, 2000.

R. L. Hough, S. Harmon, H. Tarke, S. Yamashiro, R. Quinlivan, P. Landau-Cox, M. S. Hurlburt, P. A. Wood, R. Milone, V. Renker, A. Crowell, and E. Morris. Supported independent housing: implementation issues and solutions in the San Diego McKinney Homeless Demonstration Research Project. In W. R. Breakey and J. W. Thompson, editors, *Mentally Ill and Homeless: Special Programs for Special Needs*, pages 95–117. Harwood Academic Publishers, New York, 1997.

J. Hox. *Multilevel Analysis: Techniques and Applications*. Erlbaum, Mahwah, NJ, 2002.

F. B. Hu, J. Goldberg, D. Hedeker, B. R. Flay, and M. A. Pentz. A comparison of generalized estimating equation and random-effects approaches to analyzing binary outcomes from longitudinal studies: illustrations from a smoking prevention study. *American Journal of Epidemiology*, 147:694–703, 1998.

S. L. Hui and J. O. Berger. Empirical Bayes estimation of rates in longitudinal studies. *Journal of the American Statistical Association*, 78:753–759, 1983.

T. J. Hummel and J. R. Sligo. An empirical comparison of univariate and multivariate analysis of variance procedures. *Psychological Bulletin*, 76:49–57, 1971.

K. Hur, D. Hedeker, W. Henderson, S. Khuri, and J. Daley. Modeling clustered count data with excess zeros in health care outcomes research. *Health Services and Outcomes Research Methodology*, 3:5–20, 2002.

M. S. Hurlburt, P. A. Wood, and R. L. Hough. Providing independent housing for the homeless mentally ill: a novel approach to evaluating long-term longitudinal housing patterns. *Journal of Community Psychology*, 24:291–310, 1996.

J. E. Huttenlocher, W. Haight, A. S. Bryk, and M. Seltzer. Early vocabulary growth: relation to language input and gender. *Developmental Psychology*, 27:236–248, 1991.

H. Huynh and L. S. Feldt. Estimation of the Box correction for degrees of freedom from sample data in the randomized block and split plot designs. *Journal of Educational Statistics*, 1:69–82, 1976.

D. D. Ingram and J. C. Kleinman. Empirical comparisons of proportional hazards and logistic regression models. *Statistical in Medicine*, 8:525–538, 1989.

IOM. *Institute of Medicine Report on Organ Procurement and Transplantation: Assessing Current Policies and the Potential Impact of the DHHS Final Rule*. National Academy Press, Washington, D.C., 1999.

H. Ishwaran. Univariate and multirater ordinal cumulative link regression with covariate specific cutpoints. *Canadian Journal of Statistics*, pages 715–730, 2000.

H. Ishwaran and C.A. Gatsonis. A general class of hierarchical ordinal regression models with applications to correlated ROC analysis. *Canadian Journal of Statistics*, 28:731–750, 2000.

D. R. Jacobs, R.W. Jeffery, and P. J. Hannan. Methodological issues in worksite health intervention research: II. Computation of variance in worksite data: unit of analysis. In K. Johnson, J. H. LaRosa, and C. J. Scheirer, editors, *Proceedings of the 1988 Methodological Issues in Worksite Research Conference*, pages 77–88. United States Department of Health and Human Services, Airlie, VA, 1989.

J. Jansen. On the statistical analysis of ordinal data when extravariation is present. *Applied Statistics*, 39:75–84, 1990.

R. I. Jennrich and M. D. Schluchter. Unbalanced repeated-measures models with structured covariance matrices. *Biometrics*, 42:805–820, 1986.

N. L. Johnson, S. Kotz, and A. W. Kemp. *Univariate Discrete Distributions, 2nd edition*. Wiley, New York, 1992.

R. H. Jones. *Longitudinal Data with Serial Correlation: A State-Space Approach*. Chapman and Hall, New York, 1993.

R. H. Jones. Polynomials with asymptotes for longitudinal data. *Statistics in Medicine*, 15: 61–74, 1996.

R. H. Jones and F. Boadi-Boateng. Unequally spaced longitudinal data with AR(1) serial correlations. *Biometrics*, 47:161–175, 1991.

B. Jovanovic, D. Hosmer, and J. Murray. A method for fitting Poisson regression model with extra zeros. *Proceedings of the Section on Epidemiology of the American Statistical Association*, 1994.

D. Kaplan and R. George. Evaluating latent growth models through ex post simulation. *Journal of Educational and Behavioral Statistics*, 23:216–235, 1998.

G. Kauermann and G. Tutz. Semi- and nonparametric modeling of ordinal data. *Journal of Computational and Graphical Statistics*, 12:176–196, 2003.

M. G. Kenward. Selection models for repeated measurements with non-random dropout: an illustration of sensitivity. *Statistics in Medicine*, 17:2723–2732, 1998.

S. F. Khuri, J. Daley, W. G. Henderson, K. Hur, and et al. Risk adjustment of the postoperative mortality rate for the comparative assessment of the quality of surgical care: results of the National Veterans Affairs Surgical Risk Study. *Journal of the American College of Surgeons*, 185(4):315–327, 1997.

G. King. Event count models for international relations: generalizations and applications. *International Studies Quarterly*, 33:123–147, 1989.

J. R. Kramer, H. Kitazume, W. L. Proudfit, and et al. Segmental analysis of the rate of progression in patients with progressive coronary atherosclerosis. *American Heart Journal*, 106:1427–1431, 1983.

I. Kreft and J. de Leeuw. *Introducing Multilevel Modeling*. Sage, Thousand Oaks, CA, 1998.

E. Läärä and J. N. S. Matthews. The equivalence of two models for ordinal data. *Biometrika*, 72:206–207, 1985.

N. M. Laird. Missing data in longitudinal studies. *Statistics in Medicine*, 7:305–315, 1988.

N. M. Laird and J. H. Ware. Random-effects models for longitudinal data. *Biometrics*, 38: 963–974, 1982.

D. Lambert. Zero-inflated Poisson regression with an application to defects in manufacturing. *Technometrics*, 34:1–14, 1992.

J. Lawless. Negative binomial and mixed Poisson regression. *Canadian Journal of Statistics*, 15:209–225, 1987.

J. F. Lawless and G. E. Willmot. A mixed Poisson-inverse-Gaussian regression model. *Canadian Journal of Statistics*, 17:171–181, 1989.

J. P. Leigh, M. M. Ward, and J. F. Fries. Reducing attrition bias with an instrumental variable in a regression model: results from a panel of rheumatoid arthritis patients. *Statistics in Medicine*, 12:1005–1018, 1993.

E. Lesaffre and B. Spiessens. On the effect of the number of quadrature points in a logistic random-effects model: an example. *Applied Statistics*, 50:325–335, 2001.

K.-Y. Liang and S. L. Zeger. Longitudinal data analysis using generalized linear models. *Biometrika*, 73:13–22, 1986.

K.-Y. Liang and S. L. Zeger. Regression analysis for correlated data. *Annual Review of Public Health*, 14:43–68, 1993.

H. Lin, C. E. McCulloch, and R. A. Rosenheck. Latent pattern mixture models for informative intermittent missing data in longitudinal studies. *Biometrics*, 60:295–305, 2004.

J. K. Lindsey and P. Lambert. On the appropriateness of marginal models for repeated measurements in clinical trials. *Statistics in Medicine*, 17:447–469, 1998.

R. C. Littell, R. J. Freund, and P. C. Spector. *SAS System for Linear Models, 3rd edition*. SAS Institute Inc., Cary, NC, 1991.

R. J. A. Little. A test of missing completely at random for multivariate data with missing values. *Journal of the American Statistical Association*, 83:1198–1202, 1988.

R. J. A. Little. Pattern-mixture models for multivariate incomplete data. *Journal of the American Statistical Association*, 88:125–133, 1993.

R. J. A. Little. A class of pattern-mixture models for normal incomplete data. *Biometrika*, 81:471–483, 1994.

R. J. A. Little. Modeling the drop-out mechanism in repeated-measures studies. *Journal of the American Statistical Association*, 90:1112–1121, 1995.

R. J. A. Little and D. B Rubin. Causal effects in clinical and epidemiological studies via potential outcomes: concepts and analytical approaches. *Annual Reviews of Public Health*, 21:121–145, 2000.

R. J. A. Little and D. B. Rubin. *Statistical Analysis with Missing Data, 2nd edition*. Wiley, New York, 2002.

L. C. Liu and D. Hedeker. A mixed-effects regression model for longitudinal multivariate ordinal data. *Biometrics*, 62, 2006.

Q. Liu and D. A. Pierce. A note on Gauss–Hermite quadrature. *Biometrika*, 81:624–629, 1994.

J. S. Long. *Regression Models for Categorical and Limited Dependent Variables*. Sage Publications, Thousand Oaks, CA, 1997.

N. T. Longford. A fast scoring algorithm for maximum likelihood estimation in unbalanced mixed models with nested random effects. *Biometrika*, 74:817–827, 1987.

N. T. Longford. A quasilikelihood adaptation for variance component analysis. *Proceedings of the Section on Statistical Computing of the American Statistical Association*, 1988.

N. T. Longford. *Random Coefficient Models*. Oxford University Press, New York, 1993.

N. T. Longford. Logistic regression with random coefficients. *Computational Statistics and Data Analysis*, 17:1–15, 1994.

F. M. Lord. *Applications of Item Response Theory to Practical Testing Problems*. Erlbaum, Hillside, NJ, 1980.

M. Lorr and C. J. Klett. *Inpatient Multidimensional Psychiatric Scale: Manual*. Consulting Psychologists Press, Palo Alto, CA, 1966.

T. E. MaCurdy. The use of time series processes to model the error structure of earnings in a longitudinal data analysis. *Journal of Econometrics*, 18:83–114, 1982.

G. S. Maddala. *Limited-Dependent and Qualitative Variables in Econometrics*. Cambridge University Press, Cambridge, UK, 1983.

J. R. Magnus. *Linear Structures*. Oxford University Press, New York, 1988.

H. Mansour, E. V. Nordheim, and J. J. Rutledge. Maximum likelihood estimation of variance components in repeated measures designs assuming autoregressive errors. *Biometrics*, 41:287–294, 1985.

M. M. Marini, A. R. Olsen, and D. B. Rubin. Maximum-likelihood estimation in panel studies with attrition. *Sociology Methodology, 1980*, pages 314–357, 1980.

E. Maris. Covariance adjustment versus gain scores - revisited. *Psychological Methods*, 3: 309–327, 1998.

E. C. Marshall and D. Spiegelhalter. Institutional performance. In A. H. Leyland and H. Goldstein, editors, *Multilevel Modelling of Health Statistics*, pages 127–142. Wiley, New York, 2001.

D. Martin and S. Katti. Fitting of certain contagious distributions to some available data by the maximum likelihood method. *Biometrics*, 21:34–48, 1965.

J. W. Mauchly. Significance test for sphericity of a normal n-variate distribution. *Annals of Mathematical Statistics*, 29:204–209, 1940.

J. J. McArdle and F. Hamagami. Modeling incomplete longitudinal and cross-sectional data using latent growth structural models. *Experimental Aging Research*, 18:145–166, 1992.

P. McCullagh. Regression models for ordinal data (with discussion). *Journal of the Royal Statistical Society, Series B*, 42:109–142, 1980.

P. McCullagh and J. A. Nelder. *Generalized Linear Models, 2nd edition*. Chapman and Hall, New York, 1989.

C. E. McCulloch. Symmetric matrix derivatives with applications. *Journal of the American Statistical Association*, 77:679–682, 1982.

C. E. McCulloch and S. R. Searle. *Generalized, Linear, and Mixed Models*. Wiley, New York, 2001.

C.E. McCulloch, H. Lin, E. H. Slate, and B. W. Turnbull. Discovering subpopulation structure with latent class mixed models. *Statistics in Medicine*, 21:417–429, 2002.

D. McFadden. Conditional logit analysis of qualitative choice behavior. In P. Zarembka, editor, *Frontiers in Econometrics*. Academic Press, New York, 1973.

D. McFadden. Qualitative response models. In W. Hildenbrand, editor, *Advances in Econometrics*, pages 1–37. Cambridge University Press, Cambridge, UK, 1980.

R. D. McKelvey and W. Zavoina. A statistical model for the analysis of ordinal level dependent variables. *Journal of Mathematical Sociology*, 1975.

P. D. Mehta and S. G. West. Putting the individual back into individual growth curves. *Psychological Methods*, 5:23–43, 2000.

B. Michiels, G. M. Molenberghs, L. Bijnens, T. Vangeneugden, and H. Thijs. Selection models and pattern-mixture models to analyze longitudinal quality of life data subject to drop out. *Statistics in Medicine*, 21:1023–1041, 2002.

G. M. Molenberghs, B. Michiels, M. G. Kenward, and P. J. Diggle. Missing data mechanisms and pattern-mixture models. *Statistica Neerlandica*, 52:153–161, 1998.

G. M. Molenberghs, H. Thijs, I. Jansen, C. Beunckens, M. G. Kenward, C. Mallinckrodt, and R. J. Carroll. Analyzing incomplete longitudinal clinical trial data. *Biostatistics*, 5: 445–464, 2004.

D. F. Morrison. *Multivariate Statistical Methods, 2nd edition*. Mc-Graw Hill, New York, 1976.

J. Mullahy. Specification and testing of some modified count data models. *Journal of Econometrics*, 33:341–365, 1986.

D. M. Murray. *Design and Analysis of Group-Randomized Trials*. Oxford University Press, New York, 1998.

B. Muthén, D. Kaplan, and M. Hollis. On structural equation modeling with data that are not missing completely at random. *Psychometrika*, 52:431–462, 1987.

B. Muthén and K. Masyn. Discrete-time survival mixture analysis. *Journal of Educational and Behavioral Statistics*, 30:27–58, 2005.

B. Muthén and K. Shedden. Finite mixture modeling with mixture outcomes using the EM algorithm. *Biometrics*, 55:463–469, 1999.

J. A. Nelder and R. W. M. Wedderburn. Generalized linear models. *Journal of the Royal Statistical Society, Series A*, 135:370–384, 1972.

J. M. Neuhaus and N.P. Jewell. Some comments on rosner's multiple logistic model for clustered data. *Biometrics*, 46:523–534, 1990.

J. M. Neuhaus, J. D. Kalbfleisch, and W. W. Hauck. A comparison of cluster-specific and population-averaged approaches for analyzing correlated binary data. *International Statistical Review*, 59:25–35, 1991.

R. Niaura, B. Spring, B. Borrelli, D. Hedeker, M.G. Goldstein, N. Keuthen, J. DePue, J. Kristeller, J. Ockene, A. Prochazka, J.A. Chiles, and D.B. Abrams. Multicenter trial of fluoxetine as an adjunct to behavioral smoking cessation treatment. *Journal of Consulting and Clinical Psychology*, 70:887–896, 2002.

E. C. Norton, G. S. Bieler, S. T. Ennett, and G. A. Zarkin. Analysis of prevention program effectiveness with clustered data using generalized estimating equations. *Journal of Consulting and Clinical Psychology*, 64:919–926, 1996.

V. Núñez-Antón and G. G. Woodworth. Analysis of longitudinal data with unequally spaced observations and time-dependent correlated errors. *Biometrics*, 50:445–456, 1994.

T. Park. A comparison of the generalized estimating equation approach with the maximum likelihood approach for repeated measurements. *Statistics in Medicine*, 12:1723–1732, 1993.

H. D. Patterson and R. Thompson. Recovery of inter-block information when block sizes are unequal. *Biometrika*, 58:545–554, 1971.

E. S. Pearson and H. O. Hartley. *Biometrika Tables for Statisticians, Volume I*. Biometrika Trust, London, 1976.

J. F. Pendergast, S. J. Gange, M. A. Newton, M. J. Lindstrom, M. Palta, and M. R. Fisher. A survey of methods for analyzing clustered binary response data. *International Statistical Review*, 64:89–118, 1996.

M. S. Pepe and G. A. Anderson. A cautionary note on inference for marginal regression models with longitudinal data and general correlated response data. *Communications in Statistics–Simulation*, 23:939–951, 1994.

B. Peterson and F. E. Harrell. Partial proportional odds models for ordinal response variables. *Applied Statistics*, 39:205–217, 1990.

J. C. Pinheiro and D. M. Bates. Approximations to the log-likelihood function in the non-linear mixed-effects model. *Journal of Computational and Graphical Statistics*, 4: 12–35, 1995.

H. K. Preisler. Analysis of a toxicological experiment using a generalized linear model with nested random effects. *Inter. Statist. Rev*, 57:145–159, 1989.

R. L. Prentice. Binary regression using an extended beta-binomial distribution, with discussion of correlation induces by covariate measurement errors. *Journal of the American Statistical Assocation*, 81:321–327, 1986.

R. L. Prentice and L. A. Gloeckler. Regression analysis of grouped survival data with application to breast cancer data. *Biometrics*, 34:57–67, 1978.

S. Rabe-Hesketh, A. Skrondal, and A. Pickles. Reliable estimation of generalized linear mixed models using adaptive quadrature. *The Stata Journal*, 2:1–21, 2002.

A. E. Raftery. Bayesian model selection in social research. In P. V. Marsden, editor, *Sociological Methodology 1995*, pages 111–164. Blackwell, Oxford, 1995.

R. Raman and D. Hedeker. A mixed-effects regression model for three-level ordinal response data. *Statistics in Medicine*, 24:3331–3345, 2005.

S. W. Raudenbush and A. S. Bryk. *Hierarchical Linear Models, 2nd edition*. Sage, Thousand Oaks, CA, 2002.

S. W. Raudenbush, A. S. Bryk, Y. F. Cheong, and R. Congdon. *HLM 5. Hierarchical Linear and Nonlinear Modeling*. Scientific Software International, Chicago, 2000a.

S. W. Raudenbush, M.-L. Yang, and M. Yosef. Maximum likelihood for generalized linear models with nested random effects via high-order, multivariate Laplace approximation. *Journal of Computational and Graphical Statistics*, 9:141–157, 2000b.

S. F. Reardon, R. Brennan, and S. L. Buka. Estimating multi-level discrete-time hazard models using cross-sectional data: neighborhood effects on the onset of adolescent cigarette use. *Multivariate Behavioral Research*, in press, 2002.

N. Reisby, L. F. Gram, P. Bech, A. Nagy, G. O. Petersen, J. Ortmann, I. Ibsen, S. J. Dencker, O. Jacobsen, O. Krautwald, I. Sondergaard, and J Christiansen. Imipramine: clinical effects and pharmacokinetic variability. *Psychopharmacology*, 54:263–272, 1977.

S. P. Reise. Using multilevel logistic regression to evaluate person-fit in IRT models. *Multivariate Behavioral Research*, 35:543–568, 2000.

D. Revelt and K. Train. Mixed logit with repeated choices: household's choices of appliance efficiency level. *Review of Economics and Statistics*, 80:647–657, 1998.

M. S. Ridout. Reader Reaction: Testing for random dropouts in repeated measurement data. *Biometrics*, 47:1617–1619, 1991.

J. M. Robins, A. G. Rotnitzky, and L. P. Zhao. Analysis of semiparametric regression models for repeated outcomes in the presence of missing data. *Journal of the American Statistical Association*, 90:106–121, 1995.

J. Rochon. ARMA covariance structures with time heteroscedasticity for repeated measures experiments. *Journal of the American Statistical Association*, 87:777–784, 1992.

G. Rodríguez and N. Goldman. An assessment of estimation procedures for multilevel models with binary responses. *Journal of the Royal Statistical Society, Series A*, 158: 73–89, 1995.

P. Rosenbaum and D. B. Rubin. The central role of the propensity score in observational studies for causal effects. *Biometrika*, 70:41–55, 1983.

B. Rosner. *Fundamentals of Biostatistics, 4th edition*. Wadsworth Publishing, New York, 1995.

K. J. Rothman. No adjustments are needed for multiple comparisons. *Epidemiology*, 1: 43–46, 1990.

A. G. Rotnitzky and J. M. Robins. Analysis of semi-parametric regression models with nonignorable nonresponse. *Statistics in Medicine*, 16:81–102, 1999.

J. Roy. Modeling longitudinal data with nonignorable dropouts using a latent dropout class model. *Biometrics*, 59:829–836, 2003.

R. M. Royall. Model robust confidence intervals using maximum likelihood estimators. *International Statistical Review*, 54:221–226, 1986.

D. B. Rubin. Estimating causal effects of treatments in randomized and nonrandomized studies. *Journal of Educational Psychology*, 66:688–701, 1974.

D. B. Rubin. Inference and missing data. *Biometrika*, 63:581–592, 1976.

D. B. Rubin. Formalizing subjective notions about the effect of nonrespondents in sample surveys. *Journal of the American Statistical Association*, 72, 359:538–543, 1977.

D. B. Rubin. *Multiple Imputation for Nonresponse in Surveys*. Wiley, New York, 1987.

D. B. Rubin. Multiple imputation after 18 years. *Journal of the American Statistical Association*, 91, 359:473–489., 1996.

F. Samejima. Estimation of latent ability using a response pattern of graded scores. *Psychometrika Monograph No. 17*, 1969.

D. J. Saville. Multiple comparison procedures: the practical solution. *The American Statistician*, 44:174–180, 1990.

J. L. Schafer. *Analysis of Incomplete Multivariate Data*. Chapman and Hall, New York, 1997.

J. L. Schafer and J. W. Graham. Missing data: our view of the state of the art. *Psychological Methods*, 7:147–177, 2002.

T. H. Scheike and T. K. Jensen. A discrete survival model with random effects: an application to time to pregnancy. *Biometrics*, 53:318–329, 1997.

M. D. Schluchter. Analysis of incomplete multivariate data using linear models with structured covariance matrices. *Statistics in Medicine*, 7:317–324, 1988.

M. D. Schluchter. Methods for the analysis of informatively censored longitudinal data. *Statistics in Medicine*, 11:1861–1870, 1992.

G. Schwarz. Estimating the dimension of a model. *Annals of Statistics*, 6:461–464, 1978.

L. S. Seiden and L. A. Dykstra. *Psychopharmacology: A Biochemical and Behavioral Approach*. Van Nostrand Reinhold, New York, 1977.

A. Serretti, E. Lattuada, R. Zanardi, L. Franchini, and E. Smeraldi. Patterns of symptom improvement during antidepressant treatment of delusional depression. *Psychiatry Research*, 94:185–190, 2000.

C.-F. Sheu. Regression analysis of correlated binary outcomes. *Behavior Research Methods, Instruments, and Computers*, 32:269–273, 2000.

W. J. Shih. Reader Reaction: On informative and random dropouts in longitudinal studies. *Biometrics*, 48:970–971, 1992.

O. Siddiqui. Modeling clustered count and survival data with an application to a school based smoking prevention study. *PhD Dissertation, University of Illinois at Chicago*, 1996.

O. Siddiqui, D. Hedeker, B. R. Flay, and F. B. Hu. Intraclass correlation estimates in a school-based smoking prevention study: outcome and mediating variables, by gender and ethnicity. *American Journal of Epidemiology*, 144:425–433, 1996.

J. D. Singer and J. B. Willett. It's about time: using discrete-time survival analysis to study duration and the timing of events. *Journal of Educational and Behavioral Statistics*, 18: 155–195, 1993.

J. D. Singer and J. B. Willett. *Applied Longitudinal Data Analysis*. Oxford University Press, New York, 2003.

S. Singh. A note on inflated Poisson distribution. *Journal of the Indian Statistical Association*, 1:140–144, 1963.

T. Snijders and R. Bosker. *Multilevel Analysis: An Introduction to Basic and Advanced Multilevel Modeling*. Sage, Thousand Oaks, CA, 1999.

D. J. Spiegelhalter, A. Thomas, N. G. Best, and W. R. Gilks. *BUGS: Bayesian Inference Using Gibbs Sampling, version 0.50*. Technical Report, MRC Biostatistics Unit, Cambridge, UK, 1995.

R. Stiratelli, N. M. Laird, and J. H. Ware. Random-effects models for serial observations with binary response. *Biometrics*, 40:961–971, 1984.

J. F. Strenio, H. I. Weisberg, and A. S. Bryk. Empirical Bayes estimation of individual growth curve parameters and their relationship to covariates. *Biometrics*, 39:71–86, 1983.

A. H. Stroud and D. Sechrest. *Gaussian Quadrature Formulas*. Prentice Hall, Englewood Cliffs, NJ, 1966.

Student. The probable error of the mean. *Biometrika*, 6:1–25, 1908.

T. A. Stukel. Comparison of methods for the analysis of longitudinal interval count data. *Statistics in Medicine*, 12:1339–1351, 1993.

M. A. Tanner. *Tools for Statistical Inference: Methods for the Exploration of Posterior Distributions and Likelihood Functions, 3rd edition*. Springer Verlag, New York, 1996.

T. R. Ten Have. A mixed effects model for multivariate ordinal response data including correlated discrete failure times with ordinal responses. *Biometrics*, 52:473–491, 1996.

T. R. Ten Have, A. R. Kunselman, E. P. Pulkstenis, and J. R. Landis. Mixed effects logistic regression models for longitudinal binary response data with informative dropout. *Biometrics*, 54:367–383, 1998.

T. R. Ten Have, A. R. Kunselman, and L. Tran. A comparison of mixed effects logistic regression models for binary response data with two nested levels of clustering. *Statistics in Medicine*, 18:947–960, 1999.

T. R. Ten Have, R. Landis, and J. Hartzel. Population-averaged and cluster-specific models for clustered ordinal response data. *Statistics in Medicine*, 15:2573–2588, 1996.

T. R. Ten Have, R. Landis, and S. Weaver. Association models for periodontal disease progression: a comparison of methods for clustered binary data. *Statistics in Medicine*, 14:413–429, 1993.

T. R. Ten Have and D. H. Uttal. Subject-specific and population-averaged continuation ratio logit models for multiple discrete time survival profiles. *Applied Statistics*, 43: 371–384, 1994.

J. V. Terza. Ordinal probit: a generalization. *Communications in Statistical Theory and Methods*, 14:1–11, 1985.

P. F. Thall. Mixed Poisson likelihood regression models for longitudinal interval count data. *Biometrics*, 44:197–209, 1988.

H. Thijs, G. M. Molenberghs, B. Michiels, G. Verbeke, and D. Curran. Strategies to fit pattern-mixture models. *Biostatistics*, 3:245–265, 2002.

N. Thomas, N. T. Longford, and J. Rolph. A statistical framework for severity adjustment of hospital mortality rates. *Working paper, RAND, Santa Monica, CA*, 1992.

L. L. Thurstone. Psychophysical analysis. *American Journal of Psychology*, 38:368–389, 1927.

A. N. Tosteson and C. B. Begg. A general regression methodology for ROC curve estimation. *Medical Decision Making*, 8:204–215, 1988.

G. Touloumi, A. G. Babiker, S. J. Pocock, and J. H. Darbyshire. Impact of missing data due to dropouts on estimators for rates of change in longitudinal studies: a simulation study. *Statistics in Medicine*, 20:3715–3728, 2001.

R. K. Tsutakawa. Mixed models for analyzing geographic variability in mortality rates. *Journal of the American Statistical Association*, 83:37–42, 1988.

G. Tutz and W. Hennevogl. Random effects in ordinal regression models. *Computational Statistics and Data Analysis*, 22:537–557, 1996.

UNOS. *United Network for Organ Sharing*. http://www.unos.org, 1999.

USGAO. United States General Accounting Office. Report to the Ranking Minority Member, Committee on Labor and Human Resources, U.S. Senate. *Organ Procurement Organizations: Alternatives Being Developed to More Accurately Assess Performance*, GAO/HEHS-98-26, 1997.

R. van der Leeden, K. Vrijburg, and J. de Leeuw. A review of two different approaches for the analysis of growth data using longitudinal mixed linear models. *Computational Statistics and Data Analysis*, 21:583–605, 1996.

G. Verbeke and E. Lesaffre. A linear mixed-effects model with heterogeneity in the random-effects population. *Journal of American Statistical Association*, 91:217–221, 1996.

G. Verbeke and G. Molenberghs. *Linear Mixed Models for Longitudinal Data*. Springer, New York, 2000.

E. F. Vonesh and V. M. Chinchilli. *Linear and Nonlinear Models for the Analysis of Repeated Measures*. Marcel Dekker, New York, 1997.

M. A. Wakefield, F. J. Chaloupka, N. J. Kaufman, C. T. Orleans, D. C. Barker, and E. E. Ruel. Effect of restrictions on smoking at home, at school, and in public places on teenage smoking: cross sectional study. *British Medical Journal*, 321:333–337, 2001.

A. Wald. Tests of statistical hypotheses concerning several parameters when the number of observations is large. *Transactions of the American Mathematical Society*, 54:426–482, 1943.

R. W. M. Wedderburn. Quasi-likelihood functions, generalized linear models and the Gaussian method. *Biometrika*, 61:439–447, 1974.

P. H. Westfall, R. D. Tobias, D. Rom, R. D. Wolfinger, and Y. Hochberg. *Multiple Comparisons and Multiple Tests Using SAS*. SAS Institute Inc., Cary, NC, 1999.

P. Willner. *Depression: A Psychobiological Synthesis*. Wiley, New York, 1985.

B. J. Winer. *Statistical Principles in Experimental Design, 2nd edition*. McGraw-Hill, New York, 1971.

C. Winship and R. D. Mare. Regression models with ordinal variables. *American Sociological Review*, 49:512–525, 1984.

R. D. Wolfinger. Covariance structure selection in general mixed models. *Communications in Statistics, Simulation and Computation*, 22:1079–1106, 1993.

G. Y. Wong and W. M. Mason. The hierarchical logistic regression model for multilevel analysis. *Journal of the American Statistical Association*, 80:513–524, 1985.

B. D. Wright. Solving measurement problems with the Rasch model. *Journal of Educational Measurement*, 14:97–116, 1977.

D. B. Wright. Comparing groups in a before-after design: when t-test and ANCOVA produce different results. *British Journal of Educational Psychology*, 2005.

M. C. Wu and K. R. Bailey. Estimation and comparison of changes in the presence of informative right censoring: conditional linear model. *Biometrics*, 45:939–955, 1989.

M. C. Wu and R. J. Carroll. Estimation and comparison of changes in the presence of right censoring by modeling the censoring process. *Biometrics*, 44:175–188, 1988.

H. Xie, G. McHugo, A. Sengupta, D. Hedeker, and R. Drake. An application of the thresholds of change model to the analysis of mental health data. *Mental Health Services Research*, 3:107–114, 2001.

W. Xu and D. Hedeker. A random-effects models for classifying treatment response in longitudinal clinical trials. *Journal of Biopharmaceutical Statistics*, 11:253–273, 2002.

S. L. Zeger and K. Y. Liang. Longitudinal data analysis for discrete and continuous outcomes. *Biometrics*, 42:121–130, 1986.

S. L. Zeger, K. Y. Liang, and P. S. Albert. Models for longitudinal data: a generalized estimating equation approach. *Biometrics*, 44:1049–1060, 1988.

C. J. W. Zorn. Evaluating zero-inflated and hurdle Poisson specifications. *Working Paper. Department of Political Science, Ohio State University, Columbus, OH*, 1996.

C. J. W. Zorn. Generalized estimating equation models for correlated data: a review with applications. *American Journal of Political Science*, 45:470–490, 2001.

TOPIC INDEX

WILEY SERIES IN PROBABILITY AND STATISTICS

ESTABLISHED BY WALTER A. SHEWHART AND SAMUEL S. WILKS

The ***Wiley Series in Probability and Statistics*** is well established and authoritative. It covers many topics of current research interest in both pure and applied statistics and probability theory. Written by leading statisticians and institutions, the titles span both state-of-the-art developments in the field and classical methods.

Reflecting the wide range of current research in statistics, the series encompasses applied, methodological and theoretical statistics, ranging from applications and new techniques made possible by advances in computerized practice to rigorous treatment of theoretical approaches.

This series provides essential and invaluable reading for all statisticians, whether in academia, industry, government, or research.

† ABRAHAM and LEDOLTER · Statistical Methods for Forecasting
AGRESTI · Analysis of Ordinal Categorical Data
AGRESTI · An Introduction to Categorical Data Analysis
AGRESTI · Categorical Data Analysis, *Second Edition*
ALTMAN, GILL, and McDONALD · Numerical Issues in Statistical Computing for the Social Scientist
AMARATUNGA and CABRERA · Exploration and Analysis of DNA Microarray and Protein Array Data
ANDĚL · Mathematics of Chance
ANDERSON · An Introduction to Multivariate Statistical Analysis, *Third Edition*
* ANDERSON · The Statistical Analysis of Time Series
ANDERSON, AUQUIER, HAUCK, OAKES, VANDAELE, and WEISBERG · Statistical Methods for Comparative Studies
ANDERSON and LOYNES · The Teaching of Practical Statistics
ARMITAGE and DAVID (editors) · Advances in Biometry
ARNOLD, BALAKRISHNAN, and NAGARAJA · Records
* ARTHANARI and DODGE · Mathematical Programming in Statistics
* BAILEY · The Elements of Stochastic Processes with Applications to the Natural Sciences
BALAKRISHNAN and KOUTRAS · Runs and Scans with Applications
BALAKRISHNAN and NG · Precedence-Type Tests and Applications
BARNETT · Comparative Statistical Inference, *Third Edition*
BARNETT · Environmental Statistics
BARNETT and LEWIS · Outliers in Statistical Data, *Third Edition*
BARTOSZYNSKI and NIEWIADOMSKA-BUGAJ · Probability and Statistical Inference
BASILEVSKY · Statistical Factor Analysis and Related Methods: Theory and Applications
BASU and RIGDON · Statistical Methods for the Reliability of Repairable Systems
BATES and WATTS · Nonlinear Regression Analysis and Its Applications
BECHHOFER, SANTNER, and GOLDSMAN · Design and Analysis of Experiments for Statistical Selection, Screening, and Multiple Comparisons

*Now available in a lower priced paperback edition in the Wiley Classics Library.
†Now available in a lower priced paperback edition in the Wiley–Interscience Paperback Series.

BELSLEY · Conditioning Diagnostics: Collinearity and Weak Data in Regression
† BELSLEY, KUH, and WELSCH · Regression Diagnostics: Identifying Influential Data and Sources of Collinearity
BENDAT and PIERSOL · Random Data: Analysis and Measurement Procedures, *Third Edition*
BERRY, CHALONER, and GEWEKE · Bayesian Analysis in Statistics and Econometrics: Essays in Honor of Arnold Zellner
BERNARDO and SMITH · Bayesian Theory
BHAT and MILLER · Elements of Applied Stochastic Processes, *Third Edition*
BHATTACHARYA and WAYMIRE · Stochastic Processes with Applications
† BIEMER, GROVES, LYBERG, MATHIOWETZ, and SUDMAN · Measurement Errors in Surveys
BILLINGSLEY · Convergence of Probability Measures, *Second Edition*
BILLINGSLEY · Probability and Measure, *Third Edition*
BIRKES and DODGE · Alternative Methods of Regression
BLISCHKE AND MURTHY (editors) · Case Studies in Reliability and Maintenance
BLISCHKE AND MURTHY · Reliability: Modeling, Prediction, and Optimization
BLOOMFIELD · Fourier Analysis of Time Series: An Introduction, *Second Edition*
BOLLEN · Structural Equations with Latent Variables
BOLLEN and CURRAN · Latent Curve Models: A Structural Equation Perspective
BOROVKOV · Ergodicity and Stability of Stochastic Processes
BOULEAU · Numerical Methods for Stochastic Processes
BOX · Bayesian Inference in Statistical Analysis
BOX · R. A. Fisher, the Life of a Scientist
BOX and DRAPER · Empirical Model-Building and Response Surfaces
* BOX and DRAPER · Evolutionary Operation: A Statistical Method for Process Improvement
BOX, HUNTER, and HUNTER · Statistics for Experimenters: Design, Innovation, and Discovery, *Second Editon*
BOX and LUCEÑO · Statistical Control by Monitoring and Feedback Adjustment
BRANDIMARTE · Numerical Methods in Finance: A MATLAB-Based Introduction
BROWN and HOLLANDER · Statistics: A Biomedical Introduction
BRUNNER, DOMHOF, and LANGER · Nonparametric Analysis of Longitudinal Data in Factorial Experiments
BUCKLEW · Large Deviation Techniques in Decision, Simulation, and Estimation
CAIROLI and DALANG · Sequential Stochastic Optimization
CASTILLO, HADI, BALAKRISHNAN, and SARABIA · Extreme Value and Related Models with Applications in Engineering and Science
CHAN · Time Series: Applications to Finance
CHARALAMBIDES · Combinatorial Methods in Discrete Distributions
CHATTERJEE and HADI · Sensitivity Analysis in Linear Regression
CHATTERJEE and PRICE · Regression Analysis by Example, *Third Edition*
CHERNICK · Bootstrap Methods: A Practitioner's Guide
CHERNICK and FRIIS · Introductory Biostatistics for the Health Sciences
CHILÈS and DELFINER · Geostatistics: Modeling Spatial Uncertainty
CHOW and LIU · Design and Analysis of Clinical Trials: Concepts and Methodologies, *Second Edition*
CLARKE and DISNEY · Probability and Random Processes: A First Course with Applications, *Second Edition*
* COCHRAN and COX · Experimental Designs, *Second Edition*
CONGDON · Applied Bayesian Modelling
CONGDON · Bayesian Models for Categorical Data
CONGDON · Bayesian Statistical Modelling
CONOVER · Practical Nonparametric Statistics, *Third Edition*

*Now available in a lower priced paperback edition in the Wiley Classics Library.
†Now available in a lower priced paperback edition in the Wiley–Interscience Paperback Series.

COOK · Regression Graphics
COOK and WEISBERG · Applied Regression Including Computing and Graphics
COOK and WEISBERG · An Introduction to Regression Graphics
CORNELL · Experiments with Mixtures, Designs, Models, and the Analysis of Mixture Data, *Third Edition*
COVER and THOMAS · Elements of Information Theory
COX · A Handbook of Introductory Statistical Methods
* COX · Planning of Experiments
CRESSIE · Statistics for Spatial Data, *Revised Edition*
CSÖRGŐ and HORVÁTH · Limit Theorems in Change Point Analysis
DANIEL · Applications of Statistics to Industrial Experimentation
DANIEL · Biostatistics: A Foundation for Analysis in the Health Sciences, *Eighth Edition*
* DANIEL · Fitting Equations to Data: Computer Analysis of Multifactor Data, *Second Edition*
DASU and JOHNSON · Exploratory Data Mining and Data Cleaning
DAVID and NAGARAJA · Order Statistics, *Third Edition*
* DEGROOT, FIENBERG, and KADANE · Statistics and the Law
DEL CASTILLO · Statistical Process Adjustment for Quality Control
DEMARIS · Regression with Social Data: Modeling Continuous and Limited Response Variables
DEMIDENKO · Mixed Models: Theory and Applications
DENISON, HOLMES, MALLICK and SMITH · Bayesian Methods for Nonlinear Classification and Regression
DETTE and STUDDEN · The Theory of Canonical Moments with Applications in Statistics, Probability, and Analysis
DEY and MUKERJEE · Fractional Factorial Plans
DILLON and GOLDSTEIN · Multivariate Analysis: Methods and Applications
DODGE · Alternative Methods of Regression
* DODGE and ROMIG · Sampling Inspection Tables, *Second Edition*
* DOOB · Stochastic Processes
DOWDY, WEARDEN, and CHILKO · Statistics for Research, *Third Edition*
DRAPER and SMITH · Applied Regression Analysis, *Third Edition*
DRYDEN and MARDIA · Statistical Shape Analysis
DUDEWICZ and MISHRA · Modern Mathematical Statistics
DUNN and CLARK · Basic Statistics: A Primer for the Biomedical Sciences, *Third Edition*
DUPUIS and ELLIS · A Weak Convergence Approach to the Theory of Large Deviations
EDLER and KITSOS · Recent Advances in Quantitative Methods in Cancer and Human Health Risk Assessment
* ELANDT-JOHNSON and JOHNSON · Survival Models and Data Analysis
ENDERS · Applied Econometric Time Series
† ETHIER and KURTZ · Markov Processes: Characterization and Convergence
EVANS, HASTINGS, and PEACOCK · Statistical Distributions, *Third Edition*
FELLER · An Introduction to Probability Theory and Its Applications, Volume I, *Third Edition,* Revised; Volume II, *Second Edition*
FISHER and VAN BELLE · Biostatistics: A Methodology for the Health Sciences
FITZMAURICE, LAIRD, and WARE · Applied Longitudinal Analysis
* FLEISS · The Design and Analysis of Clinical Experiments
FLEISS · Statistical Methods for Rates and Proportions, *Third Edition*
† FLEMING and HARRINGTON · Counting Processes and Survival Analysis
FULLER · Introduction to Statistical Time Series, *Second Edition*
FULLER · Measurement Error Models
GALLANT · Nonlinear Statistical Models

*Now available in a lower priced paperback edition in the Wiley Classics Library.
†Now available in a lower priced paperback edition in the Wiley–Interscience Paperback Series.

GEISSER · Modes of Parametric Statistical Inference
GELMAN and MENG · Applied Bayesian Modeling and Causal Inference from Incomplete-Data Perspectives
GEWEKE · Contemporary Bayesian Econometrics and Statistics
GHOSH, MUKHOPADHYAY, and SEN · Sequential Estimation
GIESBRECHT and GUMPERTZ · Planning, Construction, and Statistical Analysis of Comparative Experiments
GIFI · Nonlinear Multivariate Analysis
GIVENS and HOETING · Computational Statistics
GLASSERMAN and YAO · Monotone Structure in Discrete-Event Systems
GNANADESIKAN · Methods for Statistical Data Analysis of Multivariate Observations, *Second Edition*
GOLDSTEIN and LEWIS · Assessment: Problems, Development, and Statistical Issues
GREENWOOD and NIKULIN · A Guide to Chi-Squared Testing
GROSS and HARRIS · Fundamentals of Queueing Theory, *Third Edition*
* HAHN and SHAPIRO · Statistical Models in Engineering
HAHN and MEEKER · Statistical Intervals: A Guide for Practitioners
HALD · A History of Probability and Statistics and their Applications Before 1750
HALD · A History of Mathematical Statistics from 1750 to 1930
† HAMPEL · Robust Statistics: The Approach Based on Influence Functions
HANNAN and DEISTLER · The Statistical Theory of Linear Systems
HEIBERGER · Computation for the Analysis of Designed Experiments
HEDAYAT and SINHA · Design and Inference in Finite Population Sampling
HEDEKER and GIBBONS · Longitudinal Data Analysis
HELLER · MACSYMA for Statisticians
HINKELMANN and KEMPTHORNE · Design and Analysis of Experiments, Volume 1: Introduction to Experimental Design
HINKELMANN and KEMPTHORNE · Design and Analysis of Experiments, Volume 2: Advanced Experimental Design
HOAGLIN, MOSTELLER, and TUKEY · Exploratory Approach to Analysis of Variance
* HOAGLIN, MOSTELLER, and TUKEY · Exploring Data Tables, Trends and Shapes
* HOAGLIN, MOSTELLER, and TUKEY · Understanding Robust and Exploratory Data Analysis
HOCHBERG and TAMHANE · Multiple Comparison Procedures
HOCKING · Methods and Applications of Linear Models: Regression and the Analysis of Variance, *Second Edition*
HOEL · Introduction to Mathematical Statistics, *Fifth Edition*
HOGG and KLUGMAN · Loss Distributions
HOLLANDER and WOLFE · Nonparametric Statistical Methods, *Second Edition*
HOSMER and LEMESHOW · Applied Logistic Regression, *Second Edition*
HOSMER and LEMESHOW · Applied Survival Analysis: Regression Modeling of Time to Event Data
† HUBER · Robust Statistics
HUBERTY · Applied Discriminant Analysis
HUNT and KENNEDY · Financial Derivatives in Theory and Practice, *Revised Edition*
HUSKOVA, BERAN, and DUPAC · Collected Works of Jaroslav Hajek—with Commentary
HUZURBAZAR · Flowgraph Models for Multistate Time-to-Event Data
IMAN and CONOVER · A Modern Approach to Statistics
† JACKSON · A User's Guide to Principle Components
JOHN · Statistical Methods in Engineering and Quality Assurance
JOHNSON · Multivariate Statistical Simulation

*Now available in a lower priced paperback edition in the Wiley Classics Library.
†Now available in a lower priced paperback edition in the Wiley–Interscience Paperback Series.

JOHNSON and BALAKRISHNAN · Advances in the Theory and Practice of Statistics: A Volume in Honor of Samuel Kotz
JOHNSON and BHATTACHARYYA · Statistics: Principles and Methods, *Fifth Edition*
JOHNSON and KOTZ · Distributions in Statistics
JOHNSON and KOTZ (editors) · Leading Personalities in Statistical Sciences: From the Seventeenth Century to the Present
JOHNSON, KOTZ, and BALAKRISHNAN · Continuous Univariate Distributions, Volume 1, *Second Edition*
JOHNSON, KOTZ, and BALAKRISHNAN · Continuous Univariate Distributions, Volume 2, *Second Edition*
JOHNSON, KOTZ, and BALAKRISHNAN · Discrete Multivariate Distributions
JOHNSON, KEMP, and KOTZ · Univariate Discrete Distributions, *Third Edition*
JUDGE, GRIFFITHS, HILL, LÜTKEPOHL, and LEE · The Theory and Practice of Econometrics, *Second Edition*
JUREČKOVÁ and SEN · Robust Statistical Procedures: Aymptotics and Interrelations
JUREK and MASON · Operator-Limit Distributions in Probability Theory
KADANE · Bayesian Methods and Ethics in a Clinical Trial Design
KADANE AND SCHUM · A Probabilistic Analysis of the Sacco and Vanzetti Evidence
KALBFLEISCH and PRENTICE · The Statistical Analysis of Failure Time Data, *Second Edition*
KARIYA and KURATA · Generalized Least Squares
KASS and VOS · Geometrical Foundations of Asymptotic Inference
† KAUFMAN and ROUSSEEUW · Finding Groups in Data: An Introduction to Cluster Analysis
KEDEM and FOKIANOS · Regression Models for Time Series Analysis
KENDALL, BARDEN, CARNE, and LE · Shape and Shape Theory
KHURI · Advanced Calculus with Applications in Statistics, *Second Edition*
KHURI, MATHEW, and SINHA · Statistical Tests for Mixed Linear Models
* KISH · Statistical Design for Research
KLEIBER and KOTZ · Statistical Size Distributions in Economics and Actuarial Sciences
KLUGMAN, PANJER, and WILLMOT · Loss Models: From Data to Decisions, *Second Edition*
KLUGMAN, PANJER, and WILLMOT · Solutions Manual to Accompany Loss Models: From Data to Decisions, *Second Edition*
KOTZ, BALAKRISHNAN, and JOHNSON · Continuous Multivariate Distributions, Volume 1, *Second Edition*
KOTZ and JOHNSON (editors) · Encyclopedia of Statistical Sciences: Volumes 1 to 9 with Index
KOTZ and JOHNSON (editors) · Encyclopedia of Statistical Sciences: Supplement Volume
KOTZ, READ, and BANKS (editors) · Encyclopedia of Statistical Sciences: Update Volume 1
KOTZ, READ, and BANKS (editors) · Encyclopedia of Statistical Sciences: Update Volume 2
KOVALENKO, KUZNETZOV, and PEGG · Mathematical Theory of Reliability of Time-Dependent Systems with Practical Applications
LACHIN · Biostatistical Methods: The Assessment of Relative Risks
LAD · Operational Subjective Statistical Methods: A Mathematical, Philosophical, and Historical Introduction
LAMPERTI · Probability: A Survey of the Mathematical Theory, *Second Edition*
LANGE, RYAN, BILLARD, BRILLINGER, CONQUEST, and GREENHOUSE · Case Studies in Biometry
LARSON · Introduction to Probability Theory and Statistical Inference, *Third Edition*
LAWLESS · Statistical Models and Methods for Lifetime Data, *Second Edition*

*Now available in a lower priced paperback edition in the Wiley Classics Library.
†Now available in a lower priced paperback edition in the Wiley–Interscience Paperback Series.

LAWSON · Statistical Methods in Spatial Epidemiology
LE · Applied Categorical Data Analysis
LE · Applied Survival Analysis
LEE and WANG · Statistical Methods for Survival Data Analysis, *Third Edition*
LEPAGE and BILLARD · Exploring the Limits of Bootstrap
LEYLAND and GOLDSTEIN (editors) · Multilevel Modelling of Health Statistics
LIAO · Statistical Group Comparison
LINDVALL · Lectures on the Coupling Method
LIN · Introductory Stochastic Analysis for Finance and Insurance
LINHART and ZUCCHINI · Model Selection
LITTLE and RUBIN · Statistical Analysis with Missing Data, *Second Edition*
LLOYD · The Statistical Analysis of Categorical Data
LOWEN and TEICH · Fractal-Based Point Processes
MAGNUS and NEUDECKER · Matrix Differential Calculus with Applications in Statistics and Econometrics, *Revised Edition*
MALLER and ZHOU · Survival Analysis with Long Term Survivors
MALLOWS · Design, Data, and Analysis by Some Friends of Cuthbert Daniel
MANN, SCHAFER, and SINGPURWALLA · Methods for Statistical Analysis of Reliability and Life Data
MANTON, WOODBURY, and TOLLEY · Statistical Applications Using Fuzzy Sets
MARCHETTE · Random Graphs for Statistical Pattern Recognition
MARDIA and JUPP · Directional Statistics
MASON, GUNST, and HESS · Statistical Design and Analysis of Experiments with Applications to Engineering and Science, *Second Edition*
McCULLOCH and SEARLE · Generalized, Linear, and Mixed Models
McFADDEN · Management of Data in Clinical Trials
* McLACHLAN · Discriminant Analysis and Statistical Pattern Recognition
McLACHLAN, DO, and AMBROISE · Analyzing Microarray Gene Expression Data
McLACHLAN and KRISHNAN · The EM Algorithm and Extensions
McLACHLAN and PEEL · Finite Mixture Models
McNEIL · Epidemiological Research Methods
MEEKER and ESCOBAR · Statistical Methods for Reliability Data
MEERSCHAERT and SCHEFFLER · Limit Distributions for Sums of Independent Random Vectors: Heavy Tails in Theory and Practice
MICKEY, DUNN, and CLARK · Applied Statistics: Analysis of Variance and Regression, *Third Edition*
* MILLER · Survival Analysis, *Second Edition*
MONTGOMERY, PECK, and VINING · Introduction to Linear Regression Analysis, *Third Edition*
MORGENTHALER and TUKEY · Configural Polysampling: A Route to Practical Robustness
MUIRHEAD · Aspects of Multivariate Statistical Theory
MULLER and STOYAN · Comparison Methods for Stochastic Models and Risks
MURRAY · X-STAT 2.0 Statistical Experimentation, Design Data Analysis, and Nonlinear Optimization
MURTHY, XIE, and JIANG · Weibull Models
MYERS and MONTGOMERY · Response Surface Methodology: Process and Product Optimization Using Designed Experiments, *Second Edition*
MYERS, MONTGOMERY, and VINING · Generalized Linear Models. With Applications in Engineering and the Sciences
† NELSON · Accelerated Testing, Statistical Models, Test Plans, and Data Analyses
† NELSON · Applied Life Data Analysis
NEWMAN · Biostatistical Methods in Epidemiology

*Now available in a lower priced paperback edition in the Wiley Classics Library.
†Now available in a lower priced paperback edition in the Wiley–Interscience Paperback Series.

OCHI · Applied Probability and Stochastic Processes in Engineering and Physical Sciences
OKABE, BOOTS, SUGIHARA, and CHIU · Spatial Tesselations: Concepts and Applications of Voronoi Diagrams, *Second Edition*
OLIVER and SMITH · Influence Diagrams, Belief Nets and Decision Analysis
PALTA · Quantitative Methods in Population Health: Extensions of Ordinary Regressions
PANKRATZ · Forecasting with Dynamic Regression Models
PANKRATZ · Forecasting with Univariate Box-Jenkins Models: Concepts and Cases
* PARZEN · Modern Probability Theory and Its Applications
PEÑA, TIAO, and TSAY · A Course in Time Series Analysis
PIANTADOSI · Clinical Trials: A Methodologic Perspective
PORT · Theoretical Probability for Applications
POURAHMADI · Foundations of Time Series Analysis and Prediction Theory
PRESS · Bayesian Statistics: Principles, Models, and Applications
PRESS · Subjective and Objective Bayesian Statistics, *Second Edition*
PRESS and TANUR · The Subjectivity of Scientists and the Bayesian Approach
PUKELSHEIM · Optimal Experimental Design
PURI, VILAPLANA, and WERTZ · New Perspectives in Theoretical and Applied Statistics
† PUTERMAN · Markov Decision Processes: Discrete Stochastic Dynamic Programming
QIU · Image Processing and Jump Regression Analysis
* RAO · Linear Statistical Inference and Its Applications, *Second Edition*
RAUSAND and HØYLAND · System Reliability Theory: Models, Statistical Methods, and Applications, *Second Edition*
RENCHER · Linear Models in Statistics
RENCHER · Methods of Multivariate Analysis, *Second Edition*
RENCHER · Multivariate Statistical Inference with Applications
* RIPLEY · Spatial Statistics
* RIPLEY · Stochastic Simulation
ROBINSON · Practical Strategies for Experimenting
ROHATGI and SALEH · An Introduction to Probability and Statistics, *Second Edition*
ROLSKI, SCHMIDLI, SCHMIDT, and TEUGELS · Stochastic Processes for Insurance and Finance
ROSENBERGER and LACHIN · Randomization in Clinical Trials: Theory and Practice
ROSS · Introduction to Probability and Statistics for Engineers and Scientists
ROSSI, ALLENBY, and McCULLOCH · Bayesian Statistics and Marketing
† ROUSSEEUW and LEROY · Robust Regression and Outlier Detection
* RUBIN · Multiple Imputation for Nonresponse in Surveys
RUBINSTEIN · Simulation and the Monte Carlo Method
RUBINSTEIN and MELAMED · Modern Simulation and Modeling
RYAN · Modern Regression Methods
RYAN · Statistical Methods for Quality Improvement, *Second Edition*
SALEH · Theory of Preliminary Test and Stein-Type Estimation with Applications
* SCHEFFE · The Analysis of Variance
SCHIMEK · Smoothing and Regression: Approaches, Computation, and Application
SCHOTT · Matrix Analysis for Statistics, *Second Edition*
SCHOUTENS · Levy Processes in Finance: Pricing Financial Derivatives
SCHUSS · Theory and Applications of Stochastic Differential Equations
SCOTT · Multivariate Density Estimation: Theory, Practice, and Visualization
† SEARLE · Linear Models for Unbalanced Data
† SEARLE · Matrix Algebra Useful for Statistics
† SEARLE, CASELLA, and McCULLOCH · Variance Components
SEARLE and WILLETT · Matrix Algebra for Applied Economics
SEBER and LEE · Linear Regression Analysis, *Second Edition*

*Now available in a lower priced paperback edition in the Wiley Classics Library.
†Now available in a lower priced paperback edition in the Wiley–Interscience Paperback Series.

† SEBER · Multivariate Observations
† SEBER and WILD · Nonlinear Regression
SENNOTT · Stochastic Dynamic Programming and the Control of Queueing Systems
* SERFLING · Approximation Theorems of Mathematical Statistics
SHAFER and VOVK · Probability and Finance: It's Only a Game!
SILVAPULLE and SEN · Constrained Statistical Inference: Inequality, Order, and Shape Restrictions
SMALL and McLEISH · Hilbert Space Methods in Probability and Statistical Inference
SRIVASTAVA · Methods of Multivariate Statistics
STAPLETON · Linear Statistical Models
STAUDTE and SHEATHER · Robust Estimation and Testing
STOYAN, KENDALL, and MECKE · Stochastic Geometry and Its Applications, *Second Edition*
STOYAN and STOYAN · Fractals, Random Shapes and Point Fields: Methods of Geometrical Statistics
STYAN · The Collected Papers of T. W. Anderson: 1943–1985
SUTTON, ABRAMS, JONES, SHELDON, and SONG · Methods for Meta-Analysis in Medical Research
TAKEZAWA · Introduction to Nonparametric Regression
TANAKA · Time Series Analysis: Nonstationary and Noninvertible Distribution Theory
THOMPSON · Empirical Model Building
THOMPSON · Sampling, *Second Edition*
THOMPSON · Simulation: A Modeler's Approach
THOMPSON and SEBER · Adaptive Sampling
THOMPSON, WILLIAMS, and FINDLAY · Models for Investors in Real World Markets
TIAO, BISGAARD, HILL, PEÑA, and STIGLER (editors) · Box on Quality and Discovery: with Design, Control, and Robustness
TIERNEY · LISP-STAT: An Object-Oriented Environment for Statistical Computing and Dynamic Graphics
TSAY · Analysis of Financial Time Series, *Second Edition*
UPTON and FINGLETON · Spatial Data Analysis by Example, Volume II: Categorical and Directional Data
VAN BELLE · Statistical Rules of Thumb
VAN BELLE, FISHER, HEAGERTY, and LUMLEY · Biostatistics: A Methodology for the Health Sciences, *Second Edition*
VESTRUP · The Theory of Measures and Integration
VIDAKOVIC · Statistical Modeling by Wavelets
VINOD and REAGLE · Preparing for the Worst: Incorporating Downside Risk in Stock Market Investments
WALLER and GOTWAY · Applied Spatial Statistics for Public Health Data
WEERAHANDI · Generalized Inference in Repeated Measures: Exact Methods in MANOVA and Mixed Models
WEISBERG · Applied Linear Regression, *Third Edition*
WELSH · Aspects of Statistical Inference
WESTFALL and YOUNG · Resampling-Based Multiple Testing: Examples and Methods for p-Value Adjustment
WHITTAKER · Graphical Models in Applied Multivariate Statistics
WINKER · Optimization Heuristics in Economics: Applications of Threshold Accepting
WONNACOTT and WONNACOTT · Econometrics, *Second Edition*
WOODING · Planning Pharmaceutical Clinical Trials: Basic Statistical Principles
WOODWORTH · Biostatistics: A Bayesian Introduction
WOOLSON and CLARKE · Statistical Methods for the Analysis of Biomedical Data, *Second Edition*

*Now available in a lower priced paperback edition in the Wiley Classics Library.
†Now available in a lower priced paperback edition in the Wiley–Interscience Paperback Series.

WU and HAMADA · Experiments: Planning, Analysis, and Parameter Design Optimization
WU and ZHANG · Nonparametric Regression Methods for Longitudinal Data Analysis
YANG · The Construction Theory of Denumerable Markov Processes
ZELTERMAN · Discrete Distributions—Applications in the Health Sciences
* ZELLNER · An Introduction to Bayesian Inference in Econometrics
ZHOU, OBUCHOWSKI, and McCLISH · Statistical Methods in Diagnostic Medicine